Engineering Materials: Design, Properties and Fabrication

Engineering Materials: Design, Properties and Fabrication

Edited by Sally Renwick

CLANRYE
INTERNATIONAL
www.clanryeinternational.com

Clanrye International,
750 Third Avenue, 9th Floor,
New York, NY 10017, USA

ISBN: 978-1-63240-657-6

Cataloging-in-Publication Data

Engineering materials : design, properties and fabrication / edited by Sally Renwick.
    p. cm.
Includes bibliographical references and index.
ISBN 978-1-63240-657-6
1. Materials. 2. Engineering design. I. Renwick, Sally.
TA403 .E54 2018
620.11--dc23

For information on all Clanrye International publications
visit our website at www.clanryeinternational.com

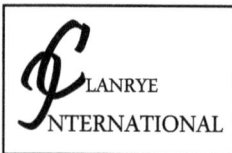

CLANRYE
INTERNATIONAL

Contents

Preface

This book has been an outcome of determined endeavour from a group of educationists in the field. The primary objective was to involve a broad spectrum of professionals from diverse cultural background involved in the field for developing new researches. The book not only targets students but also scholars pursuing higher research for further enhancement of the theoretical and practical applications of the subject.

Engineering materials involves studies the manufacture of new materials that provide solutions to a variety of problems. This field is a synthesis of various other scientific branches such as ceramics, metallurgy, chemistry and solid-state physics. Biomaterials, electronic materials such as semiconductors, optoelectronic devices, etc. are some of the widely used materials. The objective of this book is to give a general view of the different areas of engineering materials and their applications. The book is appropriate for students seeking detailed information in this area as well as for experts.

It was an honour to edit such a profound book and also a challenging task to compile and examine all the relevant data for accuracy and originality. I wish to acknowledge the efforts of the contributors for submitting such brilliant and diverse chapters in the field and for endlessly working for the completion of the book. Last, but not the least; I thank my family for being a constant source of support in all my research endeavours.

Editor

Alkyl-π engineering in state control toward versatile optoelectronic soft materials

Fengniu Lu[1] and Takashi Nakanishi[1,2,3]

[1] International Center for Materials Nanoarchitectonics (MANA), National Institute for Materials Science
(NIMS) 1-2-1 Sengen, Tsukuba 305-0047, Japan
[2] Warsaw University of Technology, Warsaw 02-507, Poland
[3] Institute for Molecular Science (IMS), 5-1 Higashiyama, Myodaiji, Okazaki 444-8787, Japan

E-mail: nakanishi.takashi@nims.go.jp

Abstract

Organic π-conjugated molecules with extremely rich and tailorable electronic and optical
properties are frequently utilized for the fabrication of optoelectronic devices. To achieve high
solubility for facile solution processing and desirable softness for flexible device fabrication, the
rigid π units were in most cases attached by alkyl chains through chemical modification.
Considerable numbers of alkylated-π molecular systems with versatile applications have been
reported. However, a profound understanding of the molecular state control through proper alkyl
chain substitution is still highly demanded because effective applications of these molecules are
closely related to their physical states. To explore the underlying rule, we review a large number
of alkylated-π molecules with emphasis on the interplay of van der Waals interactions (vdW) of
the alkyl chains and π–π interactions of the π moieties. Based on our comprehensive
investigations of the two interactions' impacts on the physical states of the molecules, a clear
guidance for state control by alkyl-π engineering is proposed. Specifically, either with proper
alkyl chain substitution or favorable additives, the vdW and π–π interactions can be adjusted,
resulting in modulation of the physical states and optoelectronic properties of the molecules. We
believe the strategy summarized here will significantly benefit the alkyl-π chemistry toward
wide-spread applications in optoelectronic devices.

Keywords: self-assembly, π-conjugated molecules, alkyl chains, optoelectronic, liquid

1. Introduction

The last few decades have witnessed the prosperous development of optoelectronic devices based on organic/polymer
materials, such as light-emitting diodes [1–3], photovoltaic
devices [1, 4], field-effect transistors [5–7] and electronic
devices [8, 9], by virtue of their extensive well-known and
potential applications. Compared with their inorganic counterparts, these organic ones are much more advantageous in
view of their light weight, low cost, flexibility, unlimited
selection of building blocks and convenience for large area
fabrications [10].

In general, π-conjugated molecules are intriguing building motifs for the design and construction of organic optoelectronic devices on account of their rich and desirable
electronic and optical properties [11, 12]. Moreover, the
optoelectronic characteristics can be facilely tailored through
chemical functionalization of the molecules. As a result,
extensive organic π-conjugated molecules with a wide range
of structures have been synthesized and fabricated into various optoelectronic devices. However, most π-conjugated
molecules, restricted by the strong π–π interactions among the
π-conjugated moieties, exist as solids at room temperature.

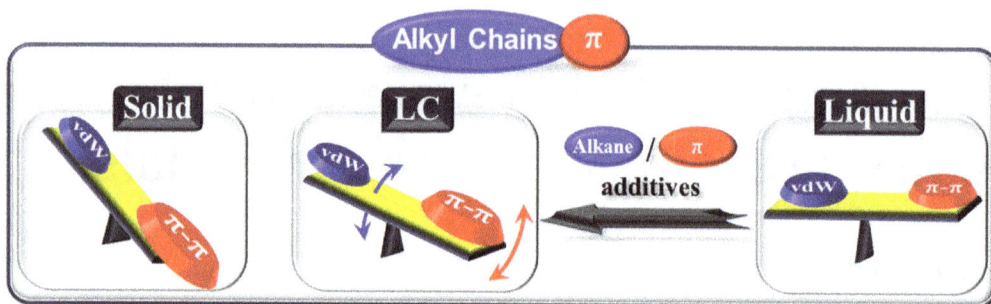

Scheme 1. State control of alkyl-π molecules by adjusting van der Waals (vdW) and π–π interactions through proper alkyl chain substitution or the introduction of additives.

These solid materials, either amorphous or crystalline, suffer poor processability for the application of flexible optoelectronic devices. In addition, large π-conjugated systems have a strong tendency to form random stacks and aggregates with extremely poor solubility in organic solvents. This limits the handling of the molecules in conventional solution-process techniques [13, 14] such as spin-coating and printing for device fabrications. As another issue, π-conjugated systems can often decompose upon oxygen attack and undergo dimerization or polymerization upon exposure to external stimuli.

The most frequently used method to overcome these issues is to attach solubilizing alkyl chain groups to the π-conjugated moieties [15, 16], which can not only soften the π-conjugated materials to some extent but also enhance the solubility and protect the π units from oxygen attack and external stimuli. Fundamentally, upon the appending of alkyl chains, van der Waals (vdW) interactions (vdW interactions mentioned in this review are specific for the vdW interactions of alkyl chains) of the chains were introduced, which can interplay with the π–π interactions of the π moieties. Based on a comprehensive investigation of various alkylated-π molecular systems reported by our group and other research teams, we find that the physical states of the molecules can be dominated by simply tuning the balance of vdW and π–π interactions (scheme 1). When the vdW interactions are far weaker than the π–π interactions, the alkyl-π molecules are in a solid state. These molecules, assisted by solvents, can assemble into various structures. With increasing vdW interactions and periodic segregation of the rigid π-conjugated moieties by the alkyl chains, thermotropic liquid crystals (LCs) can be produced. Further increasing of the vdW interactions may result in a delicate balance with the π–π interactions, which would generate a solvent-free liquid state in which both alkyl chains and π-moieties are disordered.

Because of the strong effect of alkyl chains on vdW interactions, the balance of the vdW and π–π interactions can be adjusted by diversifying the alkyl chains or modulating the chain substitution pattern. Moreover, we demonstrated that the introduction of alkane or π additives can break the balance between the vdW and π–π interactions in liquid molecules, allowing additive-directed conversion from a liquid to a highly ordered LC or a gel [17]. Previously, our group had

reviewed assembled functional materials with a focus on the smart combination of π molecules and alkyl chains [18–22]. In contrast, herein, we stress 'state control' through alkyl-π engineering based on a deep investigation of a large number of alkylated-π molecules. We aim to provide clear guidance for mastering the balance between vdW and π–π interactions in alkylated-π molecules and to direct their physical states and state-dependent optoelectronic applications.

2. Solvent-assisted solid self-assemblies of linear alkyl chain-attached π molecules

The majority of π molecules with attached linear alkyl chains appear to be solid at room temperature. Because optoelectronic device performance is strongly dependent on the precise organization of the π-conjugated moieties [23], controllable self-assembly of these solid state molecules is required for achieving excellent optoelectronic properties [24, 25]. The self-assembly behavior of these alkylated-π molecules is intrinsically affected by the alkyl chain substitution pattern due to the strong interplay of the alkyl chains' vdW interactions with the π units' π–π interactions. Moreover, within solvent systems, adjustment of external experimental conditions (solvent polarity and temperature), as well as the introduction of substrates and other interactions (electrostatic, hydrogen bonding and hydrophilic interactions), can all play significant roles in forming numerous self-assembled nano/micro structures.

2.1. Solvent polarity-modulated architectures

Our group has reported a series of linear alkyl chain-attached C_{60} derivatives **1a**–**1c** (figure 1(a)), which self-assembled into diverse well-defined 1D, 2D and 3D architectures in different organic solvents. The self-assembly of **1a** appended with 3,4,5-trishexadecyloxyl chains, prepared simply by cooling a solvent mixture from 60 °C to 20 °C, gave rise to a variety of self-assembled architectures under different solvent conditions. 1D nanofibers (figure 1(b)), 2D nanodisks (figure 1(c)) and 3D cones (figure 1(d)) were obtained in 1-propanol, 1,4-dioxane and a 1:1 tetrahydrofuran (THF)/H_2O mixture, respectively [26]. Similarly, with identical preparation

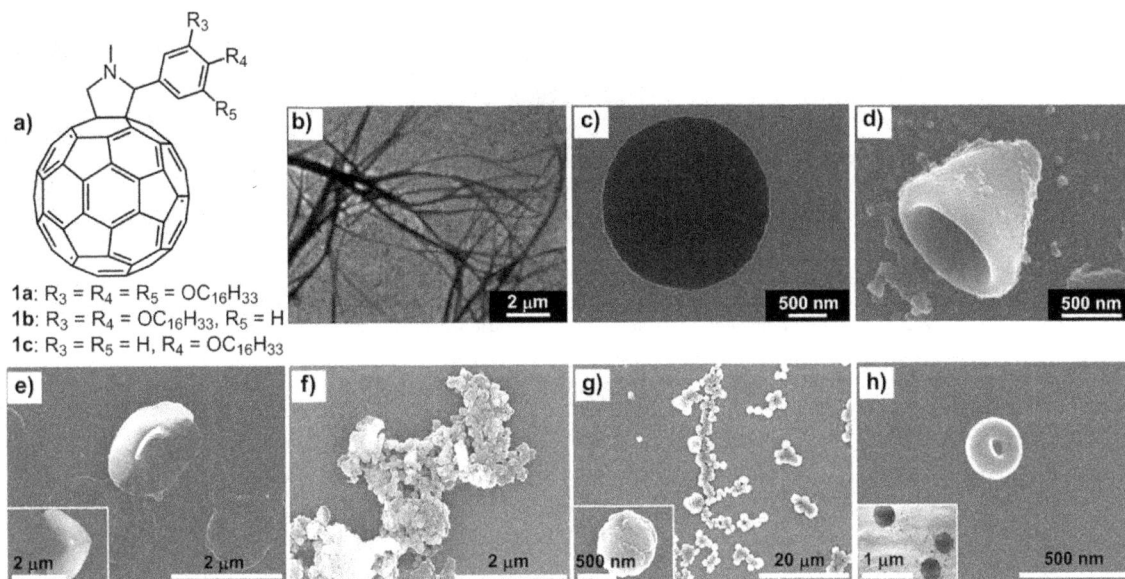

Figure 1. (a) Chemical structures of C_{60} derivatives **1a**–**1c** containing linear alkyl chains. Scanning electron microscopy (SEM) images of fibrous structures of **1a** assembled from 1-propanol (b), a nanodisk of **1a** formed from 1,4-dioxane (c) and a conical object of **1a** assembled from a 1:1 tetrahydrofuran (THF)/H_2O mixture (d). Reprinted from [26]. SEM images of disk-like sheets of **1b** formed from a 2:1 2-propanol/toluene mixture (e), SEM images of self-aggregated particles of **1b** obtained from a 1:2 THF/H_2O mixture (f), globular aggregates of **1(c)** with coarse surfaces formed from a 2:1 2-propanol/toluene mixture (g) and SEM and transmission electron microscopy (TEM) (inset) images of vesicular-spherical objects of **1c** assembled from a 1:2 THF/H_2O mixture (h). Parts (e)–(h) reprinted from T Nakanishi *et al* 2008 *Thin Solid Films* **516** 2401, © 2008 with permission from Elsevier.

procedures, 3,4-bishexadecyloxyl chains attached **1b** formed 2D disk-like sheets in a 2:1 2-propanol/toluene mixture (figure 1(e)) and rather random 3D self-aggregated particles in a 1:2 THF/H_2O mixture (figure 1(f)). The 4-hexadecyloxyl chain modified **1c** created 3D globular aggregates in a 2:1 2-propanol/toluene mixture (figure 1(g)) and 3D vesicular-spherical objects in 1:2 THF/H_2O mixtures (figure 1(h)) [27].

Such solvent-dependent self-organization originates from the amphiphilicity of the two components, C_{60} and alkyl chains, in organic solvent. Although both are hydrophobic, the sp^2-carbon rich C_{60} moiety shows higher solubility in aromatic solvents, such as toluene, while the sp^3-carbon rich alkyl chains exhibit stronger affinity to aliphatic alkanes and polar solvents such as alcohols and ethers. With these amphiphilic-like features, the two moieties go through different solvation in these solvent systems, which would remarkably influence the curvature of the ensuing assembly and result in various self-assembled structures in the nano and micrometer scales. Fine-tuning of the morphologies of these assembled objects by simply adjusting the solvent systems can generate extensive hierarchical assemblies and strikingly enrich the structures and functions of those materials.

Wang *et al* reported a linear dodecyl chain-substituted oligoarene derivative, **2**, which exhibited similar solvent-dependent self-assembly behavior as **1a**–**1c** (figure 2(a)) [28]. By drop casting solutions of **2** in different solvents onto glass substrates, three distinctive structures were obtained after evaporation of the solvents. In 1,4-dioxane, **2** self-assembled into 1D microbelts on the order of tens of micrometers in

length, several hundred nanometers in width and 50 nm in thickness (figure 2(b)). However, in THF and *n*-decane, two different 3D flower-shaped microstructures, flower-A (figure 2(c)) and flower-B (figure 2(d)), were generated. In spite of their similar diameters, around 10–20 μm, flower-A was made of hundreds of shuttle-like 1D petals, while flower-B was composed of hundreds of 1D acicula-like petals. In the light of their high surface areas, the flower-shaped objects were fabricated for explosive detection because the detection scope mainly relied on surface area. The detection speed of 3D flower-B was enhanced by more than 700 times compared with that of the 1D microbelts, providing prospects for using these self-assembled structures in chemosensing.

2.2. Temperature-influenced architectures

In addition to the solvent effect, temperature also plays a significant role in the self-assembly behaviors. Upon heating at 60 °C, the mixture of **1a** in 1,4-dioxane (1.0 mM) transformed into a transparent light-brown solution, which formed aggregates composed of nanodisks after subsequent aging at 20 °C for 24 h (figure 3(a)). The nanodisks have 0.2–1.5 μm diameters and a thickness of 4.4 nm, which is in good agreement with the thickness of the alkyl chain-interdigitated bilayers (figure 3(b)) [26]. Interestingly, further cooling of the mixture from 20 °C to 5 °C and keeping at 5 °C for 12 h resulted in precipitates comprising flower-shaped assemblies several micrometers in size (3–10 μm) with crumpled sheet-

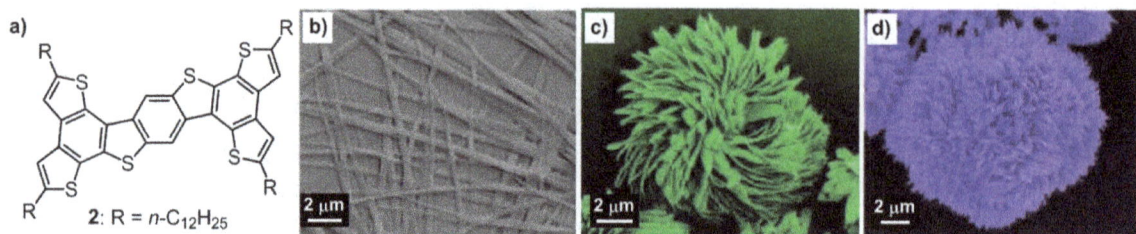

Figure 2. (a) Chemical structure of oligoarene derivative **2** containing linear alkyl chains. False-color SEM images of different nanostructures of **2**: microbelts self-assembled from 1,4-dioxane (b), flower-A formed from THF (c) and flower-B assembled from *n*-decane (d). Reprinted with permission from L Wang *et al* 2009 *Langmuir* **25** 1306, © 2009 American Chemical Society.

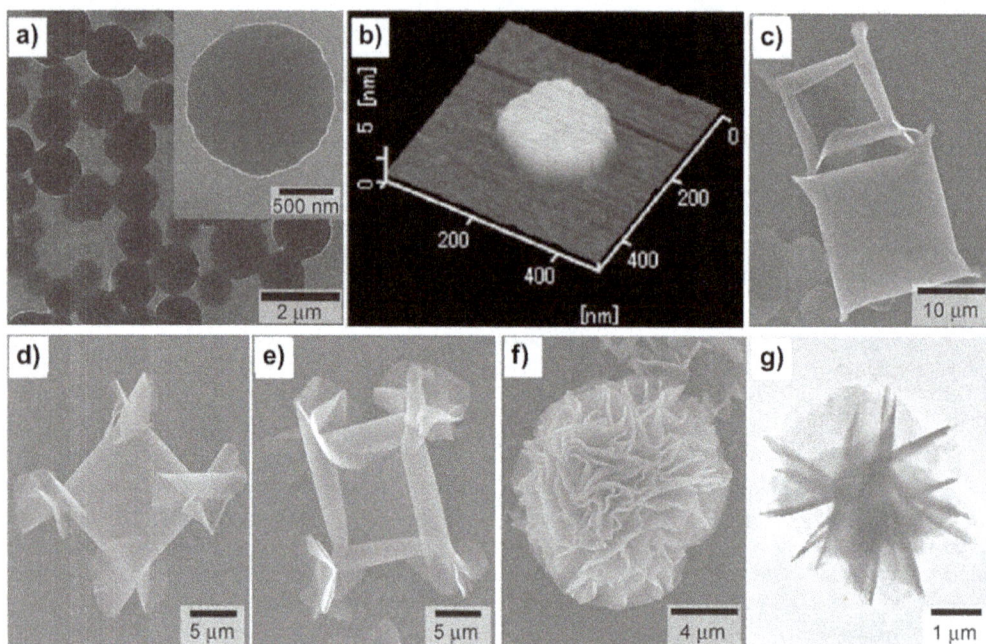

Figure 3. SEM (a) and atomic force microscopy (AFM) (b) images of nanodisks formed by cooling a 1,4-dioxane solution of **1a** from 60 °C to 20 °C. SEM (c) image of square-shaped objects loosely rolled up in each corner formed by rapid cooling of 1,4-dioxane solution of **1a** from 60 °C to 5 °C. SEM images (d), (e) of the further rolled up objects of crumpled structures at the four corners. SEM (f) and TEM (g) images of final flower-shaped assemblies of **1a** precipitated by slow aging at 5 °C. Reprinted with permission from T Nakanishi *et al* 2007 *Small* **3** 2019, © 2007 John Wiley & Sons.

or flake-like nanostructures several tens of nanometers in thickness (figures 3(f)–(g)) [29].

Such temperature-dependent morphologies have significantly promoted the understanding of the formation mechanism of the very complex flower-shaped objects. Given that bending of a thin sheet is entropically more favorable than the stretched state, the pre-formed disk objects at 20 °C have a strong tendency to roll up at the edges (figure 3(c)), which are attainable upon rapid cooling of the solution from 60 °C to 5 °C. The continual rolling up would result in spatial congestions at the four corners, leading to crumpling, bending, stretching and fracture of the disks (figures 3(d)–(e)). Once these conformations complete, the bilayer at the edges keep on growing to fix the spatial conformation of the crumpled sheets, resulting in the flower-shaped superstructures (figures 3(f)–(g)). According to this process, the slow temperature aging is indispensable for the growth from small nanodisks to microscopic flower-shaped objects. This result reveals the considerable influence of temperature on the self-organization process.

Similar self-assembly phenomenon using a heating/cooling process in 1,4-dioxane was observed for an alkylated C_{60} derivative, **3a**, bearing 3,4,5-triseicosyloxyl chains (figure 4(a)), which also formed globular objects with wrinkled nanoflake structures at the outer surface (figure 4(b)) [30]. The extraordinarily high roughness, together with the hydrophobic properties of both the C_{60} and alkyl chains, made the fabricated thin film of these globular microparticles exhibit superhydrophobicity with a water contact angle of 152° (figure 4(b), inset). The surface morphology and

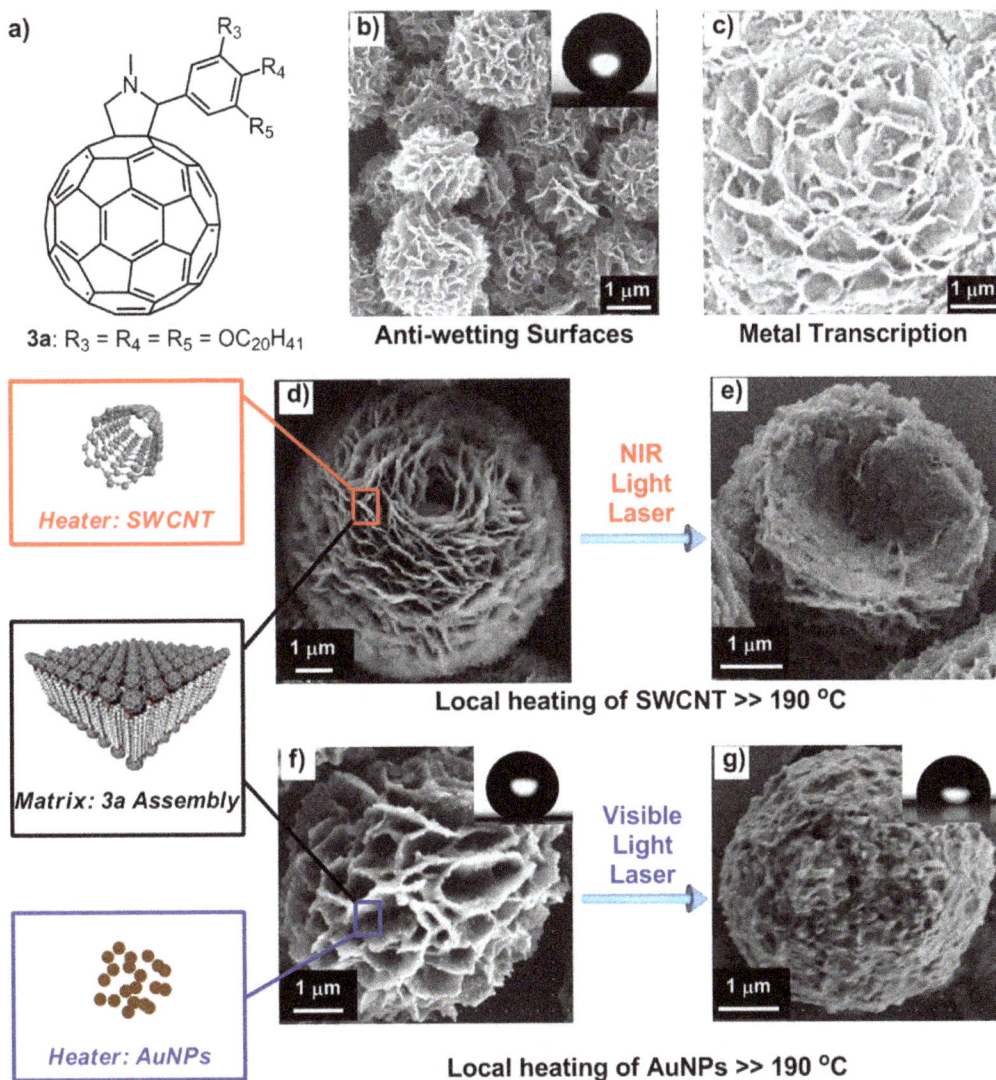

Figure 4. (a) Chemical structure of an alkylated-C$_{60}$ **3a**. (b) SEM image of globular microparticles with nanoflaked outer surfaces formed by cooling a 1,4-dioxane solution of **3a** from 70 °C to 20 °C and a photograph (inset) of a water droplet on the surface of a thin film of the globular micro-objects on a Si substrate. Reprinted with permission from T Nakanishi *et al* 2008 *Adv. Mater.* **20** 443, © 2008 John Wiley & Sons. (c) SEM image of Au nanoflakes transcribed from the nanoflake-featured microparticles of **3a**. Reprinted with permission from Y Shen *et al* 2009 *Chem. Eur. J.* **15** 2763, © 2009 John Wiley & Sons. Morphology of **3a**-SWCNT assembly before (d) and after (e) NIR light laser irradiation ($\lambda = 830$ nm). Reprinted with permission from Y Shen *et al* 2010 *J. Am. Chem. Soc.* **132** 8566, © 2010 American Chemical Society. Morphology and anti-wettability of **3a**-AuNPs before (f) and after (g) visible light laser irradiation ($\lambda = 532$ nm). Reprinted with permission from H Asanuma *et al* 2013 *Langmuir* **129** 7464, © 2013 American Chemical Society.

superhydrophobicity of the thin film is reminiscent of the self-cleaning features of the Lotus leaves.

In addition, the microparticles can also be employed as a template for nanoflaked metal surfaces by simply sputtering the desired metal directly onto a thin film of the globular objects and rinsing out of **3a** in a good solvent such as chloroform (CHCl$_3$) (figure 4(c)) [31]. The resulting Au nanoflake surfaces, retaining the high roughness features, are able to fabricate both superhydrophobic and superhydrophilic surfaces through chemical modification of hydrophobic and

hydrophilic thiol molecules. The Au nanoflake surfaces can also be applied as a surface-enhanced Raman scattering (SERS) active substrate owing to the plasmonic effect of nanostructured metal [32].

By virtue of the highly photothermally active single-walled carbon nanotubes (SWCNTs) [33, 34] and gold nanoparticles (AuNPs) [35], the nanoflake-featured micro-particles of **3a** doped with SWCNTs or AuNPs were either employed as a temperature indicator in air [36] or applied to modulate surface anti-wetting characteristics [37].

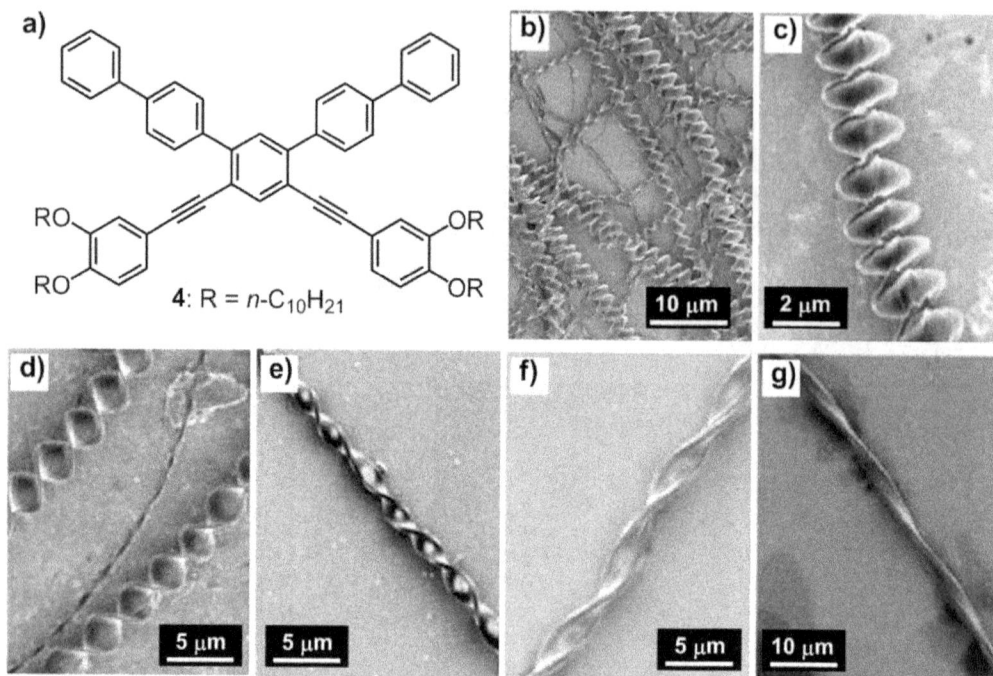

Figure 5. (a) Chemical structure of molecule **4**. (b) SEM image of self-assembled microwires from a solution of **4** in a 2:3 CHCl$_3$/ethanol (EtOH) mixture by evaporating the solvents slowly. SEM images of self-assembled **4** precipitated by cooling of its hot homogeneous solution at different temperatures: 15 °C (c), 20 °C (d), 25 °C (e), 30 °C (f) and 35 °C (g). Reprinted with permission from H-B Chen *et al* 2009 *Langmuir* **25** 5459, © 2009 American Chemical Society.

Specifically, NIR laser-induced heating of SWCNTs could generate increased temperatures, which, once reaching the melting point of the **3a**-SWCNTs assembly (190 °C), would induce morphology changes observable by various microscopy techniques (figures 4(d)–(e)). Similarly, the surface roughness of the fabricated thin films prepared with **3a**-SWCNTs or **3a**-AuNPs could be remotely controlled by NIR light laser ($\lambda = 830$ nm) or visible light laser ($\lambda = 532$ nm) irradiation, exerting significant influence on the surface-roughness-dependent anti-wetting properties (figures 4(f)–(g)).

With the same temperature-regulating self-assembly strategy, the Pei group also obtained 3D flower-shaped micro-objects using a benzothiophene derivative appended with *n*-dodecyl chains [38]. In addition, the same group reported the morphology tuning of chiral microtwists through temperature control [39]. By slowly evaporating a solution of an achiral compound **4** (figure 5(a)) in a 2:3 CHCl$_3$/ethanol (EtOH) mixture (1 mg mL^{-1}), perfectly twisted chiral structures with uniform pitch were obtained (figure 5(b)). Interestingly, with different precipitation temperatures, the pitch of the micro-twist could be easily tuned (figures 5(c)–(g)). Basically, a higher temperature led to a slower precipitation process and thus a larger pitch. This phenomenon was explained by special crystal growth kinetics, according to which the driving force for twisting derived from the imbalance of the growth

rate between the center and the edge of the self-assembled nanobelts.

2.3. Chain-substitution pattern-controlled architectures

In addition to the solvent effect on the self-assembly of **1a**–**1c**, which was briefly described in figure 1, the chain sub-stitution pattern also had significant influence on the control of organized architectures. Through the same self-assembly procedures and same temperature history within the same solvent, alkylated-C$_{60}$ derivatives, **3a**–**3c**, bearing different eicosyloxy chain numbers (figures 4(a), 6(a)), formed differ-ent self-assembled structures [40]. Specifically, compound **3a** appended with 3,4,5-triseicosyloxyl chains formed globular microparticles with a nanoflaked outer surface (figure 4(b)). While **3b** bearing 3,4-biseicosyloxyl chains generated plate-rich giant particles (figures 6(b)–(c)). The 4-monoeicosyloxyl chain-substituted **3c**, on the other hand, gave rise to sheet structures (figure 6(d)). The distinctive morphologies can be attributed to the competing π–π interactions of neighboring C$_{60}$ moieties with the vdW interactions of the alkyl chains. With fewer alkyl chains, the π–π interaction of C$_{60}$ moeties is richer and therefore induces plate-rich architectures due to the lower flexibility of the molecular organizations. On the other hand, with more alkyl chains, the π–π interaction of C$_{60}$ moieties is constrained, resulting in a suppressed planar

Figure 6. (a) Chemical structure of C_{60} derivatives **3b–3c**. SEM images of the assemblies of **3b–3c** formed by cooling their 1,4-dioxane solution from 70 °C to 20 °C (b); (c) Plate-Rich giant particles of **3b**; (d) sheet structures of **3c**. Reprinted from [40].

3b: $R_3 = R_4 = OC_{20}H_{41}$, $R_5 = H$
3c: $R_3 = R_5 = H$, $R_4 = OC_{20}H_{41}$

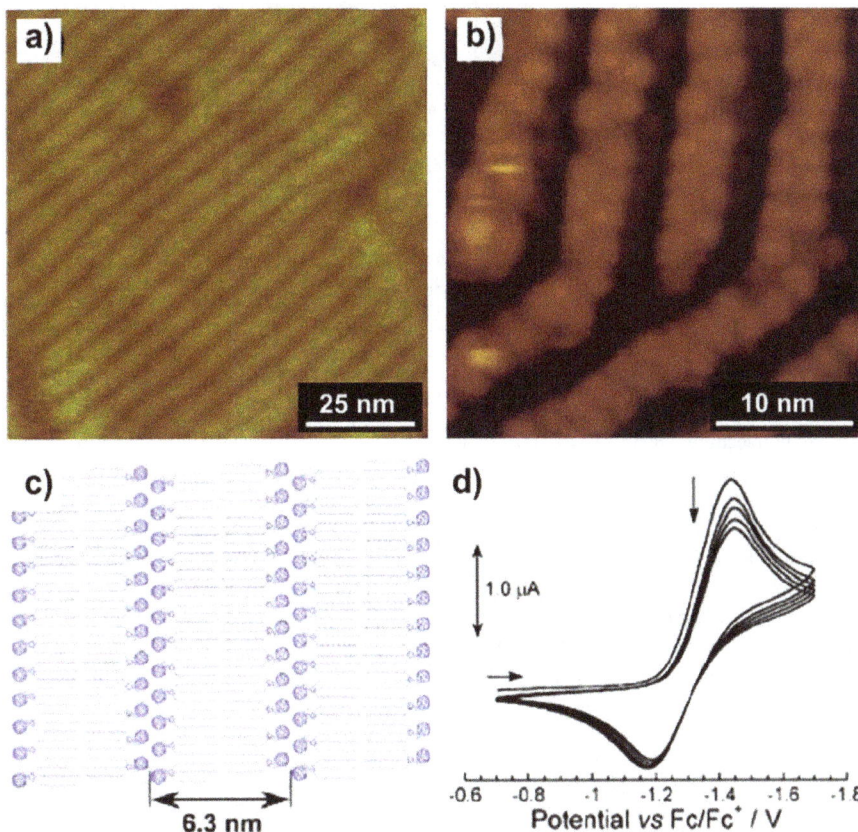

Figure 7. AFM (a) and high-resolution scanning tunneling microscopy (STM) (b) images of **1a** on HOPG spin-coated from a $CHCl_3$ solution. (c) Schematic illustration showing the molecular organization of **1a** in the lamellae. (d) Cyclic voltammogram of **3a** on HOPG (0.1 M tetra-n-butylammonium perchlorate, acetonitrile (CH_3CN), Ar atmosphere, scan rate 0.1 V s^{-1}, 20 °C). Reprinted with permission from T Nakanishi *et al* 2006 *J. Am. Chem. Soc.* **128** 6328, © 2006 American Chemical Society.

arrangement of C_{60} moeties and thus favoring globular objects with wrinkled nanoflaked outer surfaces.

2.4. Substrate-supported self-assemblies

The interaction between the appended alkyl chains of alkylated-π molecules and solid substrates, highly oriented pyrolytic graphite (HOPG) in particular, was reported to be able to drive the corresponding π molecules to form lamellae and other complicated ordered structures and therefore organize into epitaxially ordered molecular patterns [41, 42]. This technique, although widely utilized for the establishment of 2D alignments [43, 44], was seldom employed for the construction of 1D architectures [45]. Taking into consideration the significance of

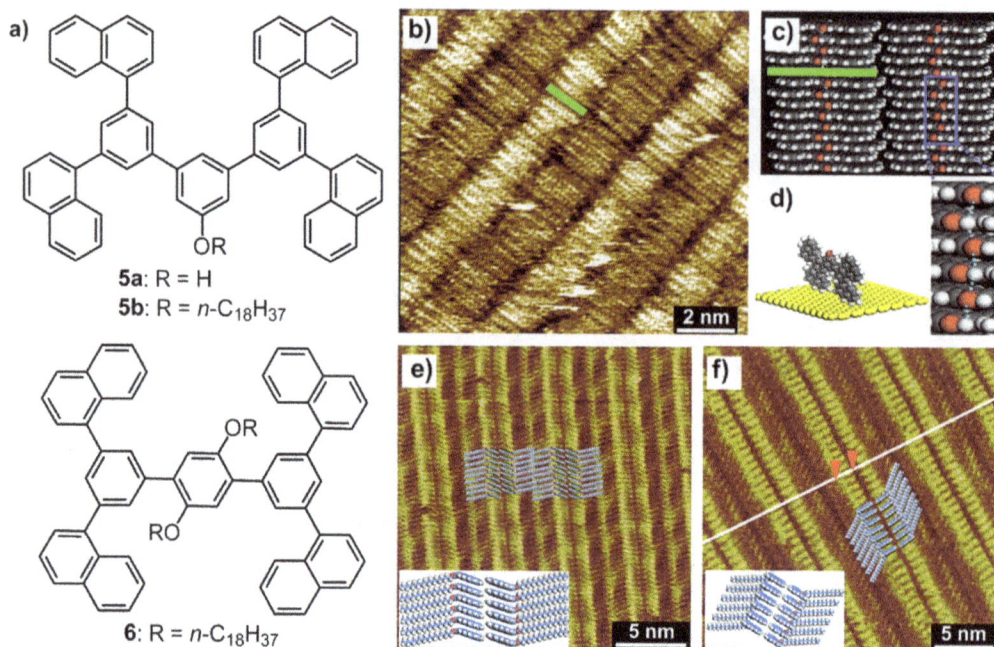

Figure 8. (a) Chemical structures of dendronized conjugated molecules **5** and **6**. (b) High-resolution STM image of the **5a** monolayer on the HOPG surface. (c) Schematic model of the **5a** monolayer adopting an edge-on stacking pattern, and an enlarged structural model showing hydrogen bonds by blue dashed lines. (d) Side-view model of an individual **5a** molecule on the HOPG surface. (e) High-resolution STM image of **5b** on the HOPG surface; inset, proposed structural model of the **5b** adlayer. (f) High-resolution STM image of the self-assembly of **6** on the HOPG surface; inset, a possible packing pattern of the molecular lamellar structure of **6**. Reprinted from Y Yang *et al* 2012 *Appl. Surf. Sci.* **263** 73, © 2012 with permission from Elsevier.

the 1D C_{60} structure in electronic nanodevices, our group applied the technique to fabricate perfectly aligned 1D C_{60} nanowires by spin-coating a dilute $CHCl_3$ solution of compound **1a** onto HOPG surfaces (figure 7(a)) [46]. The nanowires, with the C_{60} heads locating at the center in a zigzag fashion and the substituted alkyl chains stretching outward (figure 7(b)), possess lengths exceeding several hundred of nanometers. The periodicity of the nanostripes corresponds well to twice the molecular length of **1a**, revealing a perfect lamellar structure with fully extended alkyl chains of all-trans conformation (figure 7(c)).

The perfect alignment is mainly driven by the good lattice matching between the all-trans conformational alkyl chains and graphite, which force the alkyl chains to assemble along the underlying lattice axis of the basal plane of graphite. Meanwhile, the π–π interactions allow the C_{60} units to form in a zigzag fashion on the surface. This hypothesis was further supported by the similar alignment behavior of **1b** and **3a** [47].

Notably, even in the surface-confined assemblies, the C_{60} groups, i.e. **3a**, showed fully maintained electrochemical activity (figure 7(d)), suggesting these molecules as promising candidates for electronic nanodevices. Moreover, such 1D nanowires could facilely regulate the carrier transporting direction, which could be of prominent advantage for applications in semiconductors [48–51].

In addition, the aligning strategy on the substrate surface constructed here has been generalized to other functional molecules. The Miao group have investigated a series of

dendronized molecules **5a**, **5b** and **6** (figure 8(a)), which formed 2D self-organized monolayers on the HOPG surface through solution evaporation of the molecules under ambient conditions [52]. As a result of the π–π interactions, the dendronized conjugated moieties of all the compounds adopted an edge-on arrangement on the HOPG surface. Substituted by a hydroxyl group, molecule **5a** stood perpendicularly to the substrate surface, resulting from both intermolecular π–π interactions and hydrogen bonding (figures 8(b)–(d)). Molecule **5b**, however, attached by one long alkyl chain, was subjected not only to the π–π interactions but also to the vdW interactions of alkyl chains, as well as the interactions between the alkyl chains and the HOPG substrate. As a consequence, the alkyl chains laid flat on the substrate, while the conjugated units stood perpendicular to the HOPG surface (figure 8(e)). Compound **6** exhibited a similar 2D adsorbed structure to that of **5b**, except for the slightly larger intermolecular spacing, which was due to the tilted conjugated units from the HOPG surface (figure 8(f)). In addition, both **5b** and **6** displayed zigzag carbon skeletons of the alkane molecules relative to the HOPG substrate, which was due to a subtle interplay of packing and entropic effects [53].

Apart from this example, Chen *et al* have also reported a 2D self-assembly, which formed on the HOPG substrate with a long alkyl chain-substituted oligo(*p*-phenylene vinylene) (OPV) derivative [54]. Mali *et al* described a concentration-controlled 2D structural evolution of a large triangular discotic macrocycle containing an alkyl chain at the 1,2,4-

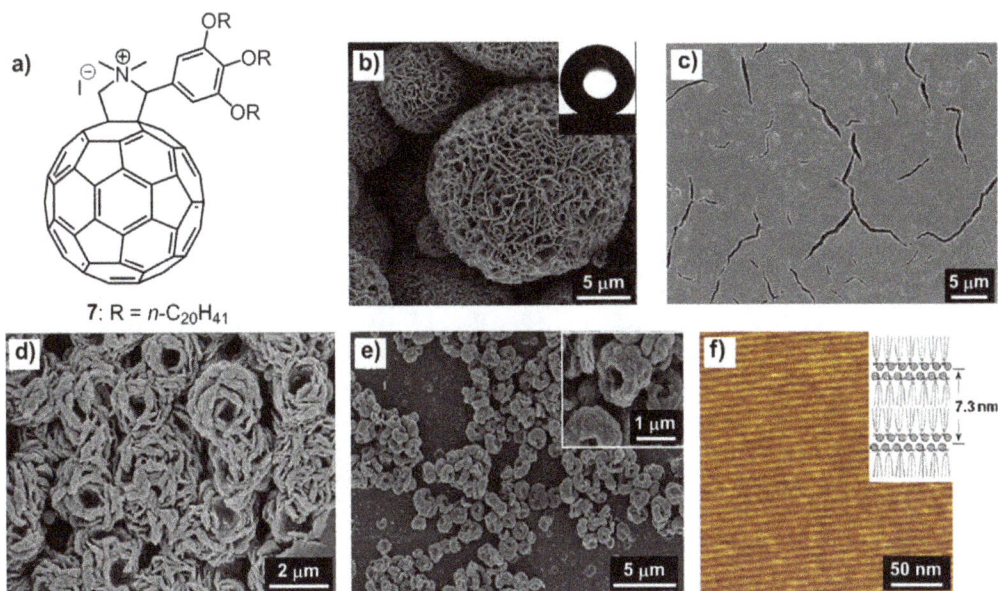

Figure 9. (a) Chemical structure of an ionic C_{60} derivative **7**. (b) SEM image of flake-like microparticles of **7** precipitated by slowly adding excess methanol (MeOH) to a concentrated CH_2Cl_2 solution of **7**; inset, a water droplet on the surface of a thin film made from the microparticles of **7**. SEM images of the self-organized structures of **7** formed on Si substrates by evaporating a 0.5 mM solution in a MeOH/CH_2Cl_2 mixed solvent with a MeOH volume content of 0 (c), 10% (d) and 30% (e). (f) AFM image of a film obtained by spin-coating a 10 μM CH_2Cl_2 solution of **7** on HOPG; inset, a schematic of lamellar form of **7** on HOPG. Reprinted with permission from H Li *et al* 2011 *Langmuir* **27** 7493, © 2011 American Chemical Society.

trichlorobenzene/HOPG interface [55]. Recently, Xu *et al* [56] and [57] Tamaki *et al* independently reported the odd–even (alkyl chain carbon number) effect on the self-assembly structures of fluorenone and anthraquinone derivatives, respectively, on HOPG. More strikingly, a 3D nanowire based 'organic radical' unit on the HOPG surface, formed from a polychlorotriphenyl radical bearing three long alkyl chains, was also reported [58].

2.5. Other interaction-assisted self-assemblies

With proper molecular design, other interactions, such as electrostatic, hydrogen bonding or hydrophilic interaction, can be introduced to alkylated-π molecules. Any of these interactions, interplaying with the π–π interaction of the π moieties and the vdW interaction of the alkyl chains, would develop versatile self-assembled structures with extensive functions.

2.5.1. Electrostatic interaction.
Our group has designed an ionic alkylated C_{60} derivative, **7** (figure 9(a)), which exhibited multiple morphologies with different processing methods [59]. Through liquid-liquid interfacial precipitation with the addition of a poor solvent, methanol (MeOH), on the top of a concentrated dichloromethane (CH_2Cl_2) solution of **7**, self-organized flake-like microparticles with high roughness were produced (figure 9(b)). However, through drop-casting of a stock solution of **7** with CH_2Cl_2, CH_2Cl_2/MeOH = 9:1 and CH_2Cl_2/MeOH < 7:3 as solvents on a Si substrate, film with some cracks (figure 9(c)), closely packed flower-like objects (figure 9(d)) and doughnut-shaped micro-objects (figure 9(e))

with rough surfaces were generated, respectively. On the other hand, the diluted CH_2Cl_2 solution of **7** (10 μM), once spin-coated on HOPG, could form perfectly straight C_{60} nanowires in which the length of nanowires exceeded 1 μm (figure 9(f)).

Such polymorphic phenomenon was benefited by the introduction of an ionic unit (pyrrolidinium iodide) on the molecule, which induced electrostatic interactions to the assemblies. Consequently, the multiple π–π, vdW and electrostatic interactions drove the formation of a variety of polymorphic self-assembled structures either from the solution or on substrates. Even with an additional ionic part, a thin film of the flake-like microparticles exhibited high water repellency with a static water contact angle of $140 \pm 3°$ (figure 9(b) inset), which was viable for the development of anti-wetting materials. In addition, the nanowires formed here, stabilized by the formation of salt bridges based on the electrostatic interactions, were the longest class of 1D C_{60} self-assemblies (>1 μm), which could be promising structures toward electronic device applications.

2.5.2. Hydrogen bonding.
The group of Yagai synthesized a series of alkylated π molecules attached by hydrogen bonding units, which gave rise to numerous functional optoelectronic materials [60–62]. For example, consider an oligo(*p*-phenylene vinylene)- (OPV) functionalized bisurea, **8** (figure 10(a)) [60], as a consequence of the cooperative π–π interactions of OPV units, vdW interactions of the alkyl chains and hydrogen bonding from the urea units, the molecule possessed a very high supramolecular

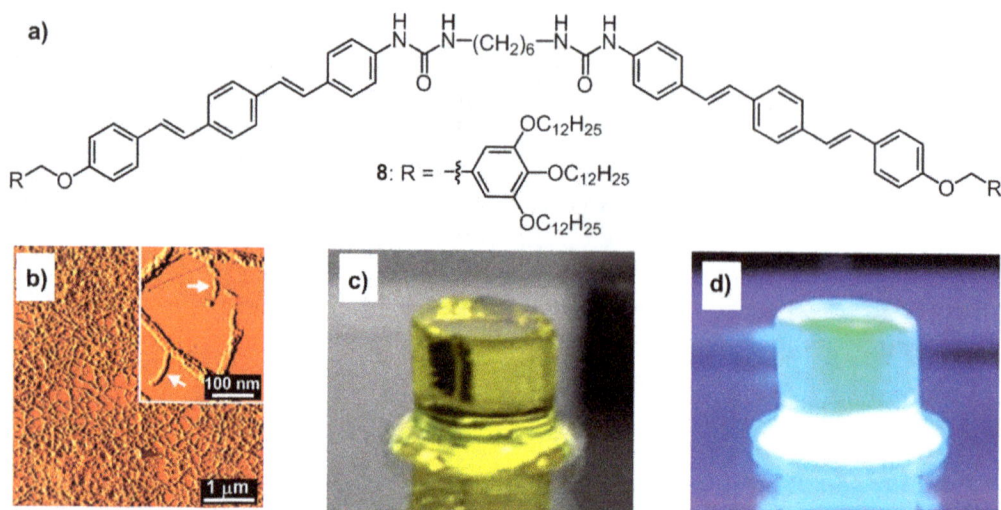

Figure 10. (a) Chemical structure of an OPV-functionalized bisurea **8**. (b) AFM phase images of **8** spin-coated on HOPG from a methylcyclohexane dispersion; inset shows a high-resolution image. Photographs of the self-supporting decane gel of **8** deposited on the glass substrate taken under visible light (c) and 365 nm UV light (d). Reprinted with permission from S Yagai *et al* 2008 *Chem. Eur. J.* **14** 5246, © 2008 John Wiley & Sons.

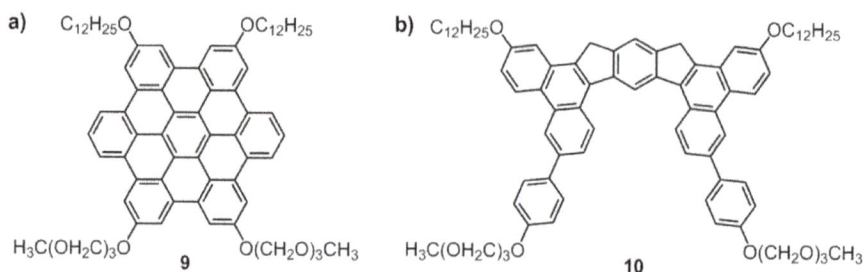

Figure 11. Chemical structures of a HBC derivative **9** (a) and a benzodithiophenen derivative **10** (b), both bearing hydrophobic *n*-dodecyl and hydrophilic TEG chains.

polymerization ability and formed noncovalent polymers with intertwined fibrous structures by simply spin-coating it in a methylcyclohexane solution on a HOPG substrate (figure 10(b)). To compare, in various organic solvents, such as *n*-decane, CH_2Cl_2 and THF, compound **8** showed a tendency to gelate, giving rise to fluorescent gels (figures 10(c)–(d)). Such a strong aggregation ability of **8** in a solution is significant for its applications as solution-incorporated soft materials.

2.5.3. Hydrophilic interaction. The Aida group reported a number of hexa-*peri*-hexabenzocoronene (HBC) derivatives substituted by both hydrophobic alkyl chains and hydrophilic triethylene glycol (TEG) chains [63, 64]. These molecules, take **9** (figure 11(a)) as an example, self-assembled into well-defined 1D nanotubes stabilized by the π–π interactions of the HBC moieties, vdW interactions of the alkyl chains and hydrophilic interactions governed by the TEG chains. The nanotube was shown to be redox active and had an electrical

conductivity of 2.5 MΩ upon oxidation, which was comparable to that of an inorganic semiconductor nanotube based on gallium nitride [65].

With the same designing strategy, the Pei group reported a butterfly-shaped benzodithiophenen derivative **10** (figure 10(b)) which self-organized into free-standing bilayer films in a solution [66]. The films, bearing both hydrophobic alkyl and hydrophilic TEG chains, could be facilely transferred onto a substrate for direct device fabrication and were employed as the active layer of organic field-effect transistors (OFETs), with the highest mobility being 0.02 cm^2 V^{-1} s^{-1}. Such substrate-independent free-standing films would be promising for producing electro-active materials with large-area deposition and solution processability.

Within solvent systems, the intrinsic balance of π–π and vdW interactions and the influence of solvent polarity and temperature, as well as the chain-substrate interaction and other interactions induced by further molecular modification, could render the linear alkyl chain-substituted π-molecules

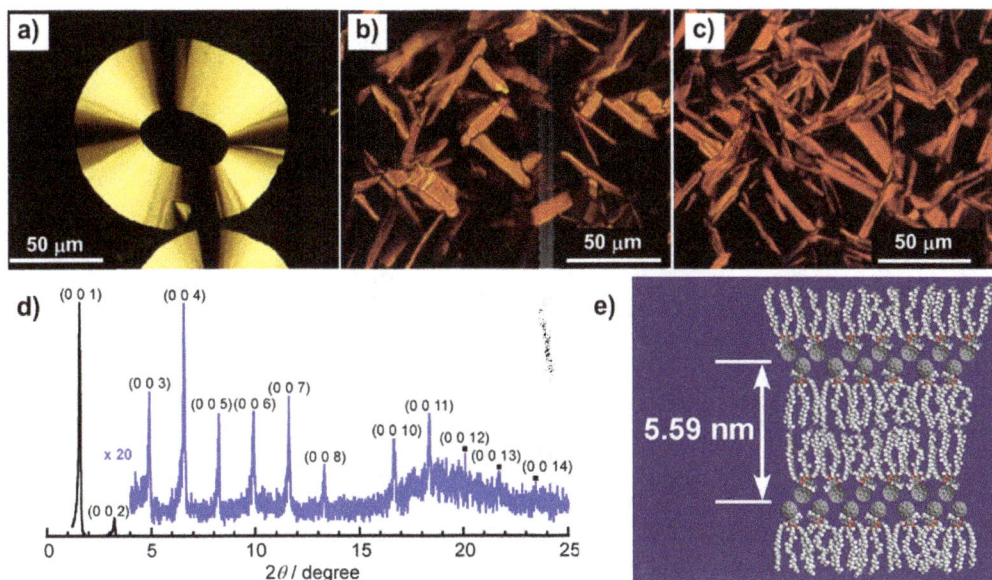

Figure 12. Polarized optical microscopy images of **3a** at 190 °C (a), **1a** at 202 °C (b) and **3b** at 200 °C (c) upon cooling from the isotropic state. (d) XRD patterns of **3a** at 185 °C. Reprinted with permission from T Nakanishi *et al* 2008 *J. Am. Chem. Soc.* **130** 9236, © 2008 American Chemical Society. (e) Proposed lamellar organization of **3a** (redrawn from [70]).

assembled into diverse 1D, 2D and 3D structures. These nano/micro architectures have turned out to be widely applicable to fabricate anti-wetting surfaces, structural templates and various optoelectronics.

3. Liquid crystals of alkyl chain-attached π molecules

Taking advantage of their efficient carrier injection from the electrode and their carrier transporting ability, as well as their softness of the state, liquid crystals (LCs) containing π-conjugated units are of particular interest for flexible optoelectronic applications [67].

3.1. Linear alkyl chain-substitution-induced thermotropic liquid crystals

In spite of the promising semiconducting properties of C_{60}, C_{60}-containing thermotropic LCs are seldom reported, not to mention their low carrier mobility due to their moderately ordered structure and low content of C_{60} in the mesophase [68]. In view of the high C_{60} content of our alkylated-C_{60}, based on the compact molecular design, the thermal properties of the above-mentioned C_{60} derivatives were investigated. **1a**, **3a** and **3b** exhibited thermotropic polymorphism and showed unprecedentedly advantageous carrier mobility in their thermotropic mesophase [69]. For all three compounds, two endothermic peaks corresponding to crystalline-to-mesomorphic and mesomorphic-to-isotropic phase transitions were observed from differential scanning calorimetry (DSC) analysis upon heating of each sample. The thermotropic

mesophase of **3a** formed in a temperature range from 62 °C to 193 °C, within which an optical texture exhibiting birefringence and fluid nature was observed under polarized optical microscopy (POM) (figure 12(a)). **1a** and **3b** showed similar LC characteristics in temperature ranges from 33 °C to 223 °C (figure 12(b)) and from 44 °C to 226 °C (figure 12(c)), respectively. Significantly, the mesophase of **3a** was able to retain the electrochemical and photoconductive properties of pristine C_{60}, featuring both reversible electrochemistry ($E_{red,1} = -0.70$ V and $E_{red,2} = -0.87$ V versus Ag/AgCl) above the crystalline-to-liquid-crystalline transition temperature (62 °C) and relatively high electron mobility of $\sim 3 \times 10^{-3}$ cm^2 V^{-1} s^{-1} at 120 °C, evaluated by a conventional time-of-flight (TOF) set-up.

The high photoconductive properties of these liquid crystalline materials are attributed not only to the high content of C_{60} (up to around 50%) but also to a suitable and dense C_{60} arrangement. As mentioned above, the C_{60} moieties and the alkyl chains act as amphiphiles in an organic solvent. The high immiscibility of the two components could significantly facilitate the segregation of C_{60} microphases into layers even under solvent-free conditions, which would naturally promote the formation of desirable molecular building blocks for liquid-crystalline organization. The densely packed C_{60} structures were confirmed by x-ray diffraction (XRD) analysis (figure 12(d)). Take **3a** as an example: at 185 °C, its XRD pattern shows a strong peak at $2\theta = 1.58°$ assigned to (0 0 1), d spacing = 5.59 nm, with a number of high-order diffraction peaks up to (0 0 14), illustrating a long-range ordered lamellar mesophase with interdigitation of only C_{60} units (figure 12(e)) [70].

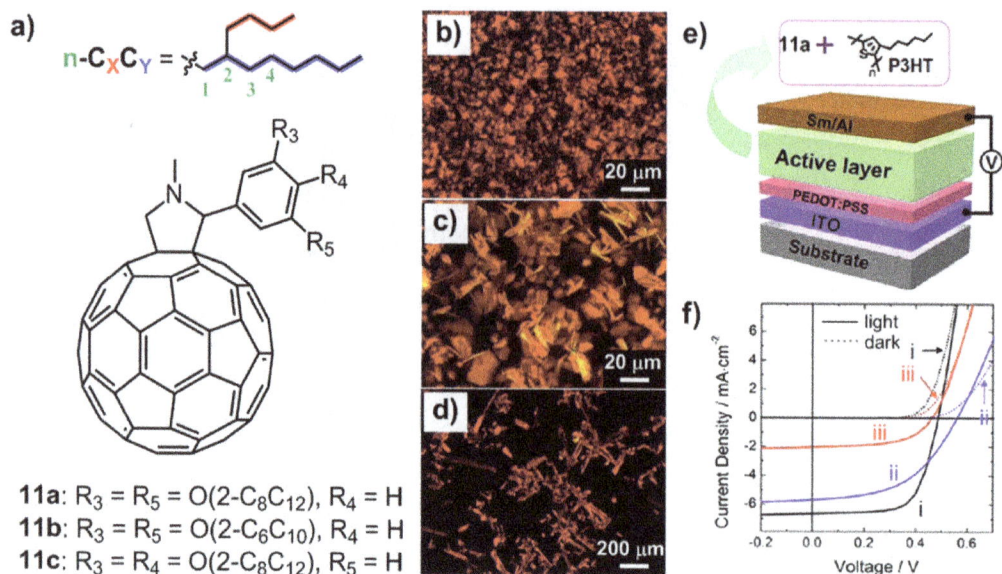

Figure 13. (a) Chemical structures of C_{60} derivatives **11a-11c** containing branched alkyl chains. POM images of **11a** (b), **11b** (c) and **11c** (d) taken in their mesophases. (e) Schematic of the BHJ cell. (f) J(V) curves of binary mixtures of PCBM/P3HT (curve i), **11a**/P3HT (curve ii) and **3a**/P3HT (curve iii), respectively. Reprinted from [71].

With satisfying optoelectronic properties guaranteed by a highly ordered structure and high content of C_{60} in the mesophase, such C_{60}-containing LCs can be employed as promising soft materials for photovoltaic applications.

3.2. Branched alkyl chain-substitution-induced thermotropic liquid crystals

Some branched alkyl chain-substituted π molecules were also found to form thermotropic liquid crystals and showed more promising properties for optoelectronic applications than those of linear ones.

Figure 13(a) shows three C_{60} derivatives **11a-11c** substituted by branched alkyl chains, all of which exhibited a thermotropic smectic phase observed by POM (figures 13(b)–(d)) [71]. Compound **11a** with branched 2-octyldodecyl (2-C_8C_{12}) chains had a mesophase to isotropic phase transition at 84 °C, which was much lower than that of compound **3a** (193 °C) attached by linear eicosyl (n-$C_{20}H_{41}$) chains, indicating the better softening effect and lower crystalline tendency of branched chains. With shorter branched chains, compound **11b** (2-C_6C_{10}) exhibited a mesophase-to-isotropic phase transition at 148 °C, which is 64 °C higher than that of **11a**. In sharp contrast, with linear chains, molecule **1a** (n-$C_{16}H_{33}$) only showed a 30 °C increase of such a phase transition compared with **3a**, proving the stronger ability of branched chains to regulate the thermotropic behavior of alkylated π molecules. Moreover, by simply changing the substitution position of **11a** from (3-, 5-) positions to (3-, 4-) positions, the resulting compound **11c** showed a drastic increase in the phase transition temperature (196.2 °C), which was attributed to the greater vdW interaction caused by

densely packed alkyl chains. As a result, chain length, substitution position and branched extent have all played significant roles in the thermotropic behavior of the alkylated C_{60} derivatives.

With the liquid-crystalline phase extended from 84 °C to room temperature, compound **11a** was selected to blend with poly(3-hexylthiophene) (P3HT) for the fabrication of bulk heterojunction (BHJ) organic solar cells (figure 13(e)). Notably, **11a**/P3HT exhibited a power conversion efficiency (PCE) of ~1.6 (figure 13(f), curve ii), which was comparable to that of [6,6]-phenyl-C_{61}-butyric acid methyl ester (PCBM)/ P3HT (figure 13(f), curve i) in our cell set-up. On the contrary, **3a**/P3HT showed a much lower PCE of ~0.5 (figure 13(f), curve iii). The dramatically enhanced PCE value of **11a**/P3HT compared with that of **3a**/P3HT was ascribed to the lowering of crystallinity and facilitating of charge transportation caused by the branched alkyl chains. In other words, too high of a crystallinity in the solar cell caused defects and reduced the performance. Therefore, ordering whilst retaining certain softness (flexibility) would play a very important role in such optoelectronic device applications.

Discotic thermotropic LCs are well known for their extensive applications in various organic devices, such as OFET [72–74] and photovoltaic cells [75], depending upon their orientation on the substrate. All these applications are derived from their fabulous ability to conduct charges on the basis of their unique molecular self-organization behaviors. In discotic LCs, the molecules stack on top of one another into columns and consequently resulted in a regular lattice, which greatly facilitates the charge carrier mobility along the 1D assembly [76–78]. A promising candidate for developing discotic LCs is HBC, which has a large aromatic core that

a)

12a: R = 2-C_2C_6
12b: R = 2-C_6C_{10}

b)

c)

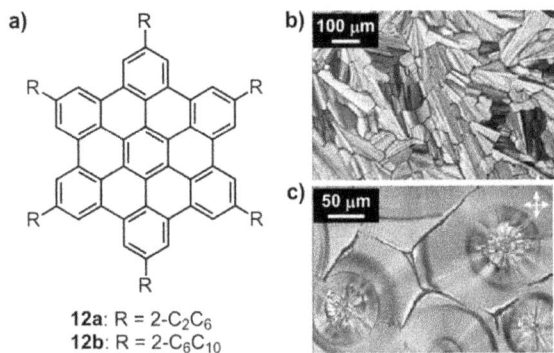

Figure 14. (a) Chemical structures of HBC derivatives **12a–12b** containing branched alkyl chains. POM images of **12a** (b) and **12b** (c) obtained after cooling from the isotropic state. Reprinted with permission from M Kastler *et al* 2006 *Adv. Mater.* **18** 2255, © 2006 John Wiley & Sons .

allows one of the highest values of intrinsic charge carrier mobility for mesogenes to be obtained [79–81].

Müllen and his coworkers have reported a number of thermotropic LCs based on HBC derivatives substituted with both linear [73, 82] and branched alkyl chains [83, 84] (figure 14). All the linear alkyl chain-substituted HBC derivatives, together with the short branched alkyl chain-equipped HBC compound **12a**, exhibited a crystalline-to-LC phase transition around 100 °C and a LC-to-isotropic phase transition around 420 °C. In dramatic contrast, the long-branched alkyl chain-modified HBC derivative, **12b**, displayed much lower phase transitions with a crystalline-to-LC phase transition at 17 °C and LC-to-isotropic phase transition at 97 °C, which further confirmed the more profound softening effect of branched chains and their stronger ability to regulate and adjust the thermotropic behavior. At room temperature, the charge carrier mobility for **12b** was determined to be 0.73 $cm^2 V^{-1} s^{-1}$, representing the highest value measured for a noncrystalline discotic LC material.

The relative softness and highly ordered phases make these alkylated-π-molecule-constituting LCs promising materials for fabrication into flexible optoelectronic devices. The thermotropic behaviors of these LC systems depend not only on the alkyl chain length and substitution positions but also on the branching. As illustrated, branched alkyl chains possess a better softening effect as well as more pronounced influence on the self-organization and carrier mobility of these alkylated-π LC systems.

4. Solvent-free liquid of alkyl chain-attached π molecules

The above-described thermotropic liquid crystals were derived from linear alkyl chain-substituted C_{60} derivatives **1**, **4** and **5** where the alkyl chains resided on the phenyl unit at (3,4,5-) or (3,4-) positions. In contrast to these LCs, nonvolatile, solvent-free room-temperature liquid C_{60} derivatives were discovered serendipitously by attaching the 2,4,6-tris

(alkyloxy)phenyl group to a C_{60} derivative where the alkyl chains were of a linear type [85]. This finding further confirmed the extraordinary impact of a chain substitution pattern on the intermolecular interactions, as depicted in part 2.3. These liquid C_{60}s, with solvent-free device processability, nonvolatility, tunable optoelectronic functions, high density of optically or electronically active π-conjugated moieties and the ability to blend with other organic or inorganic dopants, have attracted remarkable attention and were quickly extended to various functional alkylated-π systems with diverse π units substituted by both linear and branched alkyl chains.

4.1. Linear alkyl chain-substitution-induced solvent-free liquids

During the investigation of alkylated-C_{60} derivatives, our group found that compounds **14**, **15** and **17** substituted with the 2,4,6-tris(alkyloxy)phenyl group (figure 15(a)) exhibited a liquid phase at room temperature with melting points of 13.7, −36.5 and 4.5 °C, respectively (figure 15(b)) [85]. The formation of such a room-temperature liquid state could be attributed to the 2,4,6-substitution pattern, which disturbed the π–π interactions of the C_{60} units based on the independent spreading of the three chains acting as an effective steric stabilizer of individual C_{60} moieties.

Interestingly, the lower-viscous liquid C_{60}s were achieved through attachment of the longer alkyl chains due to the further weakening of C_{60}-C_{60} interactions. This viscosity trend was confirmed by a detailed study of the impact of the alkyl chain length on the rheological behavior of **14–18** [86]. Notably, in addition to viscosity, the melting points of these compounds were also strongly influenced by alkyl chain lengths (figure 15(c)). With short alkyl chains (**13**: n-C_8H_{17}), the dominant intermolecular π–π interactions of adjacent C_{60} units would lead to a melting point higher than room temperature (147–148 °C). However, with extremely long alkyl chains (**18**: n-$C_{22}H_{45}$), the vdW interactions of alkyl chains became predominant and led to increased melting point (21.5 °C) compared with **14–17**. As a consequence, only with medium alkyl chain lengths could the two interactions balance to generate a liquid phase with the melting point under room temperature (i.e. 25 °C).

Importantly, even in the liquid state, the redox properties and the high hole mobility (~3×10^{-2} $cm^2 V^{-1} s^{-1}$ at 20 °C measured by a TOF technique) of the C_{60} unit were retained, making these liquid molecules promising for developing electronic materials. Moreover, these highly fluidic liquids can act as a matrix for other optoelectronic-active dopants, resulting in various composite materials with multiple functions. For example, cadmium selenide (CdSe) nanocrystals (NCs) with tunable optical properties were successfully embedded in liquid **17** [87]. Rather than forming macroscopic aggregation, the NCs tended to organize in relatively short (~10 nm) serpentine structures (figure 15(d)), illustrating the high dispersity of CdSe NCs in liquid **17**. The favorable electronic band alignment between the NCs and the **17** matrix enabled inter-phase charge transfer, which induced a dramatic increase in the cathodic photocurrent under light illumination observed for a film of composite CdSe NCs/**17** compared to

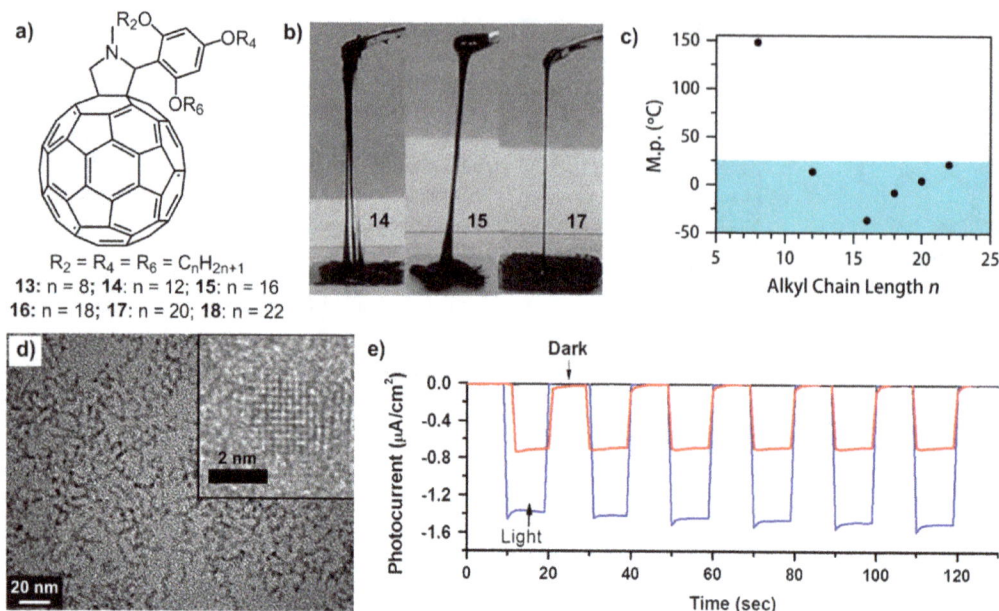

Figure 15. (a) Chemical structures of C_{60} derivatives **13–18** containing linear alkyl chains. (b) Photographs of room-temperature C_{60} liquids **14**, **15** and **17**. Reprinted with permission from T Michinobu *et al* 2006 *J. Am. Chem. Soc.* **128** 10384, © 2006 American Chemical Society. (c) Relationship between the melting point and alkyl chain length of C_{60} derivatives **13–18**, redrawn from [86]. (d) TEM image of the CdSe NCs/**17** composite; inset, high-resolution TEM image of a CdSe NC. (e) Photoelectrochemical activity of solvent-free CdSe NCs/**17** composite films on glass coated with fluorine-doped tin oxide containing **17** alone (black), 25 wt% CdSe NCs (blue) and CdSe NCs alone (red) under blue light (at 480 nm) illumination. Reprinted from [87].

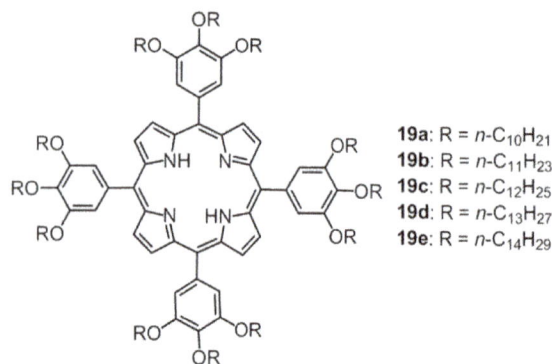

Figure 16. Chemical structures of liquid porphyrins **19a–19e** containing linear alkyl chains.

19a: R = n-$C_{10}H_{21}$
19b: R = n-$C_{11}H_{23}$
19c: R = n-$C_{12}H_{25}$
19d: R = n-$C_{13}H_{27}$
19e: R = n-$C_{14}H_{29}$

the film of pure **17** or pure CdSe NCs (figure 15(e)). This result made the composite material an excellent candidate for application in photo-sensitized photovoltaic devices.

The groups of Nowak-Król [88] and Maruyama [89] have independently reported a series of room-temperature liquid porphyrins, **19a–19e**, by introducing long linear alkyl chains to 5,10,15,20-tetraphenylporphyrin (figure 16). The thermal properties of the compounds were very sensitive to the alkyl chain lengths, with the melting point increasing proportionally to the increase of the chain length. However, a further increase of the chain length (n-$C_{15}H_{31}$ [89]) would

increase the melting point above room temperature, a behavior similar to that of liquid C_{60} derivatives [86].

Compound **19c** was employed as a suitable dispersion media for C_{60} and/or carbon nanotubes based on the intermolecular donor-acceptor and/or π–π interaction, which were confirmed by spectroscopic changes. Such dispersal may create new optical and electrical functionalities.

4.2. Branched alkyl chain-substitution-induced solvent-free liquids

As mentioned in part 3.2, branched alkyl chains possess lower crystalline tendency and a better softening ability as well as a more pronounced effect on reducing the molecular melting point than linear ones. Therefore, branched alkyl chains can be better candidates for constructing room-temperature liquid molecules containing a functional π unit. Accordingly, a series of room-temperature liquid materials by attaching branched alkyl chains to various π molecules have been reported by our group and other research groups.

Our group has synthesized a number of liquid C_{60} derivatives **20–22** attached by either swallow-tail branched alkyl chains (**20–21**) or hyperbranched alkyl chains (**22**) (figure 17) [71]. Compared with the linear alkyl chain-substituted C_{60} derivatives **14–18**, which required a 2,4,6-substitution pattern to reach a room-temperature liquid state, both **20** and **21** needed only two branched chains to generate room-temperature liquids. More significantly, **20** and **21** exhibited not only lower melting points (<−120 °C) but also less viscosity

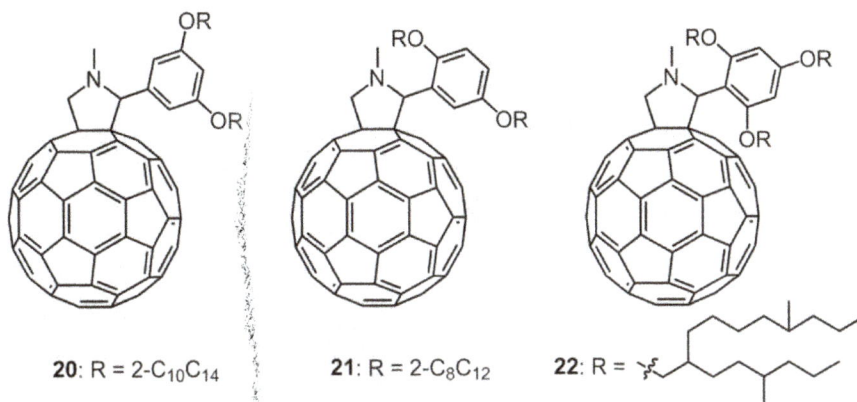

Figure 17. Chemical structures of C_{60} liquids **20–22** containing branched alkyl chains.

20: R = 2-$C_{10}C_{14}$ **21**: R = 2-C_8C_{12} **22**: R =

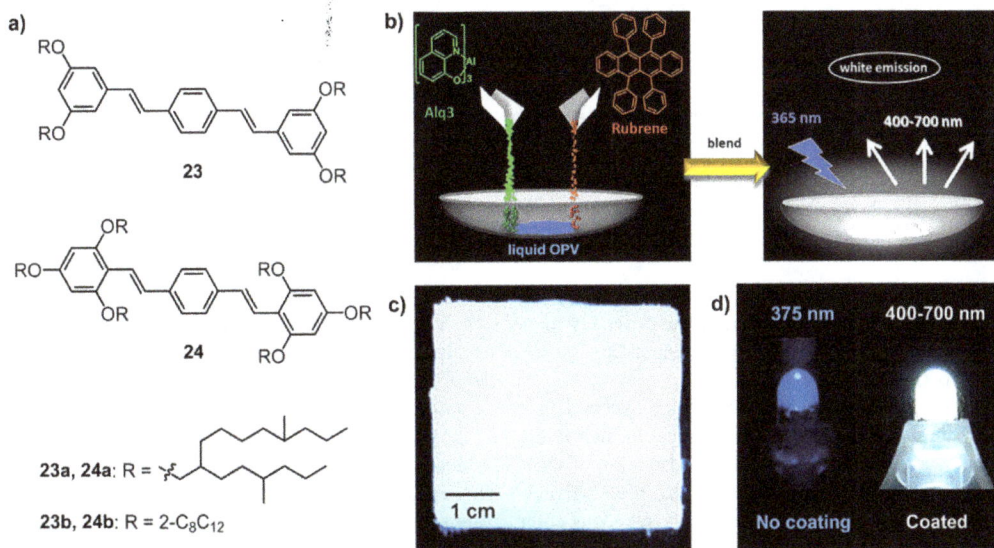

23a, 24a: R =

23b, 24b: R = 2-C_8C_{12}

Figure 18. (a) Chemical structures of OPV derivatives **23–24** containing branched alkyl chains. (b) Schematic illustration of the preparation of a solvent-free white-emitting liquid composite using liquid OPV. (c) 5×5 cm^2 area coated with the white-emitting liquid composite and exposed to UV light (365 nm). (d) Commercially available UV-LED (375 nm) before and after coating with the white-emitting liquid composite. Reprinted with permission from S S Babu *et al* 2012 *Angew. Chem. Int. Edn.* **51** 3391, © 2012 John Wiley & Sons.

than **14–18**. This was because in the case of branched chains, both the vdW interaction of the chains and the π–π interaction among the C_{60} moieties were strikingly suppressed. Interestingly, the viscosity of **21** (∼260 Pa·s) turned out to be much lower than that of **20** (∼1500 Pa·s) even though **21** had shorter branched alkyl chains. This finding emphasized the importance of substitution at the 2-position on the phenyl unit to produce less viscous C_{60} liquids.

However, a further increase of the branching extent (**22**) would lead to both an increased melting point (12 °C) and an enhanced viscosity (∼128 000 Pa·s) of the isotropic phase, which is even higher than the linear chain-substituted C_{60} liquids **14–18**. This is due to the hyperbranched-structure-induced high intra- and intermolecular friction during flowing.

Therefore, both the melting point and viscosity of alkylated C_{60} derivatives can be effectively reduced by suitable branching and proper substitution position of the alkyl chains. However, too high a branching extent would increase the melting point and viscosity again. These optoelectronically active-C_{60}-containing liquids provide opportunities for constructing flexible, printable photovoltaic devices.

Based on the designing strategy of C_{60} liquids, our group extended the π-conjugated C_{60} unit to an emissive π conjugating system and prepared OPV liquids, **23–24**, by substituting two different OPV units with branched alkyl chains (figure 18(a)) [90]. **23–24** are pale yellow fluids at room temperature with low melting points between −43 °C to −55 °C. Similar to the C_{60} liquids **20–22**, OPV liquids with

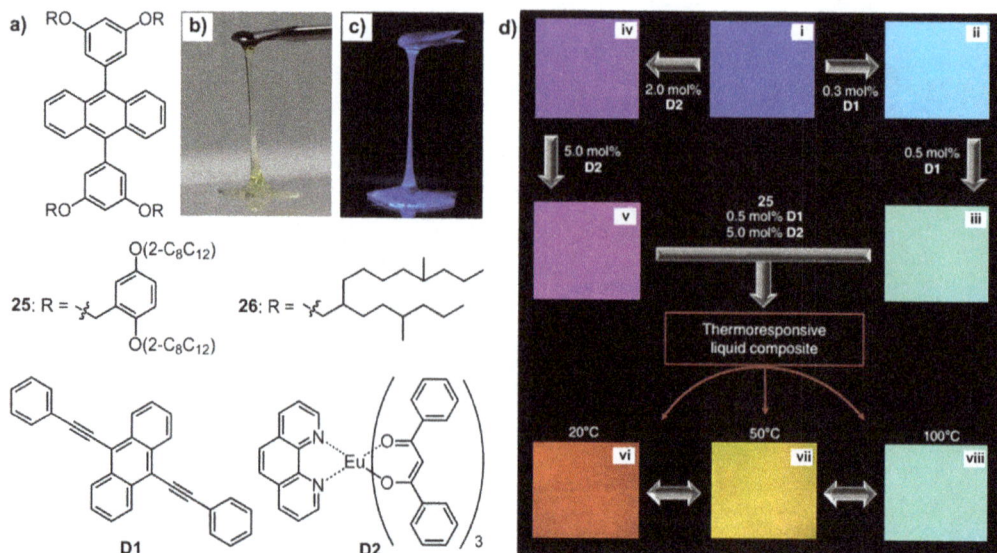

Figure 19. (a) Chemical structures of anthracene derivatives **25–26** containing branched alkyl chains and dopants **D1** and **D2**. Photographs of **26** under visible (b) and UV light (365 nm) (c). (d) Photographs of the luminescence color tunability and thermal response of the composites of **25**, **D1** and **D2**: (i) **25** alone; (ii) **25** + **D1** (0.3 mol%); (iii) **25** + **D1** (0.5 mol%); (iv) **25** + **D2** (2.0 mol%); (v) **25** + **D2** (5.0 mol%) and **25** + **D1** (0.5 mol%) + **D2** (5.0 mol%) at 20 °C (vi); 50 °C (vii) and 100 °C (viii). Adapted from [91] under a Creative Commons Attribution 3.0 Unported (CC BY) license.

swallow-tailed branched alkyl chains substituting on the (2,4, 6-) positions (i.e. **24**) possess much lower viscosity than OPVs with hyper-branched alkyl chains appending on the (3,5-) positions (i.e. **23**).

The UV–vis absorption and fluorescence spectroscopic properties of these liquids are almost identical to their dilute solution analogues, demonstrating efficient isolation of OPV units upon wrapping by the soft alkyl chains in the solvent-free liquid state. Taking into account the advantageous color purity and stability of white-emitter materials and the ability of molecular liquids to blend organic or inorganic dopants for producing functional hybrid materials, the blue-emitting **23b** or **24b** was employed as a matrix for doping green-emitting tris(8-hydroxyquinolinato)aluminium (Alq3) and orange-emitting 5,6,11,12-tetraphenylnaphthacnee (rubrene) to obtain white-emitting liquids (figure 18(b)). Simply blending the three components for just one minute would result in the composites exhibiting a white emission spanning from 400 to 700 nm. The composite maintained the strong emission features of OPV and had a quantum yield of over 35%, which, together with the low melting point of around −45 °C and low viscosity of 3.2 Pa · s, enabled writing or painting of the white-light-emitting material on various surfaces (figure 18(c)) such as the surface of a UV light-emitting diode (LED) (figure 18(d)).

Right after this achievement, our group synthesized another type of blue-emitting liquid, **25–26** (figure 19(a)), by attaching branched alkyl chains to anthracene-emitting units [91]. Both compounds are yellowish transparent viscous liquids under visible light (figure 19(b)) but show blue luminescence under UV light (figure 19(c)) due to the

reduction of π–π interactions through soft alkyl chains substitution. Compound **25**, with eight suitable branched alkyl chains, exhibits a lower melting point and viscosity than **26**, with only four hyperbranched chains, which is highly consistent with the behaviors of C_{60} and OPV liquids.

Enveloped by the alkyl chains, the anthracene cores were effectively isolated, resulting in similar absorption and fluorescence spectra of **25–26** in the solvent-free state and dilute solution. In addition, such enveloping can effectively prevent oxygen attack and dimerization of the anthracene cores, resulting in remarkably improved photostability, which was confirmed by time-dependent nuclear magnetic resonance (NMR) and emission intensity upon Xe-lamp irradiation.

Upon doping **25** with 9,10-bis(phenylethynyl)anthracene (**D1**) and tris(1,3-diphenyl-1,3-propanedionato)-(1,10-phenanthroline)europium(III) (**D2**), its luminescent color can be freely tuned based on an energy transfer mechanism, according to which **25** acts as a Förster resonance energy transfer (FRET) donor for both **D1** and **D2**. As shown in figure 19(d), with the increasing mol% of **D1**, the blue emission (i) of **25** changed to cyan (ii, 0.3 mol%) and green (iii, 0.5 mol%), while in the case of **D2**, the emission color of **25** changed to violet (iv, 2.0 mol%) and purple (v, 5.0 mol%). Moreover, by blending liquid **25** with both **D1** (0.5 mol%) and **D2** (5.0 mol%), a red-emitting composite (vi, 20 °C) was obtained. Upon increasing the temperature, the red emission changed to yellow (vii, 50 °C) and emerald green (viii, 100 °C) by virtue of the well-known temperature-dependent emission of **D2**, allowing the composite to be used as a color-indicating thermometer. Such facile luminescent color-tuning

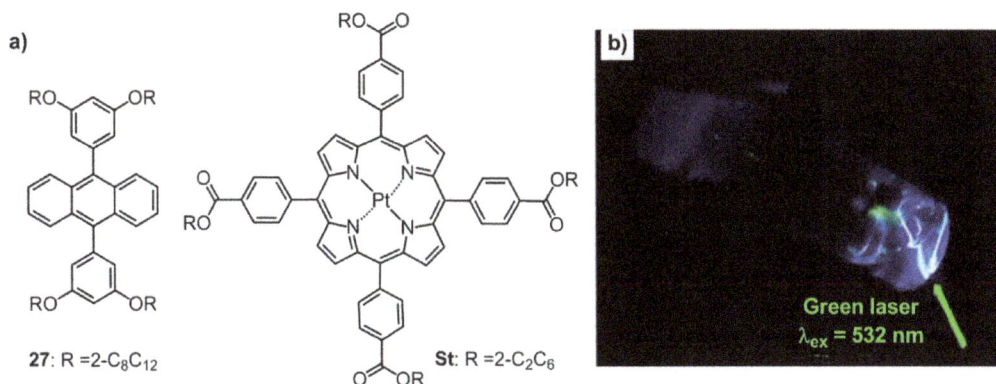

Figure 20. (a) Chemical structures of an anthracene derivative **27** and a Pt(II) porphyrin photosensitizer **St**. (b) Photograph of the doped liquid (**St/27** = 0.01 mol%) exposed to a 532 nm laser. Reprinted with permission from P Duan *et al* 2013 *J. Am. Chem. Soc.* **135** 19056, © 2013 American Chemical Society .

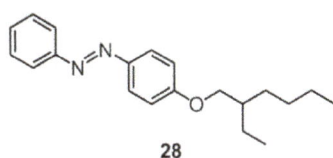

Figure 21. Chemical structure of an azobenzene derivative **28**.

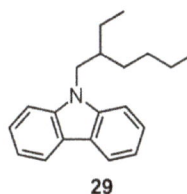

Figure 22. Chemical structure of a carbazole liquid **29**.

spanning almost the whole visible range provides technological potential for a continuous active layer in flexible materials.

Inspired by our research on photostable liquid luminophors, the Kimizuka group synthesized an anthracene liquid, **27** (figure 20(a)), and applied it to an upconversion (UC) luminescent system [92]. Liquid **27** can accommodate a Pt(II) porphyrin photosensitizer, **St**, resulting in UC properties upon light irradiation with a 532 nm green laser (figure 20(b)). Compared with a traditional UC system, which was functionalized in an organic solvent and suffered emission quenching by molecular oxygen, this new UC system exhibited strong emission features, with a high quantum yield of 28% free from the oxygen effect due to the oxygen-impermeable solvent-free liquid system.

Our group synthesized a liquid azobenzene through substitution of a branched 2-octyldodecyl (2-C_8C_{12}) chain (*vide infra*) [17]. Quite recently, Masutani *et al* [93] reported a similar liquid azobenzene **28** (figure 21) attached with a 2-ethylhexyl (2-C_2C_6) chain and explored its application as solar thermal fuel, a material in which light energy could be converted to chemical bond energy and consequently discharged as heat upon external stimuli. As a popular molecular photo-switch, trans-azobenzene is a hot candidate for solar thermal fuels because the photon energy can be stored in the photochemical generated cis isomer in the form of molecular strain energy and can be released as heat through cis-to-trans thermal isomerization. Conventional photoisomerization of azobenzene always occurs in a dilute solution, resulting in a remarkable decrease of the total volumetric energy density. Nevertheless, the azobenzene liquid, **28**, overcame this problem by facile photoisomerization even in a neat state, with a trans-to-cis rate comparable to those observed in a solution state. With such an excellent performance, liquid **28** could be promising for solar thermal storages.

Through attachment of a 2-C_2C_6 chain to the nitrogen atom, a room-temperature carbazole liquid **29** (figure 22) was obtained, which was employed as a 'solvent' in the ellipsometry measurement for the determination of electric-field-induced birefringence in photorefractive polymer composites [94]. Later, the Wada group investigated the hole mobility of **29**, which was determined to be $4 \times 10^{-6} \, cm^2 \, V^{-1} \, s^{-1}$ by TOF experiment [95]. More recently, the Adachi group applied **29** as a liquid-emitting layer in organic light-emitting diodes (OLED) [96]. Such liquid emitters, although suffering inevitable long-term degradation in an OLED, could be facilely replaced by a flow of fresh ones, which effectively solved the problem of OLED degradation resulting from the decomposition of organic materials. Moreover, even with significant device bending, the detachment between the liquid-emitting layer and electrodes could be completely avoided, allowing the realization of flexible displays. Significantly, by doping a small amount of electrolytes into the liquid-emitting layer to decrease the driving voltage and by inserting a TiO_2 hole-blocking layer to improve the carrier balance, the liquid OLED exhibited a maximum external electroluminescence (EL) quantum yield of $0.31 \pm 0.07\%$ and a maximum luminance of nearly $100 \, cd \, m^{-2}$.

Upon substitution of various π-conjugated molecules by alkyl chains with a proper substitution position or suitable chain branching extent, a number of room-temperature

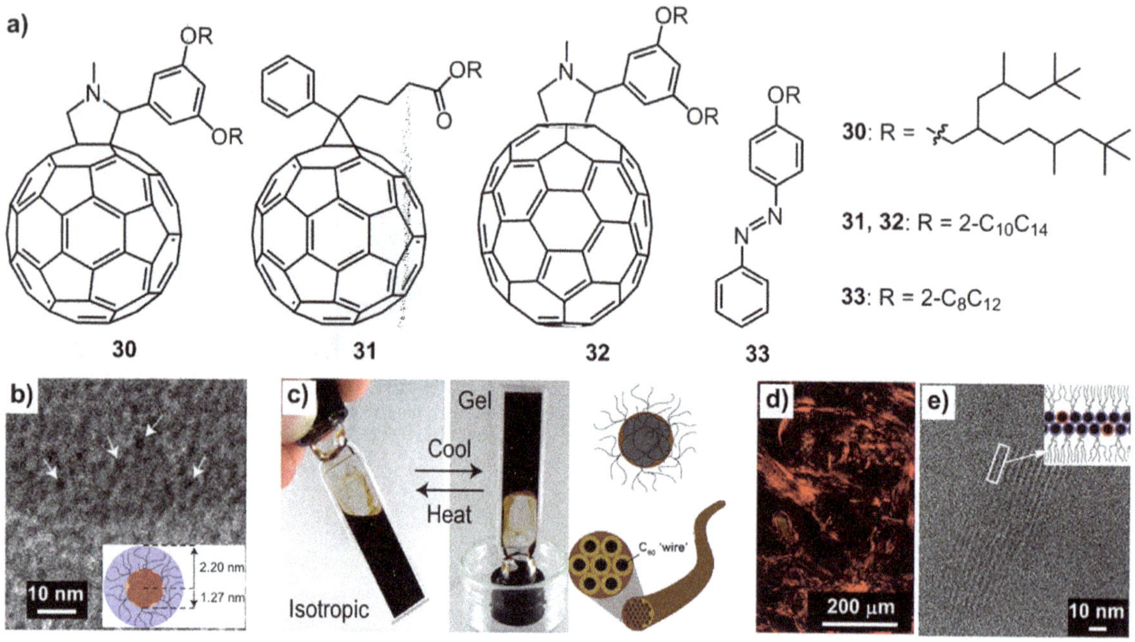

Figure 23. (a) Chemical structure of molecular liquids **30**–**33** containing branched alkyl chains. (b) Cryo-TEM image of micelles of **20** in *n*-decane; inset, proposed aggregate structure of **20**. (c) Left: photographs showing the isotropic and gelled states that arise on dissolving **30** in *n*-decane; right: schematic depictions of the assembled micelles in the isotropic state (up) and gel fibers in the gelled state (below). POM (d) and TEM (e) images for 1:10 molar ratios of C_{60} and **20** at room temperature; inset of (e) proposed lamellar structure. Reprinted with permission from M J Hollamby *et al* 2014 *Nat. Chem.* **6** 690, © 2014 Macmillan Publishers.

solvent-free liquids with high stability and distinctive applications were achieved. The physical liquid properties of these molecules are highly dependent on the substitution position, length and branching extent of the alkyl chains. In general, branched chains are proven to be more effective to reduce both melting point and viscosity than linear ones. However, further increasing the branching extent would increase the melting point and viscosity again. Based on this designing strategy, various optoelectronic liquids are expected to be constructed and applied to flexible-foldable device fabrications in the future by employing different functional π-conjugated units.

5. Directed assembly from solvent-free liquid

Compared with both solid self-assemblies and liquid crystals, the solvent-free liquid molecules, although possessing facile processability, suffered deficient orders due to the efficiently reduced π–π interactions of the π units. This shortcoming could limit their applications because high ordering of π-conjugated molecules is critical for optoelectronic devices for efficient exciton diffusion and electron transport. To address this issue, our group established a straightforward way for rebalancing the π–π and vdW interactions of solvent-free liquid alkylated-π molecules by the introduction of either alkane or π additives. As a result, the liquid materials were

directed to be various highly ordered self-assemblies upon the addition of molecular segments.

C_{60} derivatives **20** (figure 17) [71] and **30** (figure 23(a)) [17], alkylated with different branched chains, formed unstructured liquid and a disordered amorphous state at room temperature, respectively. However, upon the addition of *n*-alkanes solvents, *n*-decane for instance, the two compounds self-assembled into spherical core–shell micelles with an average diameter of 2.5 ± 0.3 nm (figure 23(b)) and into hexagonally packed gel-fibers containing insulated C_{60} nanowires with cylindrical micelles of 3.2 nm diameter (figure 23(c)), respectively.

As elucidated in part 2.1, the alkane solvents showed stronger affinity to the branched chains than to the C_{60} moieties due to the unconventional amphiphilic features of the two moieties in an organic solvent. Therefore, similar to the additive-directed assembly of conventional hydrophobic-hydrophilic surfactants in aqueous media, the introduction of alkanes would be able to direct the assembly of such alkyl-π-conjugated molecules in organic media. Compared with **20**, compound **30** with an increased branching number of alkyl chains suffered reduced interaction strength with the solvent molecules and weakening interference in π–π interactions between neighboring C_{60} units. As a result, compound **30** yields larger assemblies than compound **20**.

Similarly, additives with higher affinity to the C_{60} part rather than the alkyl chains were introduced; these additives were also capable of driving the assembly by strengthening the π–π interactions of the C_{60} moieties. For example, the addition of pristine C_{60} to **20** would direct the assembly into the lamellar mesophase (figures 23(d)–(e)).

In addition, the assembled materials, with a large fraction of optoelectronically active components, exhibit extremely high photoconductivities of a similar order as that for solid crystalline C_{60} derivatives, including PCBM [97, 98]. Therefore, the additive- (molecular segments) directed assembly strategy developed here creates a new rule to construct assembled molecular materials with distinctive functionality and complexity. Encouragingly, this strategy can be extended to another branched chain alkylated-C_{60} derivative **31** and other alkyl-π-conjugated molecules including both larger C_{70} (**32**) and smaller azobenzene (**33**) π-conjugated systems than C_{60}.

The method developed here has opened a new gateway for facile state control of alkyl-π molecules, allowing us to take full advantage of the easy processability of the liquid state and the long-range order-benefiting optoelectronic properties of the additive-directed assembled liquid-crystalline mesophase. Upon controlling the occurrence, assembled structures and functions of the assembled alkyl-π molecules from their disordered liquid state, various optoelectronic materials can be obtained which could be used for extensive practical applications in the field of flexible electronics.

6. Conclusions

This review covers a diversity of alkylated-π molecular systems with various π-conjugated units attached by different types of alkyl chains through chemical modification. The corporative vdW interactions of the alkyl chains and π–π interactions of the π-conjugated moieties affect the physical states and applications of the corresponding molecules. We have not only clarified the relationship between the balance of the two interactions and the physical states of these alkylated-π molecules but also reviewed their state-dependent optoelectronic properties toward various practical applications. With these extensive investigations, a clear guidance for molecular state control through alkyl-π engineering is provided, which can be realized by selecting proper alkyl chain types and modulating chain substitution patterns as well as by the introduction of additives. Based on this strategy, functional alkylated-π molecules with expected states and applications could be constructed such as anti-wetting surfaces fabricated from solid self-assemblies, BHJ organic solar cells with high PCE constructed from thermotropic LCs and photostable, full-color tunable luminescent systems formed from solvent-free liquid fluorophores. Hopefully, the alkyl-π engineering method summarized here would remarkably enrich the alkylated-π materials toward abundant high-performance optoelectronic applications.

Acknowledgments

The authors are grateful to all the researchers who participated in the work discussed in this paper and whose names appear in the references. FL thanks Ms A Zielinska and Dr K Okamoto for their meaningful discussions and Mr C Hassam for reading this manuscript. TN gratefully acknowledges the financial support provided by KAKENHI (25620069, 25104011) from MEXT, Japan.

References

[1] Sekine C, Tsubata Y, Yamada T, Kitano M and Doi S 2014 *Sci. Technol. Adv. Mater.* **15** 034203
[2] Sasabe H and Kido J 2013 *J. Mater. Chem.* C **1** 1699
[3] Thejo Kalyani N and Dhoble S J 2012 *Renew. Sust. Energ. Rev.* **16** 2696
[4] Lin Y, Li Y and Zhan X 2012 *Chem. Soc. Rev.* **41** 4245
[5] Yamashita Y 2009 *Sci. Technol. Adv. Mater.* **10** 024313
[6] Kumar B, Kaushik B K and Negi Y S 2014 *Polym. Rev.* **54** 33
[7] Sirringhaus H 2009 *Proc. IEEE* **97** 1570
[8] Yamamoto Y 2012 *Sci. Technol. Adv. Mater.* **13** 033001
[9] Kelley T W, Baude P F, Gerlach C, Ender D E, Muyres D, Haase M A, Vogel D E and Theiss S D 2004 *Chem. Mater.* **16** 4413
[10] Shirota Y 2000 *J. Mater. Chem.* **10** 1
[11] Maggini L and Bonifazi D 2012 *Chem. Soc. Rev.* **41** 211
[12] Würthner F and Meerholz K 2010 *Chem. Eur. J.* **16** 9366
[13] Moonen P F, Yakimets I and Huskens J 2012 *Adv. Mater.* **24** 5526
[14] Lei T and Pei J 2012 *J. Mater. Chem.* **22** 785
[15] Wang C, Dong H, Hu W, Liu Y and Zhu D 2011 *Chem. Rev.* **112** 2208
[16] Lei T, Wang J-Y and Pei J 2013 *Chem. Mater.* **26** 594
[17] Hollamby M J *et al* 2014 *Nat. Chem.* **6** 690
[18] Nakanishi T 2010 *Chem. Commun.* **46** 3425
[19] Babu S S, Möhwald H and Nakanishi T 2010 *Chem. Soc. Rev.* **39** 4021
[20] Hollamby M J and Nakanishi T 2013 *J. Mater. Chem.* C **1** 6178
[21] Li H, Choi J and Nakanishi T 2013 *Langmuir* **29** 5394
[22] Zielinska A, Leonowicz M, Li H and Nakanishi T 2014 *Curr. Opin. Colloid Interface Sci.* **19** 131
[23] Altoe V, Martin F, Katan A, Salmeron M and Aloni S 2012 *Nano Lett.* **12** 1295
[24] Li M, Ishihara S, Ji Q, Akada M, Hill J P and Ariga K 2012 *Sci. Technol. Adv. Mater.* **13** 053001
[25] Ariga K, Hill J P, Lee M V, Vinu A, Charvet R and Acharya S 2008 *Sci. Technol. Adv. Mater.* **9** 014109
[26] Nakanishi T, Schmitt W, Michinobu T, Kurth D G and Ariga K 2005 *Chem. Commun.* 5982
[27] Nakanishi T, Takahashi H, Michinobu T, Hill J P, Teranishi T and Ariga K 2008 *Thin Solid Films* **516** 2401
[28] Wang L, Zhou Y, Yan J, Wang J, Pei J and Cao Y 2009 *Langmuir* **25** 1306
[29] Nakanishi T, Ariga K, Michinobu T, Yoshida K, Takahashi H, Teranishi T, Möhwald H and Kurth D G 2007 *Small* **3** 2019
[30] Nakanishi T, Michinobu T, Yoshida K, Shirahata N, Ariga K, Möhwald H and Kurth D G 2008 *Adv. Mater.* **20** 443
[31] Shen Y, Wang J, Kuhlmann U, Hildebrandt P, Ariga K, Möhwald H, Kurth D G and Nakanishi T 2009 *Chem. Eur. J.* **15** 2763
[32] Sezer M, Feng J-J, Khoa Ly H, Shen Y, Nakanishi T, Kuhlmann U, Hildebrandt P, Möhwald H and Weidinger I M 2010 *Phys. Chem. Chem. Phys.* **12** 9822
[33] Kataura H, Kumazawa Y, Maniwa Y, Umezu I, Suzuki S, Ohtsuka Y and Achiba Y 1999 *Synth. Met.* **103** 2555
[34] Fujigaya T, Morimoto T, Niidome Y and Nakashima N 2008 *Adv. Mater.* **20** 3610

[35] Richardson H H, Carlson M T, Tandler P J, Hernandez P and Govorov A O 2009 *Nano Lett.* **9** 1139

[36] Shen Y, Skirtach A G, Seki T, Yagai S, Li H, Möhwald H and Nakanishi T 2010 *J. Am. Chem. Soc.* **132** 8566

[37] Asanuma H, Subedi P, Hartmann J, Shen Y, Möhwald H, Nakanishi T and Skirtach A 2013 *Langmuir* **29** 7464

[38] Yin J, Yan J, He M, Song Y, Xu X, Wu K and Pei J 2010 *Chem. Eur. J.* **16** 7309

[39] Chen H-B, Zhou Y, Yin J, Yan J, Ma Y, Wang L, Cao Y, Wang J and Pei J 2009 *Langmuir* **25** 5459

[40] Nakanishi T *et al* 2010 *J. Mater. Chem.* **20** 1253

[41] Rabe J P and Buchholz S 1991 *Science* **253** 424

[42] Xu S L, Zeng Q D, Wu P, Qiao Y H, Wang C and Bai C L 2003 *Appl. Phys. A* **76** 209

[43] Uemura S, Sakata M, Hirayama C and Kunitake M 2004 *Langmuir* **20** 9198

[44] Yoshimoto S, Honda Y, Murata Y, Murata M, Komatsu K, Ito O and Itaya K 2005 *J. Phys. Chem. B* **109** 8547

[45] Bonifazi D, Spillmann H, Kiebele A, de Wild M, Seiler P, Cheng F, Güntherodt H-J, Jung T and Diederich F 2004 *Angew. Chem. Int. Ed.* **43** 4759

[46] Nakanishi T, Miyashita N, Michinobu T, Wakayama Y, Tsuruoka T, Ariga K and Kurth D G 2006 *J. Am. Chem. Soc.* **128** 6328

[47] Nakanishi T, Takahashi H, Michinobu T, Takeuchi M, Teranishi T and Ariga K 2008 *Colloids Surf. A* **321** 99

[48] Severin N, Rabe J P and Kurth D G 2004 *J. Am. Chem. Soc.* **126** 3696

[49] Severin N, Barner J, Kalachev A A and Rabe J P 2004 *Nano Lett.* **4** 577

[50] Mao G, Dong W, Kurth D G and Möhwald H 2004 *Nano Lett.* **4** 249

[51] Okawa Y and Aono M 2001 *Nature* **409** 683

[52] Yang Y, Miao X, Liu G, Xu L, Wu T and Deng W 2012 *Appl. Surf. Sci.* **263** 73

[53] Hentschke R, Schürmann B L and Rabe J P 1992 *J. Chem. Phys.* **96** 6213

[54] Chen Q, Chen T, Zhang X, Wan L-J, Liu H-B, Li Y-L and Stang P 2009 *Chem. Commun.* 3765

[55] Mali K S, Schwab M G, Feng X, Müllen K and De Feyter S 2013 *Phys. Chem. Chem. Phys.* **15** 12495

[56] Xu L, Miao X, Zha B, Miao K and Deng W 2013 *J. Phys. Chem. C* **117** 12707

[57] Tamaki Y, Muto K and Miyamura K 2013 *Bull. Chem. Soc. Jpn.* **86** 354

[58] Crivillers N *et al* 2009 *J. Am. Chem. Soc.* **131** 6246

[59] Li H, Hollamby M J, Seki T, Yagai S, Möhwald H and Nakanishi T 2011 *Langmuir* **27** 7493

[60] Yagai S, Kubota S, Iwashima T, Kishikawa K, Nakanishi T, Karatsu T and Kitamura A 2008 *Chem. Eur. J.* **14** 5246

[61] Yagai S, Goto Y, Lin X, Karatsu T, Kitamura A, Kuzuhara D, Yamada H, Kikkawa Y, Saeki A and Seki S 2012 *Angew. Chem. Int. Edn* **51** 6643

[62] Yagai S, Usui M, Seki T, Murayama H, Kikkawa Y, Uemura S, Karatsu T, Kitamura A, Asano A and Seki S 2012 *J. Am. Chem. Soc.* **134** 7983

[63] Hill J P, Jin W, Kosaka A, Fukushima T, Ichihara H, Shimomura T, Ito K, Hashizume D, Ishii N and Aida T 2004 *Science* **304** 1481

[64] Jin W, Yamamoto Y, Fukushima T, Ishii N, Kim J, Kato K, Takata M and Aida T 2008 *J. Am. Chem. Soc.* **130** 9434

[65] Goldberger J, He R, Zhang Y, Lee S, Yan H, Choi H-J and Yang P 2003 *Nature* **422** 599

[66] Yin J, Zhou Y, Lei T and Pei J 2011 *Angew. Chem. Int. Edn* **50** 6320

[67] Yazaki S, Funahashi M and Kato T 2008 *J. Am. Chem. Soc.* **130** 13206

[68] Li W-S, Yamamoto Y, Fukushima T, Saeki A, Seki S, Tagawa S, Masunaga H, Sasaki S, Takata M and Aida T 2008 *J. Am. Chem. Soc.* **130** 8886

[69] Nakanishi T, Shen Y, Wang J, Yagai S, Funahashi M, Kato T, Fernandes P, Möhwald H and Kurth D G 2008 *J. Am. Chem. Soc.* **130** 9236

[70] Fernandes P A L, Yagai S, Möhwald H and Nakanishi T 2009 *Langmuir* **26** 4339

[71] Li H, Babu S S, Turner S T, Neher D, Hollamby M J, Seki T, Yagai S, Deguchi Y, Möhwald H and Nakanishi T 2013 *J. Mater. Chem. C* **1** 1943

[72] Xiao S, Myers M, Miao Q, Sanaur S, Pang K, Steigerwald M L and Nuckolls C 2005 *Angew. Chem. Int. Edn* **44** 7390

[73] Pisula W, Menon A, Stepputat M, Lieberwirth I, Kolb U, Tracz A, Sirringhaus H, Pakula T and Müllen K 2005 *Adv. Mater.* **17** 684

[74] van de Craats A M, Stutzmann N, Bunk O, Nielsen M M, Watson M, Müllen K, Chanzy H D, Sirringhaus H and Friend R H 2003 *Adv. Mater.* **15** 495

[75] Schmidt-Mende L, Fechtenkötter A, Müllen K, Moons E, Friend R H and MacKenzie J D 2001 *Science* **293** 1119

[76] Adam D, Schuhmacher P, Simmerer J, Haussling L, Siemensmeyer K, Etzbachi K H, Ringsdorf H and Haarer D 1994 *Nature* **371** 141

[77] Boden N, Bushby R J, Clements J, Jesudason M V, Knowles P F and Williams G 1988 *Chem. Phys. Lett.* **152** 94

[78] Boden N, Bushby R J, Clements J, Movaghar B, Donovan K J and Kreouzis T 1995 *Phys. Rev. B* **52** 13274

[79] van de Craats A M, Warman J M, Fechtenkötter A, Brand J D, Harbison M A and Müllen K 1999 *Adv. Mater.* **11** 1469

[80] van de Craats A M and Warman J M 2001 *Adv. Mater.* **13** 130

[81] Warman J M and van de Craats A M 2003 *Mol. Cryst. Liq. Cryst.* **396** 41

[82] Pisula W, Tomović Ž, Simpson C, Kastler M, Pakula T and Müllen K 2005 *Chem. Mater.* **17** 4296

[83] Pisula W, Kastler M, Wasserfallen D, Mondeshki M, Piris J, Schnell I and Müllen K 2006 *Chem. Mater.* **18** 3634

[84] Kastler M *et al* 2006 *Adv. Mater.* **18** 2255

[85] Michinobu T, Nakanishi T, Hill J P, Funahashi M and Ariga K 2006 *J. Am. Chem. Soc.* **128** 10384

[86] Michinobu T, Okoshi K, Murakami Y, Shigehara K, Ariga K and Nakanishi T 2013 *Langmuir* **29** 5337

[87] Kramer T J, Babu S S, Saeki A, Seki S, Aimi J and Nakanishi T 2012 *J. Mater. Chem.* **22** 22370

[88] Nowak-Król A, Gryko D and Gryko D T 2010 *Chem. Asian J.* **5** 904

[89] Maruyama S, Sato K and Iwahashi H 2010 *Chem. Lett.* **39** 714

[90] Babu S S, Aimi J, Ozawa H, Shirahata N, Saeki A, Seki S, Ajayaghosh A, Möhwald H and Nakanishi T 2012 *Angew. Chem. Int. Edn* **51** 3391

[91] Babu S S *et al* 2013 *Nat. Commun.* **4** 1969

[92] Duan P, Yanai N and Kimizuka N 2013 *J. Am. Chem. Soc.* **135** 19056

[93] Masutani K, Morikawa M and Kimizuka N 2014 *Chem. Commun.* **50** 15803

[94] Hendrickx E, Guenther B D, Zhang Y, Wang J F, Staub K, Zhang Q, Marder S R, Kippelen B and Peyghambarian N 1999 *Chem. Phys.* **245** 407

[95] Ribierre J C, Aoyama T, Muto T, Imase Y and Wada T 2008 *Org. Electron.* **9** 396

[96] Hirata S, Kubota K, Jung H H, Hirata O, Goushi K, Yahiro M and Adachi C 2011 *Adv. Mater.* **23** 889

[97] Babu S S, Saeki A, Seki S, Möhwald H and Nakanishi T 2011 *Phys. Chem. Chem. Phys.* **13** 4830

[98] Zhang X, Nakanishi T, Ogawa T, Saeki A, Seki S, Shen Y, Yamauchi Y and Takeuchi M 2010 *Chem. Commun.* **46** 8752

Probing the bulk ionic conductivity by thin film hetero-epitaxial engineering

Daniele Pergolesi[1,2], Vladimir Roddatis[3], Emiliana Fabbri[1,2], Christof W Schneider[1], Thomas Lippert[1], Enrico Traversa[4] and John A Kilner[5]

[1] Paul Scherrer Institut, Department of General Energy Research, CH-5225, Villigen-PSI, Switzerland
[2] International Center for Materials Nanoarchitectonics (WPI-MANA), National Institute for Materials Science (NIMS), 1-1 Namiki, Tsukuba, Ibaraki 305-0044, Japan
[3] CIC Energigune, Albert Einstein 48, E-01510—Miñano (Álava), Spain
[4] Physical Science and Engineering Division, King Abdullah University of Science and Technology (KAUST), Thuwal 23955-6900, Saudi Arabia
[5] Department of Materials, Imperial College London, London SW7 2BP, UK

E-mail: daniele.pergolesi@psi.ch

Abstract

Highly textured thin films with small grain boundary regions can be used as model systems to directly measure the bulk conductivity of oxygen ion conducting oxides. Ionic conducting thin films and epitaxial heterostructures are also widely used to probe the effect of strain on the oxygen ion migration in oxide materials. For the purpose of these investigations a good lattice matching between the film and the substrate is required to promote the ordered film growth. Moreover, the substrate should be a good electrical insulator at high temperature to allow a reliable electrical characterization of the deposited film. Here we report the fabrication of an epitaxial heterostructure made with a double buffer layer of $BaZrO_3$ and $SrTiO_3$ grown on MgO substrates that fulfills both requirements. Based on such template platform, highly ordered (001) epitaxially oriented thin films of 15% Sm-doped CeO_2 and 8 mol% Y_2O_3 stabilized ZrO_2 are grown. Bulk conductivities as well as activation energies are measured for both materials, confirming the success of the approach. The reported insulating template platform promises potential application also for the electrical characterization of other novel electrolyte materials that still need a thorough understanding of their ionic conductivity.

Keywords: pulsed laser deposition, high resolution transmission electron microscopy, oxygen ion conductors, impedance spectroscopy, ionic conductivity

1. Introduction

The conducting properties of oxygen ion conducting oxides, such as Sm doped CeO_2 (SDC), 8 mol% Y_2O_3 stabilized ZrO_2 (YSZ), and $La_{(1-x)}Sr_xGa_{(1-y)}Mg_yO_{(3-\delta)}$ (LSGM with typically $x = y = 0.2$), have been widely investigated for applications as electrolyte materials for solid oxide fuel cells (SOFCs) [1]. These are electrochemical devices that convert chemical energy from a fuel into electric energy with very high efficiency and small environmental impact. The importance these devices can have for the development of a power generation system to become more independent of fossil and nuclear fuel is widely recognized; however, the full exploitation of this technology is still limited by the very high operating temperatures required to achieve suitable ionic conductivity across the electrolyte [2]. The typical operating temperature of current state-of-the-art SOFCs is above 800 °C. Intense research activity is been focused on investigating the ionic conducting mechanism in oxide materials searching for the most suitable strategy to engineer materials with sufficiently high conductivity in the so-called intermediate temperature range, between 500 and 800 °C [3, 4].

In general, the micro-morphology of the conductor (average grain size, grain boundary, lattice distortions) can significantly affect the oxygen ion conduction [5]. In particular, grain boundaries appear often to block the oxygen ion migration. In the case of YSZ, the grain boundary shows conductivities up to two orders of magnitude lower than the bulk [6]. Other examples are Gd-doped CeO_2 thin polycrystalline films which showed conductivity more than one order of magnitude lower than that of highly ordered films with small grain boundary regions [7]. This effect was attributed to the development of a negatively charged space charge layer surrounding the grain boundary core [8, 9].

Electrochemical impedance spectroscopy (EIS) measurements on polycrystalline ceramic pellets are typically used to investigate the total electrical conductivity of oxygen ion conducting oxides as a function of the temperature and the gaseous environment. Thus, the measured total conductivity is composed of different conduction pathways through the grain interior (bulk) and across the grain boundary. The separation of the bulk and grain boundary contributions from EIS impedance plots is routinely performed especially at relatively low temperatures [10]. By increasing the temperature (typically above 400 °C), both the grain and grain-boundary resistances decrease, and the frequency limitations of most of the commonly available instruments restrict the determination of the two contributions. At higher temperatures, only the availability of single crystals, sintered pellets with very large average grain size or highly ordered thin films allows a direct measurement the bulk oxygen ion conductivity. Single crystals are expensive and not always easily available and sintered pellets with micro-meter grain size can be difficult to fabricate. Therefore, highly ordered thin films, typically grown by pulsed laser deposition (PLD) on insulating substrates can be a valid alternative. However, it must be taken into account that in the case of in-plane electrical measurements of thin films the EIS plot always shows only one semi-circle independent on the sample morphology due to the parallel stray capacitance of the substrate. Thus, it is possible to assign the measured resistance to the grain interior conduction pathways only if the micro-structural analysis of the sample shows no evidence of significant grain separation (high angle grain boundaries).

By carefully tuning the deposition parameters, PLD allows the growth of thin films with well-defined microstructure and morphology from almost ideal single-crystalline samples to polycrystalline samples with nanometric average grain size. However, the fabrication and the electrical characterization of such model systems strongly depend on the availability of suitable deposition substrates. The substrate material should have good lattice matching with the film material to enable an ordered growth of the conductor, and it must be a good electrical insulator at high temperatures in order not to affect the in-plane electrical characterization.

Cubic perovskite single-crystalline wafers such as $SrTiO_3$ (STO), $YAlO_3$ (YAO) or $LaAlO_3$ (LAO) are suitable substrates for a very well ordered growth of (doped-)ceria and LSGM. The latter has the same crystalline structure and a very similar lattice parameter to STO, allowing a cube-on-cube fully relaxed growth of the thin film [11, 12]. Differently, SDC has the cubic fluorite crystalline structure with lattice parameter $c = 5.44$ Å. A relatively good crystalline matching with STO can be achieved by way of the epitaxial symmetry (001)SDC//(001)STO and (100)SDC//(110)STO (i.e. with an in-plane 45° rotation of the SDC unit cell with respect to STO) [13]. In this case, SDC has an in-plane lattice mismatch of about -1.5%. In the case of YSZ, a cubic fluorite with lattice parameter of about 5.14 Å, the lattice mismatch with STO is much larger (about -7%) and the surface termination of the STO substrate seems to have a crucial role in tuning the epitaxial growth [14]. Nevertheless, highly ordered YSZ films were grown on (001)-oriented STO substrates [14–16]. Better interface quality might be achievable using (110)-oriented YAO substrates that allow a well-ordered epitaxial growth reducing significantly the lattice misfit [17].

However, it is well known that perovskite materials like STO are not good insulators at high temperatures [18]. As a consequence, the electrical conductivity of the substrate (much thicker than the film) can significantly affect the in-plane measurement of the film resistance [19].

Al_2O_3 substrates have very good insulating properties and are widely used for the fabrication of thin films of SDC and YSZ. The literature reports the growth of highly textured films of SDC on (0001)-oriented sapphire substrates [7], but also several examples of films showing multiple orientations on both (0001) and (1102)-oriented substrates [20–22]. YSZ thin films can be epitaxially grown on sapphire substrates, but these films usually show a columnar morphology consisting of parallel pillars (tens of nm in diameter) orthogonal to the substrate surface and extending through the entire film [16, 23]. Such morphology might be not appropriate for the in-plane investigation of the bulk conductivity since the conduction pathways across and along grain boundaries may dominate the conductivity.

Also MgO single-crystalline wafers offer suitable insulating properties, and indeed YSZ thin-films with very good crystalline quality have been grown on this substrate [15, 17]. However, the lattice mismatch at the interface with SDC and YSZ is as large as about 20% leading often to polycrystalline microstructures [13] and to highly defective film-substrate interfaces which may also affect in-plane conductivity measurements providing a fast ion conduction pathways along dislocation lines [15].

The problem of combining a good crystallographic matching between film and substrate, and good insulating properties of the deposition substrate has been successfully addressed in [13] and [24]. In these studies (001)-oriented MgO single-crystalline wafers were used as insulating substrates and a thin film of STO grown *in situ* on the substrate surface provided the suitable seed layer allowing the highly ordered growth of the conductor. With this design the electrical conductivity of the thin STO buffer layer, about 10 nm thick, was negligibly small compared to that of the ionic conducting films under investigation.

Both STO and MgO have a cubic symmetry but different crystalline structures, i.e. perovskite and rock salt, respectively. This in our experience often results in a sample fabrication with a limited yield probably due to small deviations of the deposition parameters that are difficult to control and typically lead to a columnar growth.

To circumvent this problem and to allow a better yield for the reproducible growth of a STO seed layer on MgO without the presence of high angle grain boundaries, a new template platform for the growth of highly textured ionic conducting thin films has been engineered. The first layer of the new template platform is $BaZrO_3$ (BZO), a cubic perovskite with a lattice parameter of about 4.19 Å, very similar to the lattice parameter of MgO (4.21 Å). The literature reports several examples of a bi-axially textured growth of doped [25, 26] and undoped [27, 28] BZO on MgO and very similar results were obtained for this work. The second layer of the template platform is STO according to the MgO + BZO + STO scheme.

2. Experimental details

A custom made PLD system (AOV Ltd) equipped with load-lock chamber and high-pressure reflection high energy electron diffraction (HP-RHEED) was used for the fabrication of the samples. Sintered ceramic pellets of $BaZrO_3$, $SrTiO_3$, CeO_2, $Ce_{0.75}Sm_{0.15}O_{2-\delta}$ and 8 mol% Y_2O_3 stabilized ZrO_2 were prepared in our laboratory and used as targets for PLD. Commercially available (100) oriented MgO single-crystalline wafers were used as deposition substrates. The substrates were ultrasonically cleaned in acetone and isopropanol before loading into the vacuum chamber. A resistive heater allowed setting the temperature of the deposition substrate up to the desired value. The thermal contact between the substrate and the heater was provided by platinum paste. The same deposition parameters were used for all materials changing only the number of laser pulses in order to grow films of the desired thickness. The temperature of the heater during the deposition was kept at 800 °C. The thermal contact between heating plate and substrates was provided by Pt paste. In this experimental condition the temperature of the substrate was about 720 °C, estimated by previous calibration with a thermocouple in good thermal contact with the heating plate.

The target-to-substrate distance was set to 65 mm. The vacuum chamber was evacuated down to the base pressure of about 10^{-6} Pa and then a high purity oxygen partial pressure of 5 Pa to vacuum was used during the film growth.

A KrF excimer laser (Coherent Lambda Physik GmbH) with a wavelength of 248 nm and a pulse width of 25 ns was focused on the target material in a spot area of about 4 mm^2 with an energy of about 80 mJ (measured at the target surface). A repetition rate between 2 and 5 Hz was used. During the ablation process, each target was rotated and oscillated, for a uniform ablation of the target surface. A stainless steel shield avoided cross-contamination of the targets. A programmable control unit allowed programming multistep deposition processes selecting for each subsequent step the target material, the target oscillation amplitude and velocity, the number of laser shots and the laser frequency for each target.

X-ray diffraction (XRD) (PANalytical X'pert Pro MPD, $\lambda = 0.1540$ nm) analysis was used for the calibration of the deposition rate by x-ray reflectivity (XRR) and to investigate the crystalline structure of the films. The growth mechanism of the thin films was monitored in situ by HP-RHEED.

High-resolution transmission electron microscopy (HR-TEM) was carried out using a Tecnai F20ST electron microscope equipped with a high-angle annular dark-field detector and operated at 200 kV. Samples for HR-TEM analysis were prepared by standard techniques, including mechanical polishing followed by ion milling using a Fischione 1010 ion mill.

For the electrical measurements, two parallel Pt electrodes about 100 nm thick were fabricated on the film surface by electron beam deposition at room temperature. 5 nm thick Ti layers were used to improve Pt adhesion. EIS measurements were performed between 100 mHz and 1 MHz in air, varying the temperature between 400 and 700 °C, using a multi-channel potentiostat VMP3 (Bio-Logic).

3. Results and discussion

To directly measure the bulk conductivity of ionic conducting oxides, thin film with high crystalline quality grown on insulating substrates can be used. Good results have been achieved for thin films of doped ceria and for CeO_2/YSZ multilayered heterostructures using STO-buffered MgO substrates [13, 24]. However, in our experience, this strategy has a limited reproducibility and tiny variation of the deposition conditions (difficult to control) and/or of the quality of the substrate surface, often result in a columnar morphology, epitaxially oriented with the substrate but with large grain boundary regions. As a consequence, the fabricated samples require an individual HR-TEM analysis to select those with no evidence of high angle grain boundary and columnar morphology. An example can be found in figure 1(a) that shows the cross section HR-TEM image of a multilayered heterostructure made of 15 CeO_2/YSZ bilayers grown epitaxially oriented with the STO-buffered MgO substrate.

The entire heterostructure is (001) epitaxially oriented with the substrate but has a clear columnar morphology. Such a columnar morphology, highlighted with the arrows in figure 1(a), is originated at the STO buffer layer that shows grains with the expected in-plane epitaxial orientation, i.e. (100)STO//(100)MgO, alternated with grains 45° tilted in-plane, i.e. (110)STO//(100)MgO. For these grains, identifiable only by HR-TEM, there cannot be any crystallographic matching at the interface between STO and MgO, thus the origin of the driving force for the nucleation of these tilted grains is not clear.

In some cases it was even possible to diagnose in situ the formation of in-plane tilted grains during the growth of the STO buffer layer. Figures 1(c) and (d) show the RHEED patterns acquired after the growth of the STO buffer layer on

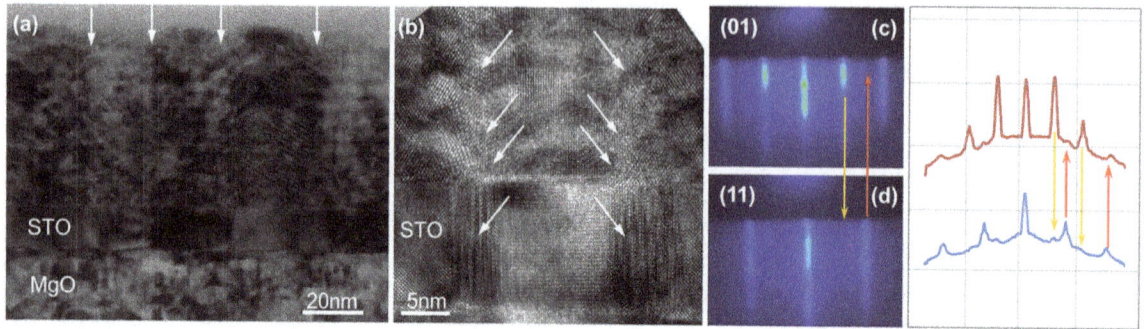

Figure 1. (a) TEM cross section micrograph of a CeO$_2$/YSZ multilayer grown on STO-buffered MgO showing the columnar morphology originating at the STO buffer layer. HR-TEM image at high magnification of the same sample is shown in (b). The arrows show the grain boundary regions. The heterostructure is epitaxially oriented with the substrate but shows two different in-plane orientations 45° tilted one another, as evidenced in (c) and (d): RHEED patterns and intensity profiles acquired after the growth of the STO layer along the (01) crystallographic direction of the substrate (c) and along the (11) direction (d).

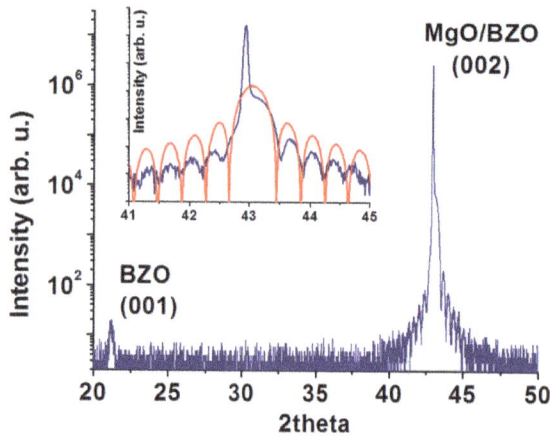

Figure 2. 2θ/θ scan plot of a 250 Å thick film of BZO on MgO. The inset shows a magnification of the angular region around the (002) peak and the fit (red line) of the interference fringes calculated for a film thickness of 57 unit cells.

MgO along the (01) and (11) in-plane directions, respectively. The analysis of the distribution of the intensity along the images shows that the in-plane (11) RHEED reflections are present also when the electron beam was aligned towards the (01) crystalline direction, and vice versa. In the case of an (almost ideal) ordered bi-axially textured growth, no overlap of the two RHEED patterns has to be expected.

To circumvent this problem, an additional seed layer of BZO was deposited between the MgO substrate and the STO buffer layer according to the design: MgO + BZO + STO.

Figure 2 shows the XRD plot of a BZO film grown on MgO. The film is about 250 Å thick, as measured by XRR. Size effect interference fringes are visible around the (002) reflection line of BZO/MgO indicating a good interface quality. The relative spacing between the interference fringes is consistent with a film thickness of about 57 unit cells ($\approx$239 Å), in good agreement with the XRR estimation of the thickness.

BZO does not reduce significantly the large lattice mismatch between MgO and STO but the formation of the previously described 45° in-plane tilted grains was never observed when the additional BZO layer was used. We believe that the different crystalline structure, cubic perovskite for BZO instead of rock salt for MgO, could be the reason that allows a more ordered and reproducible growth of the STO seed layer.

The growth and the electrical properties of the MgO + BZO + STO template platform were studied using a multilayered heterostructure of the two materials. Twenty STO/BZO bilayers were grown on BZO-buffered MgO according to the scheme: MgO + BZO + (STO + BZO) × 20. The thickness of each layer of BZO and STO was about 2 nm. Figure 3 shows the XRD, RHEED and HR-TEM analysis of this sample.

The BZO/STO heterostructure has a very high crystalline quality with no evidence of grain separation and a very small degree of out-of-plane misalignment as revealed by selected area electron diffraction (SAED) patterns. RHEED showed streaky patterns suggesting a Frank–van der Merwe (layer-by-layer) growth mechanism. The RHEED patterns did not show any significant change after the deposition of the first BZO film indicating that a well-oriented perovskite structure was preserved up to the 21st layer of the heterostructure. However, Fourier-transformed HR-TEM images revealed in some parts of the heterostructure a precession of the [100] direction of STO or BZO within a range of 4–5° and the average linear spacing between low angle grain boundary at the interface with the substrate was about 5 nm. A detailed characterization of the interfaces of BZO/STO heterostructures is also reported in reference [27] and the growth of BZO/STO superlattices on STO substrates is reported in reference [28]. The microstructural properties of the BZO/STO double buffer layer were investigated to promote the ordered growth of YBa$_2$Cu$_3$O$_{7-\delta}$ films on MgO and to avoid chemical interaction at the MgO/(Ca,Ba)Nb$_2$O$_6$ interface [27]. Antiphase boundaries and dislocations were found as lattice defects along the BZO layer, while the majority defects at the STO/

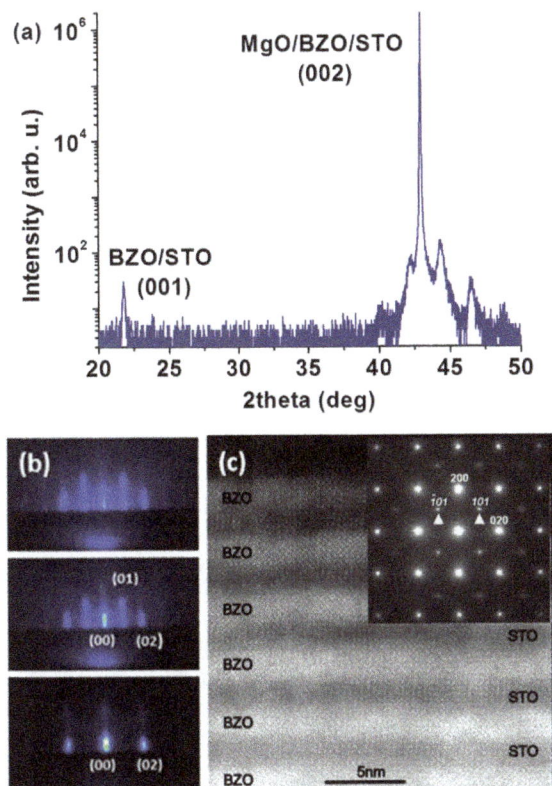

Figure 3. MgO + BZO + (STO + BZO) × 20 superlattice. (a) The XRD analysis shows the epitaxial orientation and the satellite peaks of the superlattice. (b) The RHEED patterns (from bottom to top MgO → BZO → STO) reveal an almost ideal layer-by-layer growth of the whole heterostructure. (c) The HR-STEM analysis shows a highly ordered growth of the complete structure with a very small degree of out-of-plane misalignment, as revealed by the SAED pattern shown in the inset. Spots from the MgO substrate are marked in bold while weak spots marked with arrowheads stem from the BZO/STO heterostructure.

BZO interface were identified as misfit dislocations introduced to release the excess strain. It was also found that these dislocations contribute to terminate the propagation of the BZO lattice defects toward the STO layer. This effect leads to a high crystalline quality of the STO surface [27].

The BZO/STO multi-layered heterostructure described in figure 3 was used to check the insulating properties of the BZO/STO double buffer layer used for the fabrication of the template platform, as will be discussed later.

Highly ordered SDC films and SDC/YSZ heterostructures were grown on the MgO + BZO + STO template platform, as demonstrated by figure 4(a) showing the HR-TEM cross section image of an SDC/YSZ heterostructure in the region of the double buffer layer, and by figure 5(a) showing the XRD analysis of one of the SDC films, about 200 nm thick, used for the EIS characterization. The thicknesses of the BZO and STO buffer layers of the template platform used for the electrical characterizations were about 5 nm.

From the angular position of the (002) diffraction peak of the SDC film a value of about 5.443 Å can be estimated for the out-of-plane lattice parameter revealing a relaxed crystalline structure as expected for a 200 nm thick film.

In the case of YSZ, it was not possible to reproduce the film quality reported in [15] and [16] for YSZ films grown on STO single-crystalline substrates. The growth on the STO surface of the template platform resulted in thin films mostly (001)-epitaxially oriented but with evidence of (111) minor orientation. This could be expected due to the larger lattice mismatch between YSZ film and STO substrate with respect to SDC, which may explain the different results obtained with the two materials. Moreover, the literature reports that orientation and morphology of YSZ thin films grown on STO single-crystalline wafers strongly depend on the surface termination of the substrate. In reference [14] for example, epitaxially oriented and highly ordered YSZ films could be obtained only on SrO-terminated surfaces. It might be possible to attain a SrO-terminated surface by depositing an ultra-thin layer of SrO on the STO buffer layer but this approach requires atomically flat surfaces that were not obtained for our STO films. Alternatively, a thin layer of doped or undoped CeO_2 can be used to promote a bi-axially textured growth of the YSZ film free from high angle grain boundary as shown in [24]. As an example, figure 4(b) shows the HR-TEM image of the YSZ film grown on the SDC film shown in figure 4(a).

Following this second approach, a thin layer of undoped CeO_2, about 5 nm thick, was grown *in situ* on the template platform before the deposition of the YSZ films used for the electrical characterization. These YSZ films were about 200 nm thick and showed an epitaxial orientation with the template platform as revealed by the XRD analysis reported in figure 5(b). The out-of-plane lattice parameter measured for the YSZ film is about 5.136 Å consistent with a relaxed crystalline structure. Also in this case the thicknesses of the BZO and STO buffer layers of the template platform were about 5 nm.

For comparison and for a clear identification of the small features visible on the right side of the peak of the substrate, figure 5(c) shows the XRD plot of the BZO/STO heterostructure described above.

Figure 6(a) shows the in-plane electrical conductivities of the SDC and YSZ films that were measured by EIS in the temperature range between 400 and 700 °C in air using two Pt electrodes at a distance of about 1 mm deposited onto the surface of the films. The SDC film showed activation energy of about 0.69 eV while for the YSZ film an activation energy of about 1.02 eV was measured.

These values together with the measured conductivities are in very good agreement with the value of bulk conductivities and activation energies reported in the literature for the two materials. In particular, the measured conductivity of the SDC film matches very well those measured for almost ideal single-crystalline 10% Gd-doped CeO_2 films grown (111)-oriented on sapphire [7] and for (001)-oriented SDC films grown on STO-buffered MgO [12]. Concerning YSZ, the measured conductivity and activation energy are in very

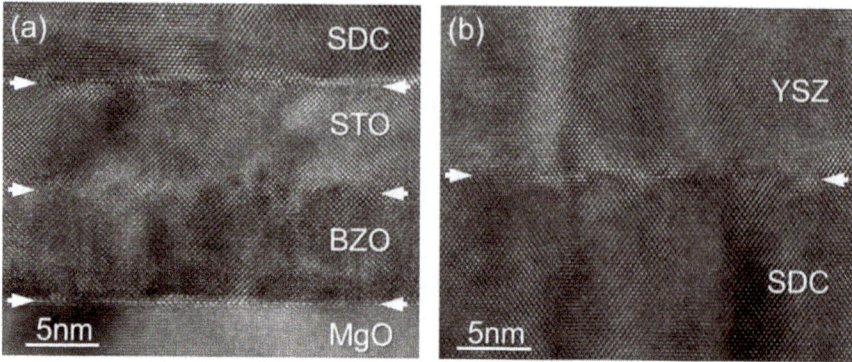

Figure 4. HR-TEM image of an SDC film grown on the MgO + BZO + STO template platform (a). The SDC layer can be used for the highly ordered growth of YSZ thin films (b).

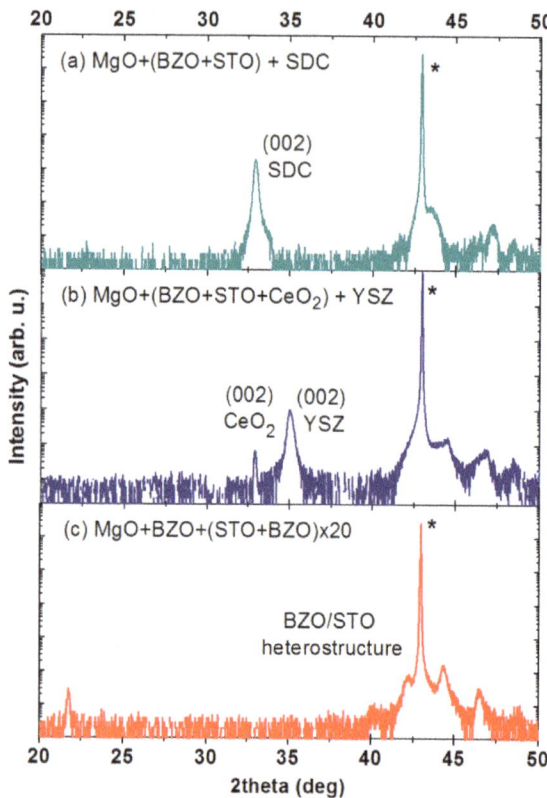

Figure 5. XRD analysis of the SDC (a) and YSZ (b) films grown on the MgO + BZO + STO template platform. For the YSZ film an additional thin layer of CeO$_2$ was used. For comparison the XRD plot of the BZO/STO heterostructure is reported (c). The asterisk indicates the (002) reflection of the MgO substrate.

range below ~500 °C. An example can be found in [30] where the bulk conductivity of yttria doped ceria was measured using single crystals prepared by inductive melting and using sintered ceramic pellets of the same material. A very good agreement was found between the two methods up to a maximum temperature of about 500 °C. However, in the temperature range of interest for practical applications, i.e. between 500 and 800 °C, only a sample morphology characterized by large average grain size and small grain boundary regions allows the direct measurement of the conducting properties of the grain interior. Such a morphology can be difficult to achieve with polycrystalline sintered samples, and highly ordered thin films grown on insulating substrates can be a valid alternative. Moreover, the described template platform could be a suitable substrate for the highly ordered and grain-boundary-free growth of ionic conducting thin films or multi-layered heterostructures used to investigate the effect of the lattice distortion on the ionic conductivity [31].

Concerning the contribution to the measured conductivity of the BZO/STO double buffer layer, figure 6(b) shows the comparison of the EIS plots measured for the YSZ film at the lowest temperature of 400 °C (the largest resistance) and for the heterostructure MgO + BZO + (STO +BZO) × 20 made with twenty STO/BZO bilayers at the highest temperature of 700 °C (the smallest resistance). The Pt electrodes were patterned with the same geometry on both samples allowing a direct comparison of the measured resistances. Taking into account that for the electrical characterizations of the SDC and YSZ films only one BZO/STO bilayer was used, these measurements clearly show that the electrical conductivity of the template platform can be neglected. Unfortunately, due to limitations of the EIS setup used for this study (in particular the geometry and dimension of the electrodes) the large electrical resistance of the BZO/STO bilayer, as well as of the heterostructure consisting on 20 BZO/STO bilayers, precluded the accurate conductivity measurement in the investigated temperature range. The use of interdigitated micro-electrodes or micropatterned side-electrodes as those reported in reference [32] has to be considered for potential future applications of this template platform; As an example for the investigation of the strain-

good agreement with values measured for 9.5 mol% YSZ single crystals [29].

It is worth highlighting here, that using sintered ceramic pellets with a typical average grain size in the range of few hundreds of nm, the EIS measurements allow to discriminate the bulk conductivity very effectively but often, depending on sample fabrication and morphology, only in the temperature

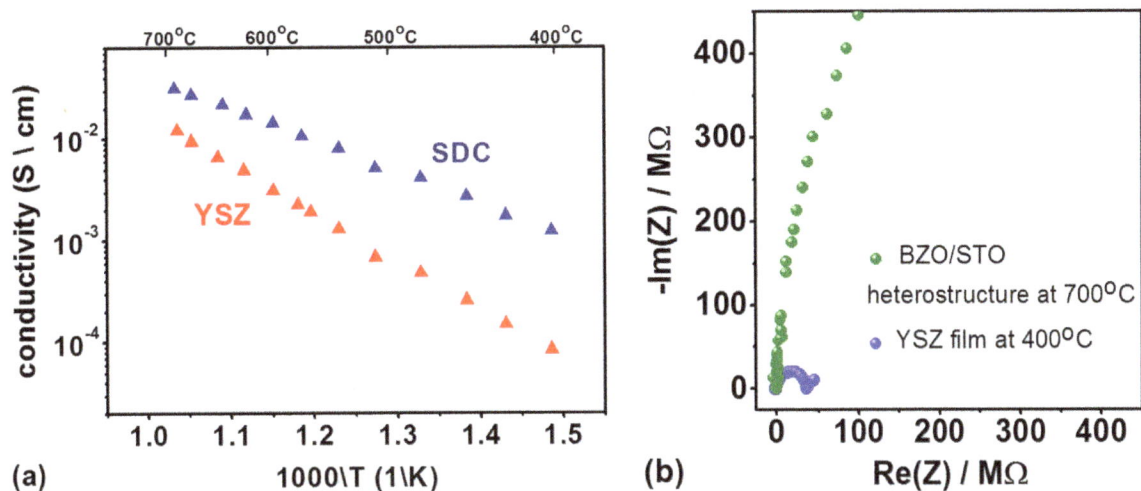

Figure 6. (a) Temperature dependence of the total electrical conductivity of SDC and YSZ thin films grown on the MgO + BZO + STO template platform. The calculated activation energies are 0.69 and 1.02 eV for SDC and YSZ, respectively. (b) Comparison of the Nyquist plots of the YSZ film at the lowest measured temperature (the larger value of resistance) and the BZO/STO superlattice (20 bilayers) at the highest temperature (the smaller value of resistance).

related effects on the ionic conductivity using ionic conducting multilayered structures or single layers whose thickness is in the range of few (tens) nm.

Finally, the measured values of conductivities and activation energies strongly suggest that surface and/or interface effects (segregation, depletion, strain), whose investigation is beyond of the scope of the present study, are not noticeable and the conductivity of the relaxed bulk volume dominates the transport properties as expected for film as thick as 200 nm. Moreover, as previously observed with similar samples [24], EIS measurements showed no evidence of chemical interdiffusion between adjacent layers.

4. Conclusions

An electrically insulating template platform allowing a bi-axially textured grain-boundary-free growth of SDC and YSZ thin films along the (001) crystallographic orientation was fabricated by PLD. The template platform consists of an epitaxial thin film heterostructure made of a double buffer layer of BZO and STO grown on MgO single-crystalline substrates. The complete growth of the template platform and of the ionic conductors was performed *in situ* and analyzed by RHEED, XRD and HR-TEM. The introduction of the BZO layer allowed a more reproducible growth of the STO seed layer that provides the suitable crystallographic matching for the ordered growth of the conductor. It is worth pointing out, that the overall crystalline quality of the fabricated samples in terms of the presence of local defects and interface roughness was found to be very similar to values reported in the literature for similar heterostructures [24]. EIS characterization showed that the presence of these local defects does not affect the conductivity measurements.

EIS analysis showed that the electrical conductivity of the template platform can be considered negligible compared to the conductivity of the ionic conducting thin films. In other words, the use of such platform provides a distinct advantage because it allows studying the ionic conductivity of highly ordered thin films not affected by the electrical conductivity of the substrate. In fact, the two ionic conducting thin films of SDC and YSZ showed the typical bulk conductivities and activation energies of the two materials.

This template platform is expected to allow the highly ordered growth and the direct measurement of the bulk conducting properties of other oxygen-ion conductors like LSGM (cubic perovskite with lattice parameter of about 391 Å), perovskite Li-ion conductors like lithium lanthanum titanate (LLTO) with lattice parameter between 3.87 and 3.88 Å, and the new family of oxygen ion conductors based on the ferroelectric perovskite $Na_{0.5}Bi_{0.5}TiO_3$ [33] whose pseudocubic unit cell possesses a lattice parameter of about 3.88 Å. Moreover, the same template platform can be used for the fabrication of highly ordered multi-layered heterostructures used to investigate the effects of the lattice distortions on the ionic conductivity [24, 31, 34].

Acknowledgements

The research leading to these results was partly supported by the World Premier International Research Centre Initiative of MEXT, Japan, and has received funding from the European Community's Seventh Framework Programme (FP7/ 2007–2013) under grant agreement n.° 290605 (COFUND: PSI-FELLOW). The authors would like to thank Dr A Tebano for useful discussion and help in crystallographic data analysis.

References

[1] Kharton V V, Marques F M B and Atkinson A 2004 *Solid State Ion.* **174** 135

[2] Wachsman E D and Lee K T 2011 *Science* **334** 935

[3] Brett D J L, Atkinson A, Brandon N P and Skinner S J 2008 *Chem. Soc. Rev.* **37** 1568

[4] Fabbri E, Bi L, Pergolesi D and Traversa E 2012 *Adv. Mater.* **24** 195

[5] Durá O J, López de la Torre M A, Vázquez L, Chaboy J, Boada R, Rivera-Calzada A, Santamaria J and Leon C 2010 *Phys. Rev.* B **81** 184301

[6] Verkerk M J, Middelhuis B J and Burggraaf A J 1982 *Solid State Ion.* **6** 159

[7] Göbel M C, Gregori G, Guo X and Maier J 2010 *Phys. Chem. Chem. Phys.* **12** 14351

[8] Göbel M C, Gregori G, Guo X and Maier J 2011 *Phys. Chem. Chem. Phys.* **13** 10940

[9] Kim S and Maier J 2002 *J. Electrochem. Soc.* **149** J73

[10] Irvine J T S, Sinclair D C and West A R 1990 *Adv. Mater.* **2** 132

[11] Matsunaga A, Kitanaka Y, Inoue R, Noguchi Y, Miyayama M and Itaka K 2014 *Key Eng. Mater.* **582** 153

[12] Joseph M, Manoravi P, Tabata H and Kawai T 2002 *J. Appl. Phys.* **92** 997

[13] Sanna S, Esposito V, Pergolesi D, Orsini A, Tebano A, Licoccia S, Balestrino G and Traversa E 2009 *Adv. Funct. Mater.* **19** 1713

[14] Cavallaro A, Ballesteros B, Bachelet R and Santiso J 2011 *Cryst. Eng. Comm.* **13** 1625

[15] Sillassen M, Eklund P, Pryds N, Johnson E, Helmersson U and Bøttiger J 2010 *Adv. Funct. Mater.* **20** 2071

[16] Gerstl M, Friedbacher G, Kubel F, Hutter H and Fleig J 2013 *Phys. Chem. Chem. Phys.* **15** 1097

[17] Pergolesi D, Fronzi M, Fabbri E, Tebano A and Traversa E 2013 *Mater. Renew. Sustain. Energy* **2** 6

[18] Gregori G, Heinze S, Lupetin P, Habermeier H U and Maier J 2013 *J. Mater. Sci.* **48** 2790

[19] Guo X 2009 *Science* **324** 465a

[20] Zaitsev A G, Ockenfuss G, Guggi D, Wördenweber R and Kruger U 1997 *J. Appl. Phys.* **81** 3069

[21] Kurian J and Naito M 2004 *Physica* C **492** 31

[22] Nandasiri M I, Nachimuthu P, Varga T, Shutthanandan V, Jiang W, Kuchibhatla S V N T, Thevuthasan S, Seal S and Kayani A N 2011 *J. Appl. Phys.* **109** 013525

[23] Scherrer B, Schlupp M V F, Stender D, Martynczuk J, Grolig J G, Ma H, Kocher P, Lippert T, Prestat M and Gauckler L J 2013 *Adv. Funct. Mater.* **23** 1957

[24] Pergolesi D, Fabbri E, Cook S N, Roddatis V, Traversa E and Kilner J A 2012 *ACS Nano* **6** 10524

[25] Pergolesi D, Fabbri E, D'Epifanio A, Di Bartolomeo E, Tebano A, Sanna S, Licoccia S, Balestrino G and Traversa E 2010 *Nat. Mater.* **9** 946

[26] Shim J H, Gür T M and Prinz F B 2008 *Appl. Phys. Lett.* **92** 253115

[27] Choudhury P R and Krupanidhi S B 2008 *J. Appl. Phys.* **104** 114105

[28] Mi S B, Jia C L, Faley M I, Poppe U and Urban K 2007 *J. Cryst. Growth* **300** 478–82

[29] Manning P S, Sirman J D, De Souza R A and Kilner J A 1997 *Solid State Ion.* **100** 1

[30] Ruiz-Trejo E, Sirman J D, Baikov Y M and Kilner J A 1998 *Solid State Ion.* **113** 565

[31] Yildiz B 2014 *MRS Bull.* **39** 147–56

[32] Schweiger S, Kubicek M, Messerschmitt F, Murer C and Rupp J L M 2014 *ACS Nano* **8** 5032

[33] Li M, Pietrowski M J, De Souza R A, Zhang H, Reaney I M, Cook S N, Kilner J A and Sinclair D C 2013 *Nat. Mater.* **13** 31

[34] Fabbri E, Pergolesi D and Traversa E 2010 *Sci. Technol. Adv. Mater.* **11** 054503

3

Non-reciprocity and topology in optics: one-way road for light via surface magnon polariton

Tetsuyuki Ochiai

Photonic Materials Unit, National Institute for Materials Science (NIMS), Tsukuba 305-0044, Japan

E-mail: OCHIAI.Tetsuyuki@nims.go.jp

Abstract

We show how non-reciprocity and topology are used to construct an optical one-way waveguide in the Voigt geometry. First, we present a traditional approach of the one-way waveguide of light using surface polaritons under a static magnetic field. Second, we explain a recent discovery of a topological approach using photonic crystals with the magneto-optical coupling. Third, we present a combination of the two approaches, toward a broadband one-way waveguide in the microwave range.

Keywords: non-reciprocity, one-way, photonic crystal, ferrite, surface magnon polariton

1. Introduction

Optical components that exclude undesirable light are indispensable in optical communication technology. Optical isolators, gyrators and circulators are among such components [1]. They commonly break reciprocal light transport. The violation of the reciprocity, that is, light transport in one direction behaves differently from that in the opposite direction, is crucial. This non-reciprocity is closely related to time-reversal symmetry (TRS) and is controlled by a static magnetic field.

In fact, conventional non-reciprocal components use the Faraday effect in which light polarization rotates along its trajectory owing to the magneto-optical coupling. Here, either a bias magnetic field or a spontaneous magnetization exists along the trajectory. To obtain a sufficient rotation angle for an isolator to work, for instance, a long optical path is necessary. This is because the rotation angle is proportional to the path length. This fact results in the rather large size of nonreciprocal optical devices.

To overcome the size demerit, the optical non-reciprocity in the Voigt geometry is desirable. In the Voigt geometry, light propagates in a direction perpendicular to the magnetic field. It enables us to construct compact non-reciprocal components. Furthermore, if the non-reciprocity works in a certain band width, we do not need to fine-tune system parameters, often employed in conventional designs using wave-interference effects.

An ideal situation is such that in a certain band width a light propagation channel exists in one direction, but does not in the opposite direction. Without any dissipation and interference, such a mode results in a non-reciprocal and one-way waveguide in a finite frequency range. In addition, the absence of the counter-propagating mode prohibits the back scattering by structural defects. Thus, the light propagation is robust against disorder.

The design of such a one-way waveguide has a long history [2]. Until recently, however, very limited systems are known: surface plasmon polaritons of metal under a magnetic field, and surface magnon polaritons of magnetic material. The polaritonic wave of the edge magneto-plasmon in a two-dimensional electron system [3] should have also a finite-band non-reciprocity below the cyclotron frequency.

In the last decade, a completely different approach to the finite-band one-way waveguide has been proposed theoretically [4–7]. Some of them are also demonstrated experimentally in the microwave range [8, 9]. This new approach utilizes a topological nature of the radiation field in photonic crystals, namely, artificial periodic structures made of different optical substances. With this approach, operating frequencies and their band width are controllable via the photonic-band engineering of underlying photonic crystals. Since there are many degrees of freedom in photonic crystals, e.g., optical substances, their geometry, and lattice constants, operating frequencies and their band width are not limited as in surface polaritons. A minimum requirement is a non-zero magneto-optical coupling (imaginary off-diagonal components of the permittivity or permeability). The resulting band width is roughly proportional to the magneto-optical coupling. Therefore, a challenge in material science is to make optical substances with large magneto-optical couplings. Although a large magneto-optical coupling of the same order as the diagonal components is available for permeability in the microwave range using ferromagnetic resonance, it is still challenging to realize a large coupling above this frequency range for permittivity or permeability.

In this paper, we propose a combination of these two approaches toward a broad-band one-way light waveguide using modulated surface magnon polaritons. A semi-infinite ferrite material with a periodic hole array in the vicinity of the sample edge strongly modulates the surface magnon polariton without the hole array. In addition, photonic bands with multiple band gaps are formed in the hole array. As a result, in a vast frequency range, a one-way light waveguide can be realized, though the frequency range is multiply divided by the photonic bands.

The paper is organized as follows. In sections 2 and 3, we briefly summarize the traditional approach using surface polaritons and the topological approach using photonic crystals, respectively. In section 4, a combined approach of these two is presented. Finally in section 5, we summarize the results obtained in this paper.

2. Non-reciprocity in surface polaritons

In optics non-reciprocity, in a narrow sense, is defined by the frequency dispersion of eigenmodes that satisfies $\omega(-k) \neq \omega(k)$. In free space, the photon dispersion is given by $\omega = c|k|/n$ with refractive index n. Obviously, it is reciprocal.

To realize the non-reciprocity, we need to break both the TRS and the space-inversion symmetry (SIS: $x \to -x$). Either the TRS or SIS alone results in $\omega(-k) = \omega(k)$. The SIS is easily broken by geometry. For instance, a flat interface between two different optical substances breaks the SIS. The TRS is broken typically by applying a static magnetic field, because the magnetic field is odd under the time reversal. However, a non-zero magnetic field does not always induce detectable non-reciprocity. In optics, effects of the TRS breaking emerge through imaginary off-diagonal components

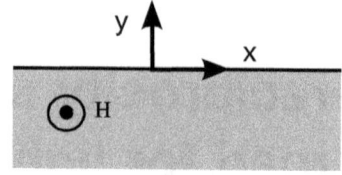

Figure 1. The geometry under consideration. The magnetic field is applied in the lower part.

in permittivity or permeability tensors. By the Onsager relation, the permittivity satisfies

$$\epsilon_{ik}(\mathbf{H}) = \epsilon_{ki}(-\mathbf{H}), \tag{1}$$

where $\mathbf{H}$ stands for a static magnetic field of either applied one $\mathbf{B}/\mu_0$ or spontaneous magnetization [10]. The same equation is satisfied for permeability. This equation along with the hermicity of the permittivity in dissipation-less systems implies that the imaginary off-diagonal components are odd functions of $\mathbf{H}$. Typically, the permittivity is written as

$$\epsilon_{ik}(\mathbf{H}) = \epsilon_s \delta_{ik} + \sqrt{-1}\,\epsilon_a \epsilon_{ijk} \hat{H}_j, \tag{2}$$

where $\hat{H}$ is the unit vector orienting $\mathbf{H}$. The ratio ϵ_a/ϵ_s is less than 10^{-3} for most transparent materials [11]. Therefore, if the imaginary off-diagonal components, ϵ_a in this case, are very small, effects of the TRS breaking are very limited. Thus, a challenge in material science is to make compound materials with large magneto-optical couplings.

In this section, we consider a flat interface between magnetic and non-magnetic media shown in figure 1 as a system without the TRS and SIS. For simplicity, we consider light propagation in the xy plane, and a static magnetic field is applied in the z direction. Obviously, the parity symmetry $\hat{\sigma}_y$ ($y \to -y$) is broken. The parity symmetry $\hat{\sigma}_x$ is also broken by the magnetic field. The parity symmetry $\hat{\sigma}_z$ is preserved irrespective of the magnetic field. As a result, the SIS $\hat{\sigma}_x \hat{\sigma}_y \hat{\sigma}_z$ is broken.

To realize the one-way road for light, we utilize a surface mode localized near the interface. Usually, such modes exist if one of the two media screens the radiation field and the other does not. Therefore, if one of the two media is taken to be a normal dielectric, the other must be a screened medium. Here, we consider two cases, screened by negative permittivity (metal) and negative permeability (ferrite).

2.1. Surface plasmon polaritons under magnetic field

One example of a finite-band non-reciprocal light waveguide is found in the surface plasmon polariton under a magnetic field. In a free-electron metal under a magnetic field along the z direction, the permittivity tensor becomes [12]

$$\overleftrightarrow{\epsilon} = \begin{pmatrix} \epsilon_\parallel & i\alpha & 0 \\ -i\alpha & \epsilon_\parallel & 0 \\ 0 & 0 & \epsilon_z \end{pmatrix}, \tag{3}$$

$$\epsilon_\parallel = 1 - \frac{\omega_p^2}{\omega^2 - \omega_c^2}, \tag{4}$$

$$\alpha = -\frac{\omega_p^2 \omega_c}{\omega\left(\omega^2 - \omega_c^2\right)}, \tag{5}$$

$$\epsilon_z = 1 - \frac{\omega_p^2}{\omega^2}, \tag{6}$$

where, ω_p and ω_c are the plasma and cyclotron frequencies, respectively. Dissipation is neglected for simplicity.

We should point out that the plasma frequency is generally in the visible-to-ultraviolet range, whereas the cyclotron frequency is in the microwave range depending on applied magnetic field. Therefore, a scale difference of order 10^6 exists between ω_p and ω_c. This difference gives very small magneto-optical coupling α in comparison to the diagonal permittivity $\epsilon_\parallel$, and results in a narrow band width of the one-way light waveguide.

Let us consider the surface plasmon polariton localized around the flat interface between metal and dielectric. This geometry supports a localized mode of the radiation field near the surface, irrespective of the applied magnetic field. That is, the surface plasmon polariton. The dispersion relation of the surface plasmon polariton under the magnetic field is given by

$$\frac{\epsilon_{a\parallel}\gamma_a^{TE} - \alpha_a k_x}{\epsilon_{a\parallel}^2 - \alpha_a^2} + \frac{\gamma_b}{\epsilon_b} = 0, \tag{7}$$

$$\gamma_a^{TE} = \sqrt{k_x^2 - \frac{\omega^2}{c^2}\frac{\epsilon_{a\parallel}^2 - \alpha_a^2}{\epsilon_{a\parallel}}\mu_a}, \tag{8}$$

$$\gamma_b = \sqrt{k_x^2 - \frac{\omega^2}{c^2}\epsilon_b\mu_b}, \tag{9}$$

where we assume the metal has the off-diagonal permittivity of equation (3) with $\epsilon_{a\parallel}$ and α_a, and the diagonal permeability μ_a. The dielectric has diagonal permittivity ϵ_b and permeability μ_b. The dynamical magnetic field of the surface plasmon polariton is polarized in the z direction. Figure 2 shows the dispersion relation of the surface plasmon polariton under a static magnetic field.

Obviously, the dispersion is non-reciprocal, and has the one-way band in the narrow frequency interval of $\omega_p/\sqrt{2} - \omega_c/2 < \omega < \omega_p/\sqrt{2} + \omega_c/2$. The band width is just ω_c, while its mid-band frequency is $\omega_p/\sqrt{2}$, namely, the surface plasmon frequency in the non-retardation limit. Therefore, the one-way band in the surface plasmon polariton is very limited.

2.2. Surface magnon polariton

As is well known, permeability in the visible and higher frequency range is almost equal to one. Dynamical magnetization can not follow applied AC magnetic field. If we consider much lower frequencies (the microwave range), however, we can enjoy large diagonal permeability along with the off-diagonal permeability (the magneto-optical

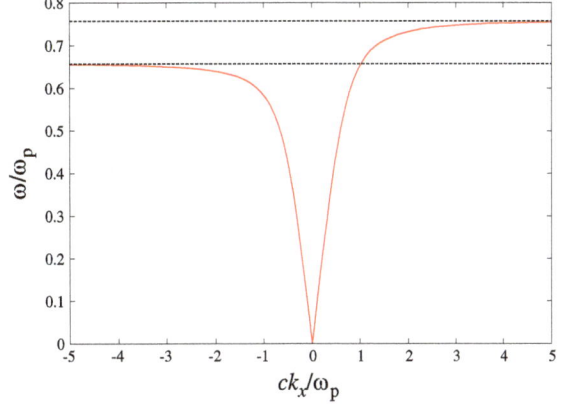

Figure 2. The dispersion relation of the surface plasmon polariton under a static magnetic field along the z direction. Just to visualize the one-way band, whose upper and lower frequencies are indicated by the dotted lines ($\omega = \omega_p/\sqrt{2} \pm \omega_c/2$), we assume $\omega_c = 0.1\omega_p$. Actually, $\omega_c \sim 10^{-6}\omega_p$, so that the one-way band width is almost invisible, if plotted in actual scale.

coupling) near the ferromagnetic resonance in ferrite materials. There, a magnetic counterpart of the surface plasmon polariton, namely, the surface magnon polariton, emerges. In a ferrite material, the spin wave is the dominating low-energy excitation. Its long-wavelength approximation is the magneto-static wave. The surface magnon is a localized version near the surface of the ferrite material. It can couple with light, forming the surface magnon polariton.

In the free-spin model of ferrite materials, the permeability is given by [1]

$$\overset{\leftrightarrow}{\mu} = \begin{pmatrix} \mu_\parallel & i\kappa & 0 \\ -i\kappa & \mu_\parallel & 0 \\ 0 & 0 & 1 \end{pmatrix}, \tag{10}$$

$$\mu_\parallel = 1 - \frac{\omega_0\omega_m}{\omega^2 - \omega_0^2}, \tag{11}$$

$$\kappa = -\frac{\omega\omega_m}{\omega^2 - \omega_0^2}, \tag{12}$$

where ω_0 is the spin precession frequency and ω_m is the saturation magnetization frequency. The precession and magnetization frequencies are given by $\mu_0\gamma H$ and $\mu_0\gamma M_s$, respectively, where μ_0 is the vacuum permeability, γ is the gyromagnetic ratio, H is the applied magnetic field in the z direction, and M_s is the saturation magnetization in the same direction. The model has the polaritonic gap in $\sqrt{\omega_0(\omega_0 + \omega_m)} < \omega < \omega_0 + \omega_m$, where the effective permeability $\mu_{eff} = (\mu_\parallel^2 - \kappa^2)/\mu_\parallel$ is negative and thus light propagation is not allowed.

The dispersion relation of the surface magnon polariton is obtained by solving the secular equation

$$\frac{\mu_{a\parallel}\gamma_a^{TM} - \kappa_a k_x}{\mu_{a\parallel}^2 - \kappa_a^2} + \frac{\gamma_b}{\mu_b} = 0, \tag{13}$$

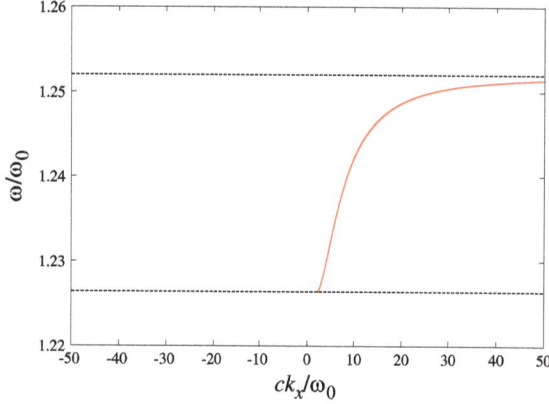

Figure 3. The dispersion relation of the surface magnon polariton under a static magnetic field along the z direction. The lower and upper frequencies of the one-way band are indicated by the dotted lines ($\omega = \sqrt{\omega_0(\omega_0 + \omega_m)}$ and $\omega = \omega_0 + \omega_m/2$, respectively). We assume $\omega_m = 0.504\omega_0$.

$$\gamma_a^{\mathrm{TM}} = \sqrt{k_x^2 - \frac{\omega^2}{c^2}\frac{\mu_{a\parallel}^2 - \kappa_a^2}{\mu_{a\parallel}}\epsilon_a}, \qquad (14)$$

where we assume the ferrite has the non-diagonal permeability of equation (10) with $\mu_{a\parallel}$ and κ_a, and the diagonal permittivity ϵ_a. The dynamical electric field of the surface magnon polariton is polarized in the z direction. In the non-retardation limit $c \to \infty$, we obtain the surface magnon (magneto-static Damon–Eshbach wave [13]) whose frequency is $\omega = \omega_0 + \omega_m/2$ available only for $k_x > 0$. For $k_x < 0$ the surface magnon is not allowed to exist. The dispersion relation of the surface magnon polariton is shown in figure 3. The dispersion relation is clearly non-reciprocal and exhibits a one-way propagation in the frequency range $\sqrt{\omega_0(\omega_0 + \omega_m)} < \omega < \omega_0 + \omega_m/2$.

A typical ferrite material is the yttrium iron garnet. It has the saturation magnetization of ~1800 Gauss. At a bias magnetic field of 3570 Oersted ($f_0 = \omega_0/2\pi = 10$ GHz), we have $\omega_m/\omega_0 = 0.504$. In this case the band-width/mid-band ratio of the surface magnon polariton is about 0.02, much larger than that in a one-way band of the surface plasmon polariton which is of order 10^{-6}.

3. Topological construction via photonic crystal

In 2008, Haldane and Raghu proposed a completely different approach to a one-way light waveguide [4]. They focused on the so-called Dirac cone found in the photonic band structure of a triangular-lattice photonic crystal. If the gap is introduced in the Dirac cone by breaking the TRS, a finite-band one-way light waveguide can be realized near the edge of the photonic crystal. Its one-way band coincides with the band gap in the Dirac cone. This is an optical analogy of chiral edge states in quantum Hall systems [14].

In condensed matter physics, it is well-established that the emergence of the Dirac cone in the electronic band structure is a critical signature in topology [15]. It is sometimes the case that the two electronic bands touch conically at a certain k point ($k = k_c$) in the Brillouin zone at a certain parameter $g = g_c$, when we sweep a physical parameter g. Before ($g < g_c$) and after ($g > g_c$) the touching, the momentum space topology, characterized by the so-called Chern number, changes abruptly [16]. If a non-zero Chern number emerges, chiral edge states are accompanied, according to the bulk-edge correspondence [17]. These chiral edge states are nothing but the one-way waveguide modes in optics. The number of the chiral edge states is given by the sum of the Chern numbers of the bands below the gap concerned.

In the proposal by Haldane and Raghu, the band touching and the separation are caused by the magneto-optical coupling. If the coupling is zero, the (gapless) Dirac cones are found at the corners of the first Brillouin zone of the triangular lattice. If it becomes nonzero, the gap is found in each Dirac cone, and the Chern number of the two bands becomes nonzero. Soon after this proposal, it is recognized that the scenario is not limited in the conical band touching (the Dirac cone), and that a quadratic band touching, which is usually found in the Brillouin zone center of the square and triangular lattices and in the Brillouin zone corner of the square lattice, is also a source of nontrivial topology [18]. In fact, chiral edge states emerge in the gap opened in the quadratic band touching, by breaking the TRS.

These theoretical proposals were confirmed experimentally in the microwave frequency range. For instance, in reference [8] the authors employed a square array of ferrite rods made of the vanadium-doped calciumirongarnet under a bias magnetic field of 0.20 Tesla. They observed a photonic band gap with the gap-width/mid-gap ratio of 6% around 4.5 GHz. In accordance with a theoretical simulation, they found unidirectional backscattering-immune edge states against metal obstacles, in the gap.

To elucidate the topological approach, let us consider photonic crystal made of ferrite rods arranged in the triangular lattice. We consider a frequency region in which the permeability tensor is not far from $\overleftrightarrow{\mu} = 1$, and the magneto-optical coupling can be viewed as a perturbation. In such frequencies, we can clearly see that the gap-opening in the Dirac cone is due to the magneto-optical coupling.

Figure 4 shows the photonic band structure of the system before and after introducing the magneto-optical coupling. We can see the band gap opening of the Dirac cone, suggesting a nontrivial topology in the gapped system. In fact, the Chern numbers of the bands are evaluated as 0 (lowest, not shown), −1 (2nd), and 2 (3rd).

Figure 5 shows the dispersion relation of the edge states in the gapped system. We can see that in the band gap around $\omega a/2\pi c = 0.62$, the dispersion curve of the edge states traverses the gap and connects the K and K' valleys of the gaped Dirac cones. As a result, the one-way transmission band is formed in the gap. We should recall that the gap opening is

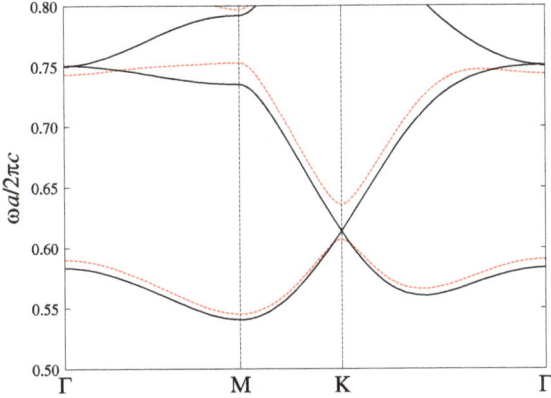

Figure 4. The photonic band structure of the transverse-magnetic polarization in the triangular array of circular ferrite rods with (solid curve) and without (dashed curve) the applied static magnetic field parallel to the rods. The background medium is air. The rod permittivity is taken to be $\epsilon = 14$, and the rod radius is $r = 0.13a$, where a is the lattice constant. As for the rod permeability, we assume $\mu_{\parallel} = 1$ and $\kappa = 0$ for the system without the magnetic field. For the system with the magnetic field, we assume dispersion-free values of $\mu_{\parallel} = 0.937$, $\kappa = -0.189$ as an approximation. These values are taken from those of equations (11) and (12) at $\omega = 3\omega_0$, provided $\omega_m = 0.504\omega_0$.

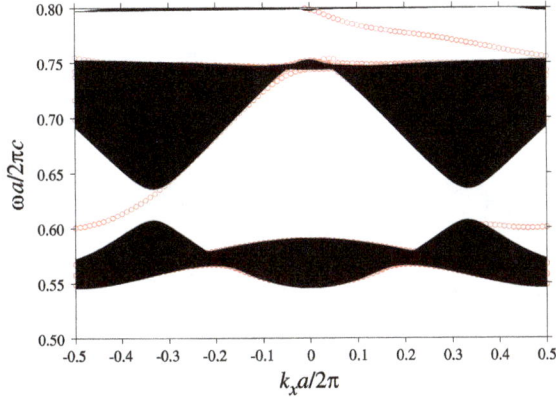

Figure 5. The dispersion relation of the edge states found in the photonic crystal edge normal to the Γ–M direction. The projected band diagram in the bulk is also plotted. The gapped photonic crystal with the same parameters as in figure 4 is employed. To clarify the edge states, we place a (perfect conductor) metal cladding away from the boundary layer with distance $(\sqrt{3}/4)a$.

due to the nonzero magneto-optical coupling κ. The gap width is proportional to κ in the degenerate perturbation around the Dirac point. The proportional constant reflects the properties of the degenerate mode at the Dirac point. Thus, the one-way band width there can be controlled by a photonic-band-mode design.

It is also remarkable that the topological approach can yield both right- and left-going one-way modes without inverting the magnetic field. For comparison, in the surface polaritons, the chirality is fixed depending on the sign of the

magnetic field. In addition, more than one edge mode can emerge in a given gap. This is because the number of the edge states is equal to the sum of the Chern numbers of the bands below the gap, and the sum is not limited in ± 1. Actually, we have the right-going mode in the Dirac cone gap around $\omega a/2\pi c = 0.62$. At the same time, the left-going mode is found in the gap around $\omega a/2\pi c = 0.78$, without inverting the magnetic field. The numbers and the chiralities of the edge states in the gaps are fully consistent with the Chern numbers of the photonic bands.

Finally, we should note that the topological approach to the one-way transport is quite universal. Basic ingredients are a periodic structure, a TRS breaking, and a band touching/ separation. Therefore, this approach can be extended to various wave phenomena on periodic structures, for instance, spin wave in magnonic crystal [19]. There, chiral edge modes of spin waves are predicted. We also note that, recently, the topological approach is extended to time-reversal invariant photonic systems inspired by the physics of topological insulators. A theoretical prediction indicates that a meta-crystal composed of a certain type of bi-anisotropic cylinders holds a helical-edge state [20]. There, spin-dependent one-way light transport can be realized. These extensions enable us to control various waves in unprecedented manners.

4. Combined structure of surface magnon polaritons and photonic crystals

In the previous section, we assumed a photonic crystal composed of a magneto-optical rod array. The one-way waveguide is obtained via the gap opening due to nonzero κ, irrespective of polaritonic effects. It is purely a topological effect, because a non-dispersive $\overset{\leftrightarrow}{\mu}$ is assumed for the rods. Here, we present a combined approach using both the polaritonic effect and topological effect. With this approach, we can enlarge the band width of the one-way waveguide.

The idea is as follows. Suppose we have a periodic hole array in a semi-infinite ferrite material with the flat interface at the edge as shown in figure 6. Depending on the geometry of the hole array, the photonic band structure is formed. Photonic band gaps can be found irrespective of the frequencies of the polariton gap in the ferrite material. These gaps act as new polariton gaps, in which light propagation is not allowed in the bulk. Therefore, the gaps can support 'new' surface magnon polaritons around the edge. The new polaritons are also affected by the topology of the photonic bands below the gap, according to the bulk-edge correspondence. Thus, the two approaches discussed in the previous sections are combined, giving us a chance to enhance the bandwidth of the one-way road for light.

Let us consider the hole array of the triangular lattice embedded in a ferrite material. The lattice constant is chosen such that $\omega_0 a/2\pi c = 0.3$ with $\omega_m = 0.504\omega_0$. The photonic band structure of the bulk hole-array system is shown in figure 7. Since the periodic modulation by the hole array is very strong, many photonic-band gaps are formed particularly above $\omega a/2\pi c > 0.4$. The photonic bands in this region are

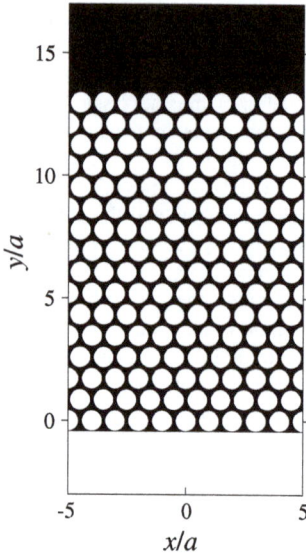

Figure 6. Triangular hole array in a semi-infinite ferrite material. The region outside the material is air.

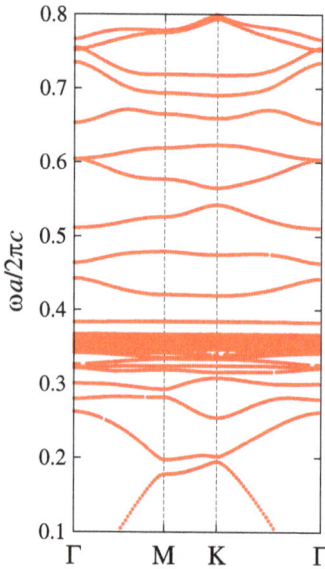

Figure 7. The photonic band structure of the transverse-magnetic polarization in the triangular array of circular holes with radius $0.4a$ in a ferrite material. The following parameters are assumed for the ferrite material: $\epsilon = 14$, $\omega_0 a/2\pi c = 0.3$ and $\omega_m a/2\pi c = 0.1512$. The polariton gap of the background ferrite material ranges from $0.3678 < \omega a/2\pi c < 0.4512$.

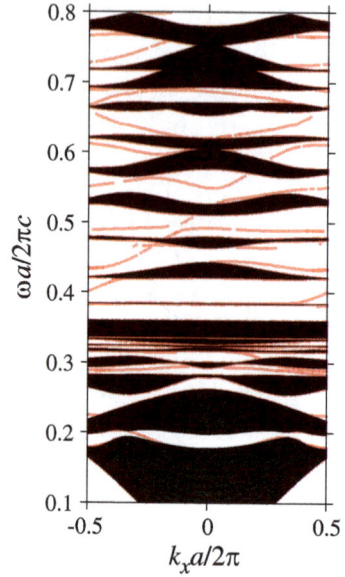

Figure 8. The dispersion relation of the edge states in the ferrite photonic crystal. The same parameters as in figure 7 are employed. As in figure 6, we assume a semi-infinite ferrite material with the periodic hole array of the triangular lattice, in the vicinity of the material edge. The region outside the ferrite material is air. The hole array has 16 layers along the Γ–M direction. The distance between the boundary hole layer and the flat interface is $(\sqrt{3}/4)a$. The edge states shown are localized around the flat interface. Those localized near the opposite-boundary hole layer are not observed.

index $n = \sqrt{\epsilon\left(\mu_\parallel^2 - \kappa^2\right)/\mu_\parallel}$ of the ferrite becomes so large there.

The dispersion curve of the edge states is shown in figure 8. Each dot represents a peak in the optical density of states as a function of parallel momentum k_x to the boundary. All peaks are plotted irrespective of their peak widths. Most peaks are inside the light cone of either the air side or ferrite side. Therefore, the peaks correspond to quasi-guided edge states, which can couple with incident light. The peaks form the dispersion curves of the edge states. We can find several vacancies in the dispersion curves, because the relevant peaks become broad, or merge with the singularity of the Wood anomaly (found in the diffraction thresholds).

A striking feature is found above $\omega a/2\pi c \sim 0.4$. There, photonic-band gaps are relatively wide and the dispersion curves of the edge states are clearly visible. It is remarkable that the edge-state dispersion curves are almost continuous and chiral (having positive slopes) in a wide frequency range or $0.4 < \omega a/2\pi c < 0.55$, taking account of the periodicity in the surface Brillouin zone. Although the curves are separated by the photonic band regions, the band-width/mid-band ratio of the one-way band is about 0.3 if the intercepting photonic band regions are neglected. This value is much larger than that in the homogeneous ferrite material, which is just 0.02.

Therefore, the periodic hole array near the edge of the semi-infinite ferrite material supports the broad-band one-way waveguide of the modulated surface magnon polariton.

not necessarily topological, because the bands are separated not simply by the magneto-optical coupling. Rather, the periodic modulation in the permittivity with a large contrast seems to play a major role in the gap formation. For comparison, the polariton gap of the homogeneous ferrite material ranges from $0.3678 < \omega a/2\pi c < 0.4512$. Just below the gap, a dense band region is found. This is because the refractive

5. Conclusions

We have presented constructions of optical one-way waveguides using the surface polaritons and topological effect. They use the magneto-optical coupling in the permittivity or permeability tensors in common. The construction via the polariton utilizes the polaritonic gap, and the resulting band of the one-way waveguide is very limited both in its frequency range and band width. The topological construction utilizes photonic crystal, and is free from these limitations. However, the resulting one-way band depends strongly on the magneto-optical coupling of the optical substances used in the photonic crystal. We propose a combined design of these two approaches, using a periodic hole array in a semi-infinite ferrite material. It enables us to construct a broad-band one-way waveguide, in the vicinity of the edge of the semi-infinite specimen.

Acknowledgment

This work was supported by JSPS KAKENHI (Grant No. 23540380).

References

[1] Pozar D M 2005 *Microwave Engineering* (New York: Wiley)
[2] Camley R 1987 *Surf. Sci. Rep.* **7** 103–87
[3] Volkov V and Mikhailov S 1988 *Sov. Phys. JETP* **67** 1639–53
[4] Haldane F D M and Raghu S 2008 *Phys. Rev. Lett.* **100** 013904
[5] Wang Z, Chong Y D, Joannopoulos J D and Soljačić M 2008 *Phys. Rev. Lett.* **100** 013905
[6] Ao X, Lin Z and Chan C T 2009 *Phys. Rev.* B **80** 033105
[7] Ochiai T and Onoda M 2009 *Phys. Rev.* B **80** 155103
[8] Wang Z, Chong Y D, Joannopoulos J D and Soljačić M 2009 *Nature* **461** 772–5
[9] Poo Y, Wu R X, Lin Z, Yang Y and Chan C T 2011 *Phys. Rev. Lett.* **106** 093903
[10] Landau L D, Lifshitz E M and Pitaevskii L P 1985 *Electrodynamics of Continuous Media* (Oxford: Butterworth-Heinemann)
[11] Zvezdin A K and Kotov V A 1997 *Modern Magnetooptics and Magnetooptical Materials* (New York: CRC Press)
[12] Dresselhaus G, Kip A F and Kittel C 1955 *Phys. Rev.* **100** 618–25
[13] Damon R W and Eshbach J 1961 *J. Phys. Chem. Solids* **19** 308–20
[14] Girvin S M 1999 Course 2: the quantum Hall effect: novel excitations and broken symmetries *Topological Aspects of Low Dimensional Systems* ed A Comtet, T Jolicoeur, S Ouvry and F David (Berlin: Springer) p 53
[15] Oshikawa M 1994 *Phys. Rev.* B **50** 17357–63
[16] Avron J E, Seiler R and Simon B 1983 *Phys. Rev. Lett.* **51** 51–53
[17] Hatsugai Y 1993 *Phys. Rev. Lett.* **71** 3697–700
[18] Chong Y D, Wen X G and Soljačić M 2008 *Phys. Rev.* B **77** 235125
[19] Shindou R, Matsumoto R, Murakami S and Ohe J I 2013 *Phys. Rev.* B **87** 174427
[20] Khanikaev A B, Mousavi S H, Tse W K, Kargarian M, MacDonald A H and Shvets G 2013 *Nat. Mater.* **12** 233–9

4

Effects of SO$_2$ on selective catalytic reduction of NO with NH$_3$ over a TiO$_2$ photocatalyst

Akira Yamamoto[1], Kentaro Teramura[1,2,3], Saburo Hosokawa[1,2] and Tsunehiro Tanaka[1,2]

[1] Department of Molecular Engineering, Graduate School of Engineering, Kyoto University, Kyotodaigaku Katsura, Nishikyo-ku, Kyoto 615-8510, Japan
[2] Elements Strategy Initiative for Catalysts & Batteries (ESICB), Kyoto University, Kyotodaigaku Katsura, Nishikyo-ku, Kyoto 615-8520, Japan
[3] Precursory Research for Embryonic Science and Technology (PRESTO), Japan Science and Technology Agency (JST), 4-1-8 Honcho, Kawaguchi, Saitama 332-0012, Japan

E-mail: teramura@moleng.kyoto-u.ac.jp

Abstract

The effect of SO$_2$ gas was investigated on the activity of the photo-assisted selective catalytic reduction of nitrogen monoxide (NO) with ammonia (NH$_3$) over a TiO$_2$ photocatalyst in the presence of excess oxygen (photo-SCR). The introduction of SO$_2$ (300 ppm) greatly decreased the activity of the photo-SCR at 373 K. The increment of the reaction temperature enhanced the resistance to SO$_2$ gas, and at 553 K the conversion of NO was stable for at least 300 min of the reaction. X-ray diffraction, FTIR spectroscopy, thermogravimetry and differential thermal analysis, x-ray photoelectron spectroscopy (XPS), elemental analysis and N$_2$ adsorption measurement revealed that the ammonium sulfate species were generated after the reaction. There was a strong negative correlation between the deposition amount of the ammonium sulfate species and the specific surface area. Based on the above relationship, we concluded that the deposition of the ammonium sulfate species decreased the specific surface area by plugging the pore structure of the catalyst, and the decrease of the specific surface area resulted in the deactivation of the catalyst.

Keywords: photocatalyst, NH$_3$-SCR, SO$_2$

1. Introduction

The emission of nitrogen oxides (NO$_x$) causes air pollution problems such as acid rain and photochemical smog. Selective catalytic reduction (SCR) of NO$_x$ with ammonia (NH$_3$) is a commercial de-NO$_x$ process in the stationary and mobile NO$_x$ emission sources. The main reaction is as follows:

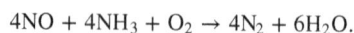

$$4NO + 4NH_3 + O_2 \rightarrow 4N_2 + 6H_2O.$$

In stationary emission sources, the industrial catalyst for the process is vanadium-oxide-based catalysts such as V$_2$O$_5$–WO$_3$/TiO$_2$ [1, 2]. The catalysts show high activity and high selectivity to N$_2$ in the temperature window of 523–673 K. This type of catalyst has to be installed upstream to the particulate collector and to the flue-gas desulfurization unit in order to meet the optimum working temperature. However, the catalysts are not available in diesel engines because of the wide temperature window (423–773 K) of the exhaust gas. A promising catalyst which shows high activity at low temperatures is strongly desired for diesel engines.

Mn-based catalysts were eagerly investigated due to their high activity at low temperatures [3–8]. Mn-based catalysts show almost 100% of the conversion in the temperature range of 373–473 K, although the selectivity to N_2 is slightly lower in some reaction conditions [3, 4]. The problem with Mn-based catalysts is low resistance to SO_2 poisoning. Kijlstra et al [9] proposed that the formation of $MnSO_4$ is the main reason for the deactivation of MnO_x/Al_2O_3 catalysts. The formed sulfates decomposed at 1020 K, which means regeneration of the catalysts is only possible at much higher temperatures than the reaction temperature. On this point, the simultaneous pursuit of the operation at low temperatures and the high resistance to SO_2 gas is a challenging and important topic in de-NO_x technology.

We have previously reported the photo-assisted SCR of NO with NH_3 over TiO_2 photocatalysts in the presence of excess oxygen (O_2) at ambient temperature. In the photocatalytic system, high conversion and high selectivity were achieved at the gas hourly space velocity (GHSV) of $25\,000\,h^{-1}$ (conversion of NO > 90%, selectivity to N_2 > 95%) [10]. We have carried out various investigations: the elucidation of the reaction mechanism using spectroscopic methods [11] and kinetic analysis [12], the improvement of the activity by metal doping [13] and combining of the temperature effect with the photo-SCR [14], and the development of a visible-light-sensitive photocatalyst with dye-sensitization [15]. Other research groups also carried out a theoretical study [16], which verified the proposed reaction mechanism by our group [11, 17], and a kinetic study using an annular fixed-film photoreactor [18]. However, the potential of the photocatalyst for the resistance to SO_2 gas has not been investigated yet, although the resistance is essential to make the photo-SCR system practicable. Thus, our objective is to investigate the effect of SO_2 gas on the performance of photo-SCR over the TiO_2 photocatalyst. In addition, the systematic characterization of the catalyst after the reaction was carried out to elucidate the deactivation mechanism of the TiO_2 photocatalyst.

2. Experimental details

2.1. Materials

The TiO_2 (ST-01) powder was purchased from Ishihara Sangyo Kaisha, Ltd. Before use, the TiO_2 power was hydrated in distilled water for 2 h at 353 K and then evaporated to dryness at 353 K. The powder after the hydration was tabletted (diameter 20 mm) and calcined at 673 K for 3 h in a furnace under a dry air flow. The tablet was granulated using 25 and 50 mesh sieves to obtain the granules with a diameter of 300–600 μm. Ammonium sulfate (wako) and sodium

sulfate (wako) were used as reference samples without further purification.

2.2. Photocatalytic reaction

Photo-SCR was carried out using a conventional fixed bed flow reactor at an atmospheric pressure in the same way as in our previous reports [14]. A quartz reactor was used for the reaction, and the reactor volume was 0.12 mL (12 × 10 × 1 mm). 110 mg of TiO_2 granules with a diameter of 300–600 μm were introduced to the reactor and pretreated at 673 K in a 10% O_2/He gas at a flow rate of 50 mL min^{-1} for 60 min. The reaction gas composition was as follows: NO (1000 ppm), NH_3 (1000 ppm), O_2 (2%), SO_2 (300 ppm, if present), He balance. A 200 W Hg–Xe lamp equipped with fiber optics, collective lens and a mirror (San-Ei Electric Co., Ltd, UVF-204 S type B) was used as a light source. The measured light irradiance was 360 mW cm^{-2}. N_2 and N_2O products were analyzed by a SHIMADSU GC-8A TCD gas chromatograph with MS-5A and Porapak Q columns, respectively.

2.3. Characterization

The crystalline phase of the TiO_2 powder was determined by the x-ray diffraction (XRD) technique using a Rigaku Ultima IV x-ray diffractometer with Cu-Kα radiation ($\lambda = 1.5406$ Å). The crystallite size was determined from the full width at half maximum (FWHM) of the diffraction peak of the anatase TiO_2 (101) plane ($2\theta = 25.2°$) using the Scherrer equation. The Fourier transform infrared (FTIR) transmission spectra were recorded on a JASCO FT/IR-4200 spectrometer at room temperature at a spectral resolution of 4 cm^{-1}, accumulating 16 scans. The background spectrum was measured without any sample in air and was subtracted from the sample spectra. The catalysts before and after the reaction and the reference samples (ammonium sulfate and sodium sulfate) were diluted with KBr by the sample-to-KBr ratio of 1.5:98.5 and 0.2:99.8 (w/w), respectively, and then pressed into pellets. Thermogravimetric (TG) analysis and differential thermal analysis (DTA) were performed on a Rigaku Thermo plus TG 8120 apparatus at a heating rate of 5 K min^{-1} under a dry air flow condition at a flow rate of 80 mL min^{-1} in the range of 298–1173 K using Al_2O_3 pans. The XPS measurement was conducted on a Shimadzu ESCA-3400 spectrometer. Samples were mounted on a silver sample holder using a conductive carbon tape and were analyzed using Mg Kα radiation in a vacuum chamber in 0.1 eV steps. The position of the carbon peak (284.6 eV) for C1s was used to calibrate the binding energy for all the samples. The surface composition was estimated by the band areas of the XP spectrum of S2p, N1s, Ti2p and O1s and the corresponding relative sensitivity factors [19]. Elemental analyses (EA) were performed on two CHN analyzers (MT-5, Yanaco Co., Ltd and JM10, J-Science Lab Co., Ltd) to analyze the contents of C, H and N, and combustion ion chromatography (Dionex ICS-1500, Mitsubishi Chemical Analytech AQF-2100H) was used to analyze the S content. The N_2 adsorption/desorption isotherm was

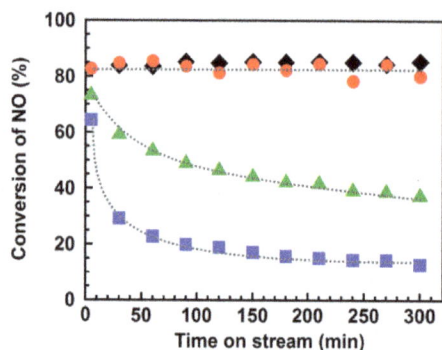

Figure 1. Time course of the photo-SCR in the presence or absence of SO_2 gas at various temperatures. ($\blacklozenge$) SO_2: 0 ppm, 433 K, ($\blacktriangle$) SO_2: 300 ppm, 373 K, ($\blacksquare$) SO_2: 300 ppm, 433 K, ($\bullet$) SO_2: 300 ppm, 553 K. NO: 1000 ppm, NH_3: 1000 ppm, O_2: 2%, He: balance gas, flow rate: 200 mL min^{-1}, GHSV: 100 000 h^{-1}, light source: 200 W Hg–Xe lamp.

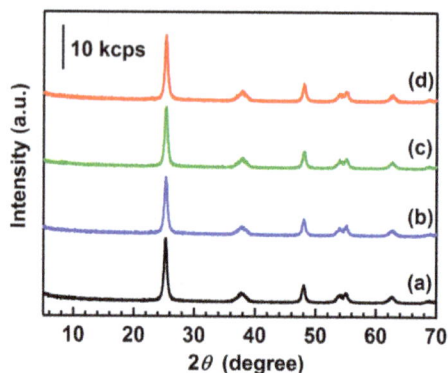

Figure 2. XRD patterns of the catalysts before and after the reaction. (a) BR, (b) AR-373K, (c) AR-433K and (d) AR-533K. Patterns are offset for clarity.

measured at 77 K using liquid nitrogen. The Brunauer–Emmett–Teller (BET) method was utilized to calculate the specific surface areas (S_{BET}). The total specific surface area (S_{tot}) and the external specific surface area (S_{ext}) were calculated from the linear fitting of the V–t plots (fitting range: 0–0.6 nm for the S_{tot} and 3.6–4.6 nm for the S_{ext}). The internal specific surface (S_{int}) area for the mesoporous materials was obtained by subtracting the S_{ext} from the S_{tot}. The pore-sized distribution was calculated from the Barrett–Joyner–Halenda (BJH) plots.

3. Results and discussion

3.1. Effect of SO_2 addition on the activity of the photo-SCR

Figure 1 shows the time course of the photo-SCR over the TiO_2 photocatalyst under illumination in the presence or absence of SO_2 (300 ppm) gas at various temperatures. The conversion of NO was stable for 300 min at 433 K in the absence of SO_2 gas, as we reported previously [14]. In the presence of SO_2 at 433 K, the conversion of NO decreased with the reaction time, which indicates that the SO_2 poisoned the catalyst, as in the case of the Mn-based catalysts [9]. The reaction temperature had a significant effect on the deactivation rate of the TiO_2 photocatalyst. At 373 K, the conversion decreased more rapidly than at 433 K, and the conversion was almost stable for at least 300 min at 553 K. For simplification, the sample before the reaction was abbreviated as BR, and the samples after the reactions at 373 K, 433 K and 553 K were abbreviated as AR-373K, AR-433K and AR-553K, respectively.

3.2. XRD patterns

The XRD patterns of the catalysts are shown in figure 2. In all the catalysts, only the diffraction pattern of anatase TiO_2 was observed. Crystalline sizes of anatase TiO_2 were estimated by the Scherrer equation using the diffraction peaks of (101), and

Table 1. Crystalline size and specific surface area of catalysts.

Sample	d^a (nm)	S_{BET} [b] (m^2 g^{-1})	S_{tot} [c] (m^2 g^{-1})	S_{ext} [d] (m^2 g^{-1})	S_{int} [e] (m^2 g^{-1})
BR	15.2	121	123	10.6	112
AR-373K	15.0	84.3	87.0	10.8	76.2
AR-433K	15.2	96.6	99.6	9.0	90.6
AR-553K	15.3	115	118	9.0	109

[a] Crystalline size calculated from the FWHM of the diffraction peak of the anatase TiO_2 (101) plane.
[b] Specific surface area determined by N_2 adsorption isotherm at 77 K using the BET method.
[c] Total specific surface area calculated from the V–t plots.
[d] External specific surface area calculated from the V–t plots.
[e] Internal specific surface area calculated by subtracting the external specific surface area from the total specific surface area.

the results are listed in table 1. The crystalline size did not change (about 15 nm) after the reaction, revealing that the aggregation of TiO_2 particles did not occur under the reaction conditions at all the reaction temperatures.

3.3. N_2 adsorption/desorption experiments

In N_2 adsorption/desorption experiments at 77 K (figure S1), all the samples exhibited a typical IV-type isotherm and had a vertically long hysteresis loop in the relative pressure (P/P_0) range of 0.8–1.0, suggesting that all the samples had a porous structure. The pore-sized distribution of the catalyst before the reaction had a sharp peak at 10 nm (figure S2(A)), and the size was the same order of the crystalline size of TiO_2 (15 nm). These results indicate that the mesopores were formed by the gaps between primary TiO_2 particles. The peak became smaller after the reaction in the presence of SO_2 gas. The specific surface area calculated from the V–t plots (figure S3(B)) is summarized in table 1. The S_{tot} calculated from V–t plots was in good agreement in the S_{BET}. The S_{ext} of all the samples was estimated from the V–t plots as 9.0–10.8 m^2 g^{-1}. The S_{int} was calculated by subtracting the S_{ext} from the S_{tot}. The S_{int} was 76.2–112 m^2 g^{-1}, which shows

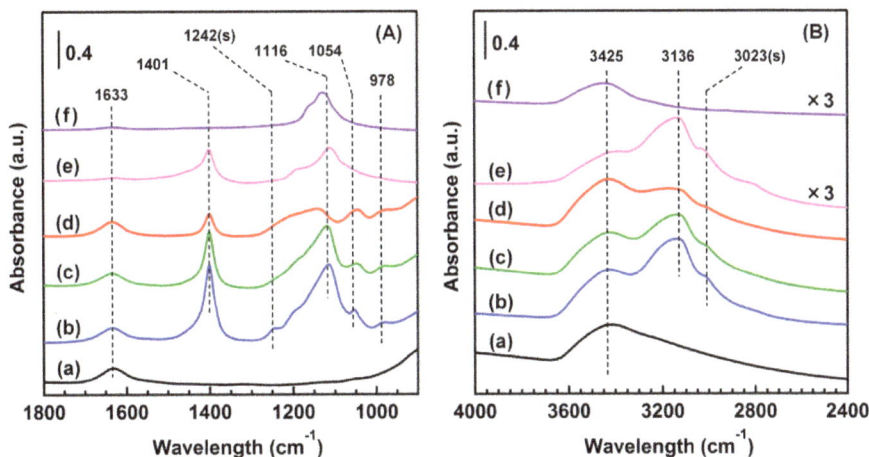

Figure 3. FTIR spectra of the catalysts and the reference samples in the region of (A) 900–1800 cm^{-1} and (B) 2400–4000 cm^{-1}. (a) BR, (b) AR-373K, (c) AR-433K, (d) AR-533K, (e) (NH$_4$)$_2$SO$_4$ and (f) Na$_2$SO$_4$. Spectra are offset for clarity.

that the largest part of the S_{tot} was derived from the mesopores.

3.4. FTIR spectra

Figure 3(A) shows the FTIR spectra of the catalysts in the region of 900–1800 cm^{-1}. In all the catalysts, a band at 1633 cm^{-1} was observed and was attributed to the deformation vibration of water molecules adsorbed on the TiO$_2$ surface. New bands at 1401, 1242, 1116, 1054 and 978 cm^{-1} appeared after the reaction in the presence of SO$_2$ at 373 K. The sharp band at 1401 cm^{-1} was assigned to the bending vibration of ammonium (NH$_4^+$) ions [20] and was also observed in the case of the reference (NH$_4$)$_2$SO$_4$ powder. Free sulfate ions (SO$_4^{2-}$, T_d symmetry) show two infrared peaks at 1104 (ν_3) and 613 (ν_4) cm^{-1} [20]. The band at 1116 cm^{-1} was due to ν_3 vibration of free sulfate ions and was also observed in both the cases of (NH$_4$)$_2$SO$_4$ and Na$_2$SO$_4$. When a SO$_4^{2-}$ ion is bound to the TiO$_2$ surface, the symmetry can be lowered to either C_{3v} or C_{2v}. The lowering symmetry causes the split of the ν_3 vibration band into two peaks for a C_{3v} symmetry and splits into three peaks for a C_{2v} symmetry [20]. Thus, the bands at 1242, 1054 and 978 cm^{-1} are assigned to the surface-coordinated SO$_4^{2-}$ ions. The surface-coordinated SO$_4^{2-}$ ions could have the C_{2v} symmetry based on the number of bands. The SO$_4^{2-}$ ion with a C_{2v} configuration is either chelating bidentate or bridge bidentate [20]. In AR-433K, the shape of the spectrum was similar to that of AR-373K, although the absorbance of the bands at 1401 and 1116 cm^{-1} were slightly weaker than those of AR-373K. In AR-553K, the band at 1116 cm^{-1} disappeared, which means that the deposition of free sulfates was inhibited in the reaction at 553 K, although the other bands at 1401, 1054 and 978 cm^{-1} remained. In the region of 2400–4000 cm^{-1} (figure 3(B)), three adsorption bands at 3425, 3136 and 3023 cm^{-1} were observed after the reaction at 373 K. The broad band between 3600–2800 cm^{-1} is the stretching

vibration of OH groups derived from surface hydroxyl groups and adsorbed water molecules, which also appeared in the sample before the reaction. The other two bands were observed in the catalysts after the reaction. The bands at 3136 and 3023 cm^{-1} are attributed to the asymmetric stretching vibration (ν_3) and symmetric stretching vibration (ν_1) of the NH$_4^+$ ions, respectively. The two bands also appeared in the case of (NH$_4$)$_2$SO$_4$, which strongly advocated the generation of NH$_4^+$ ions after the reaction. The peak intensity at 3136 and 3023 cm^{-1} decreased as the reaction temperature increased, which was consistent with the decrease of the band at 1401 cm^{-1} in figure 3(A). The FTIR results clearly revealed the generation of the free and surface-coordinated SO$_4^{2-}$ ions and NH$_4^+$ ions on the TiO$_2$ surface after the reaction. The conversion of NO after 300 min of the reaction decreased in the following order: AR-553K (80.1%) > AR-433K (37.2%) > AR-373K (12.9%). The order was consistent with that of the peak intensities of the free SO$_4^{2-}$ ions and NH$_4^+$ ions, which suggests that the generation of ammonium sulfate species (e.g. (NH$_4$)$_2$SO$_4$ and (NH$_4$)HSO$_4$) induced the deactivation of the catalyst.

3.5. TG-DTA analysis

TG profiles of the catalysts are shown in figure 4(A). Several steps of the weight loss were observed in all the catalysts after the reaction, although the TiO$_2$ has only one step of the weight loss around 330 K. From room temperature to 1173 K, the weights of BR, AR-373K, AR-433K and AR-553K decreased by 3.9, 16.1, 12.5 and 8.1%, respectively. Figure 4(B) shows the DTA profiles of the catalysts. All the profiles had a strong exothermic band around 1150 K without the weight loss, which was derived from the phase transition of TiO$_2$ from anatase to rutile. In the DTA profile of BR (see the inset of figure 4(B)), a broad exothermic band was observed around 820 K without the weight loss. The peak was observed in all the catalysts before and after the reaction and was possibly due to the crystallization of TiO$_2$ [21]. In

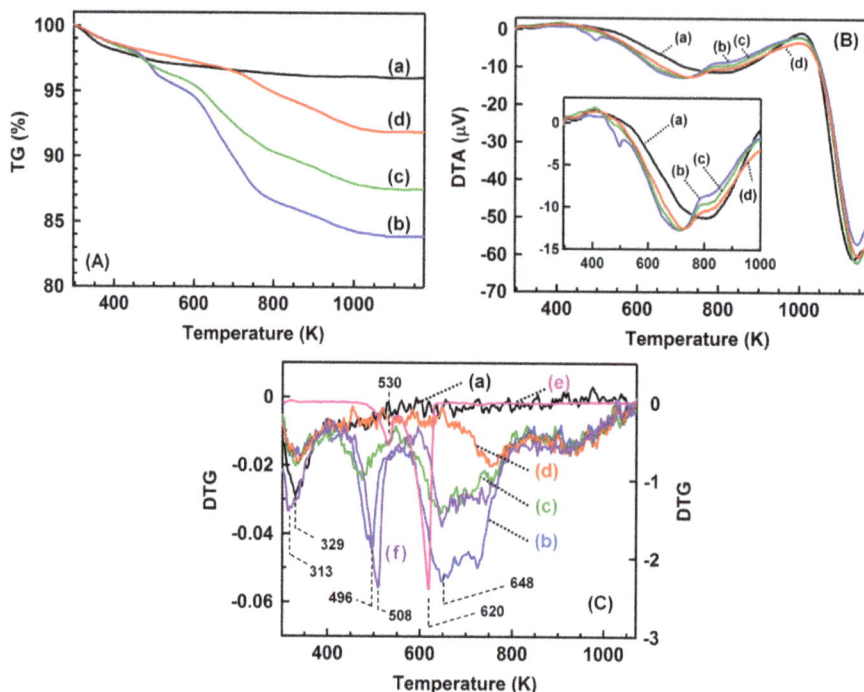

Figure 4. (A) TG profiles, (B) DTA profiles and (C) derivative thermogravimetry (DTG) profiles of the catalysts and the reference samples. (a) BR, (b) AR-373K, (c) AR-433K, (d) AR-553K, (e) $(NH_4)_2SO_4$ and (f) physical mixture of $(NH_4)_2SO_4$ and TiO_2 ($(NH_4)_2SO_4$: 10% by weight).

addition, other exothermic bands were observed around 500 and 700 K in the case of the catalysts after the reaction.

Figure 4(C) shows the first derivatives of the TG profiles (DTG) in figure 4(A). The negative band around 330 K was observed in all the catalysts and was assigned to desorption of molecular water. Other weight loss peaks were observed around 496 K, 648 K, 740 and 940 K. DTG profiles of the $(NH_4)_2SO_4$ powder and a physical mixture sample of the $(NH_4)_2SO_4$ powder and the TiO_2 powder are also shown in figure 4(C). In the $(NH_4)_2SO_4$ powder, two peaks were observed at 530 K and 620 K, which were attributed to the following reactions: (1) and (2), respectively [22]

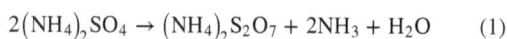

$$2(NH_4)_2SO_4 \rightarrow (NH_4)_2S_2O_7 + 2NH_3 + H_2O \quad (1)$$

$$3(NH_4)_2S_2O_7 \rightarrow 6SO_2 + 2NH_3 + 2N_2 + 9H_2O. \quad (2)$$

In the physical mixture sample of the $(NH_4)_2SO_4$ powder and the TiO_2 powder, the weight loss profile had negative peaks around 313 K, 508 K, 648 K, 760 K and 940 K, and the peak positions were in good agreement with those of the profile of AR-373 K. The analogy of the profiles strongly supports the generation of the $(NH_4)_2SO_4$ species on the TiO_2 surface after the reaction at 373 K. The bands around 500 K and 650 K decreased as the reaction temperature increased from 373 K to 553 K, which suggests that the increase of the reaction temperature inhibits the deposition of the $(NH_4)_2SO_4$ species on the TiO_2 surface.

3.6. XPS

In the S 2p XP spectra of BR (figure 5(A)), no band was observed in the range of 165–175 eV. In all the catalysts after the reaction, asymmetric bands were observed at the peak position of 168.6–168.8 eV. The asymmetry of the bands is because of the overlap of the split sublevels of the $2p_{3/2}$ and $2p_{1/2}$ states of S atoms (separation of bands: 1.2 eV) by spin–orbit coupling [23]. The spectra of the catalysts after the reaction were reasonably fitted using two Gaussian functions (figure S3 in the supplementary data and table S1). In all the catalysts after the reaction, the peak positions were 168.5–168.6 and 169.7–169.9 eV, which corresponded to the peak positions of S $2p_{3/2}$ and S $2p_{1/2}$ of the sulfate (SO_4^{2-}) species, respectively [20]. The ratios of the peak areas of S $2p_{3/2}$ to those of S $2p_{1/2}$ were 1.96–1.99 in all the catalysts after the reaction, and the values were in good agreement with the theoretical value of 2. The FWHM of the each peak was the same in all the catalysts (about 1.6 eV), which suggests that the S 2p XP spectra of the catalysts are derived from a single sulfur species of SO_4^{2-} (S^{6+}).

The N 1s XP spectra are shown in figure 5(B). The peak positions of the N bands of the NO_2^- and NO_3^- species on the TiO_2 surface were reported to be 403.5 eV and 407.0 eV, respectively [24, 25]. There was no band in the region of 403.5–407.0 eV, indicating that the NO_2^- and NO_3^- species did not exist in all the catalysts after the reaction. In AR-373K, there were two bands with the peak positions at 399.9 and 401.8 eV. The shoulder band at 399.9 eV was attributed to N

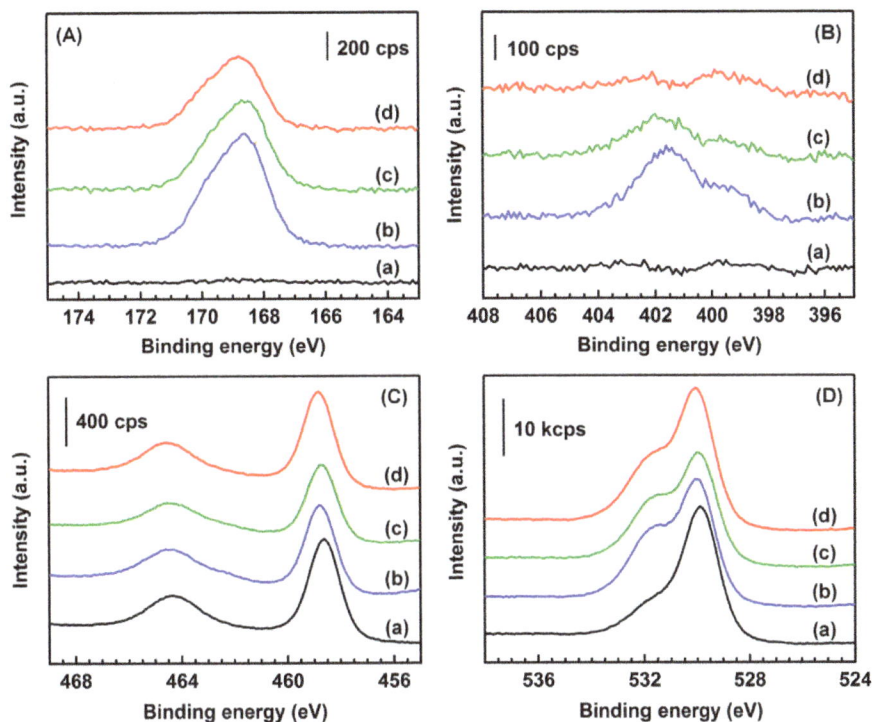

Figure 5. XP spectra of (A) S 2p, (B) N 1s, (C) Ti 2p and (D) O 1s of the catalysts. (a) BR, (b) AR-373K, (c) AR-433K and (d) AR-533K. Patterns are offset for clarity.

atoms of NH_3 adsorbed on Lewis acid sites of TiO_2, and the band at 401.8 eV was attributed to N atoms of the NH_4^+ species adsorbed on Brønsted acid sites and/or ammonium salts [20]. The band at 401.8 eV decreased with increasing the reaction temperature and disappeared in the spectrum of AR-533K.

In the Ti 2p XP spectra (figure 5(C)), the binding energies of the band of Ti $2p_{1/2}$ and $2p_{3/2}$ were 458.6–558.8 and 464.4–464.6 eV in all the catalysts before and after the reaction (table S2 in the supplementary data). The values were consistent with those reported in the literature [20]. The O 1s XPS bands (figure 5(D)) of BR were fitted by two Gaussian functions. The peak positions of the two bands were 530.9 eV and 529.8 eV. The band with a higher binding energy of 530.9 eV was due to the O atom of water molecules adsorbed on the surface and surface hydroxyl groups, and the band at 529.8 eV corresponded to the lattice oxygen on the TiO_2 surface [26]. The spectra of the samples after the reaction were fitted using two Gaussians. The result of the fitting is shown in table S3. The relative intensities of the higher energy band to the lower energy band increased after the reaction, which was interpreted by the generation of sulfate species because the oxygen atoms in sulfate have a band at 532.4 eV, as shown in figure S4 in the supplementary data.

Table 2 shows the surface composition estimated by the XPS analysis. The ratio of S atoms to Ti atoms (S/Ti) decreased in the order of AR-373K > AR-433K > AR-553K. The N atoms to Ti atoms (N/Ti) ratio also decreased with

Table 2. Surface composition estimated by XPS.

Sample	S/Ti	N/Ti	N/S
BR	0	0	—
AR-373K	0.28	0.11	0.41
AR-433K	0.27	0.07	0.26
AR-553K	0.14	0.01	0.10
$(NH_4)_2SO_4$	—	—	0.85

increasing the reaction temperature. The decreases of the S/Ti and N/Ti ratio should be mainly due to the decomposition of the $(NH_4)_2SO_4$ species based on the FTIR and TG-DTA analyses. The ratio of N atoms to S atoms (N/S) decreased from 0.41 (373 K) to 0.10 (553 K) with the reaction temperature. The N/S values of the catalysts were lower than that of $(NH_4)_2SO_4$ powder (0.85). The low N/S values of the catalysts suggests the existence of sulfur species other than $(NH_4)_2SO_4$. From the XPS analysis, SO_3^{2-} species were not detected, and the valence of all the sulfur species was +6. The FTIR spectroscopy revealed the generation of the surface-coordinated SO_4^{2-} species with the C_{2v} and/or C_{3v} symmetries. Thus, the low N/S values were due to the generation of the surface-coordinated SO_4^{2-} species. Based on the above discussion, the decrease of the N/S values with the increase of the reaction temperature is interpreted by the preferential decrease of the $(NH_4)_2SO_4$ species compared to the surface-coordinated SO_4^{2-} species with increasing the reaction temperature.

Table 3. Results of elemental analysis.

Sample	Concentration (wt%)		Surface density (nm^{-2})		N/S (atom/atom)
	S	N	S	N	
BR	0.06	0	0.09	0	0
AR-373K	3.5	2.6	5.5	9.3	1.7
AR-433K	2.6	1.6	4.1	5.6	1.4
AR-553K	1.7	0.64	2.7	2.3	0.85

3.7. Elemental analysis (EA)

Concentrations of sulfur atoms and nitrogen atoms were estimated by EA (table 3). The concentration of S atoms and N atoms increased after the reaction and decreased with increasing the reaction temperature. The tendency corresponded to the results of XPS analysis. Surface densities of each atom were calculated by dividing the concentrations by the BET specific surface area of BR (table 1). The surface density of the Ti atoms was calculated to be 7.0 nm^{-2} using a (100) plane of anatase TiO$_2$ [27] and the BET specific surface area of BR. In AR-373K, the surface densities of sulfur atoms and nitrogen atoms were 5.5 and 9.3 nm^{-2}, respectively, and the sum of the values were almost twice as much as that of Ti atoms. The result suggests the generation of a bulk (NH$_4$)$_2$SO$_4$ species. The N/S ratio estimated by EA (table 3) had the same tendency as that evaluated by XPS: the N/S ratio decreases with increasing the reaction temperature. In AR-373K, the N/S ratio was 1.7 and was close to 2, which was the theoretical value of the chemical composition of (NH$_4$)$_2$SO$_4$. Thus, the biggest part of the SO$_4$$^{2-}$ species is present as a (NH$_4$)$_2$SO$_4$ salt in AR-373K. However, the N/S ratio by the XPS analysis is 0.41, which was almost half of the experimental value of the reference (NH$_4$)$_2$SO$_4$ powder (table 2). The discrepancy between N/S ratios estimated by EA and XPS could be interpreted by considering the generation of the bulk (NH$_4$)$_2$SO$_4$ species. XPS analysis is more sensitive to

surface-coordinated SO$_4$$^{2-}$ species than bulk (NH$_4$)$_2$SO$_4$ species. The N/S ratio estimated by XPS should be lower than the real amount of the (NH$_4$)$_2$SO$_4$ species, which was estimated by the EA, when the (NH$_4$)$_2$SO$_4$ species has a bulk structure. Thus, the lower N/S ratio by XPS than that by EA also implies the generation of the bulk (NH$_4$)$_2$SO$_4$ species.

3.8. Structure of the ammonium sulfate species and deactivation mechanism

N$_2$ adsorption/desorption experiments revealed the decrease of the S_{tot} after the reaction, which was mainly due to the decrease of the S_{int}. The decrease of the S_{int} is not due to the aggregation of the TiO$_2$ particles during the reaction because the crystalline size of the TiO$_2$ particles did not change after the reaction (table 1). FTIR, TG-DTA, XPS and EA revealed the generation of the bulk (NH$_4$)$_2$SO$_4$ species on the TiO$_2$ surface after the reaction. The amount of (NH$_4$)$_2$SO$_4$ in AR-373K, which contains the largest amount of S and N atoms among the three catalysts after the reaction, is calculated to be 12 wt% assuming that all the N atoms, which were estimated by EA, exist in the (NH$_4$)$_2$SO$_4$ form. However, the XRD diffraction peak of the bulk (NH$_4$)$_2$SO$_4$ species was not observed in AR-373K, which implies that the generated bulk (NH$_4$)$_2$SO$_4$ species has an amorphous structure. Thus, the generated bulk (NH$_4$)$_2$SO$_4$ species plugged the pores of the catalysts, which resulted in the decrease of the S_{int} (figure 6).

The amount of the (NH$_4$)$_2$SO$_4$ species decreased with increasing the reaction temperature. Figures 7(A) and (B) show a correlation between the S_{tot} and the contents of N and S estimated by the EA. The increment of the contents of S and N drastically decreased the S_{tot}. FTIR analysis revealed two SO$_4$$^{2-}$ species: one is the (NH$_4$)$_2$SO$_4$ species and the other is the surface-coordinated SO$_4$$^{2-}$ species. The negative linear correlation for the N content strongly suggests that the decrease of the S_{tot} is not because of the generation of the surface-coordinated SO$_4$$^{2-}$ species but is because of the bulk (NH$_4$)$_2$SO$_4$ species. The conversion of NO after 300 min of the reaction was plotted vs. the S_{tot} (figure 8). The strong

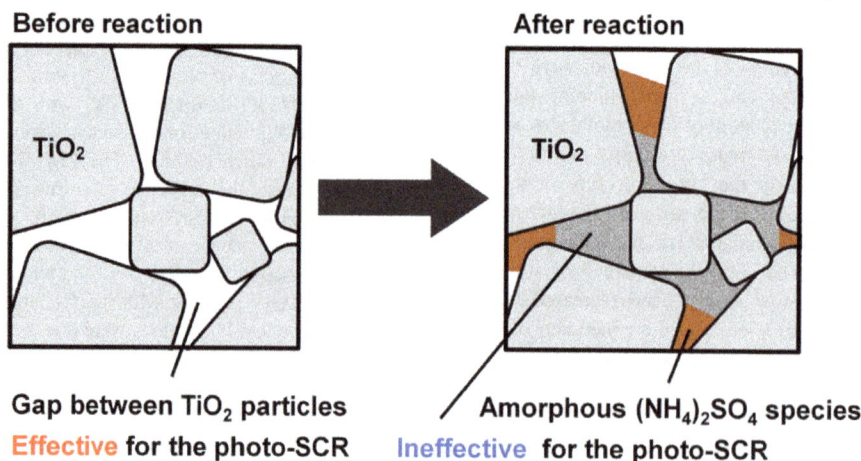

Before reaction

TiO$_2$

After reaction

TiO$_2$

Gap between TiO$_2$ particles
Effective for the photo-SCR

Amorphous (NH$_4$)$_2$SO$_4$ species
Ineffective for the photo-SCR

Figure 6. Deposition model of sulfate species of the catalysts.

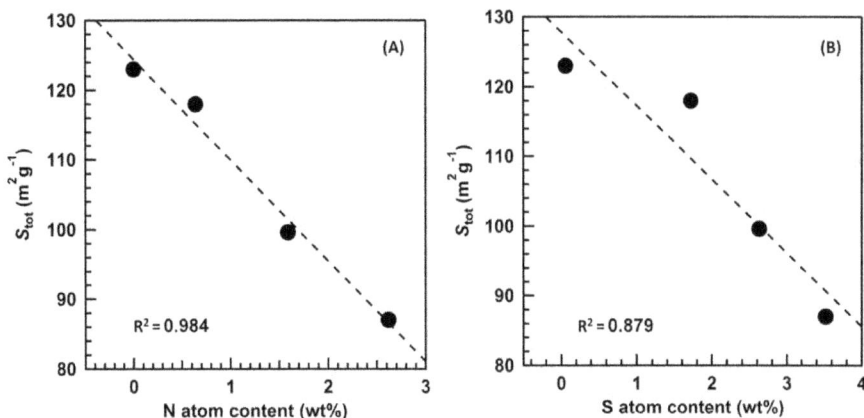

Figure 7. Effect of the contents of N atoms (A) and S atoms (B) on the S_{tot} estimated from the V–t plots.

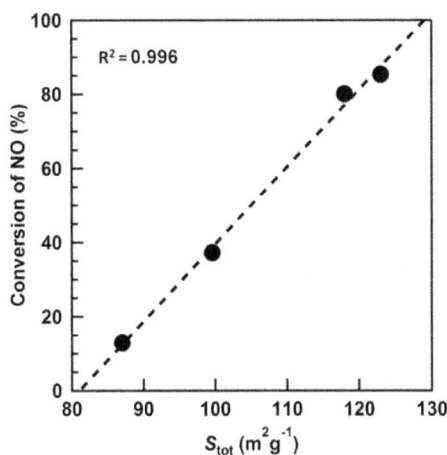

Figure 8. Effect of the S_{tot} estimated by the V–t plots on the conversion of NO after 300 min of the photo-SCR at various temperatures.

positive and linear correlation was obtained, which indicates that the decrease of the S_{tot} results in the decrease of the conversion of NO. Based on the above discussion, we concluded that the generation of the $(NH_4)_2SO_4$ species plugged a part of the mesopores derived from the gap between the TiO_2 particles, which resulted in the decrease of the S_{tot} and the deactivation of the catalyst.

4. Conclusions

In this study, tolerance to SO_2 gas was investigated in the photo-SCR of NO with NH_3 over the TiO_2 photocatalyst. The introduction of SO_2 drastically decreased the conversion of NO at 373 K. The increment of the reaction temperature drastically improved the stability of the catalyst, and at 553 K, the deactivation was not observed for 300 min of the reaction. FTIR and XPS results suggest that two SO_4^{2-} species exist on the TiO_2 surface: the surface-coordinated SO_4^{2-} species and

the free SO_4^{2-} species as a bulk $(NH_4)_2SO_4$ formed after the photocatalytic reaction. The deactivation occurs due to a pore plugging by the deposition of the $(NH_4)_2SO_4$ species on the TiO_2 surface on the basis of the correlation among the contents of N and S atoms after reaction, the S_{tot} estimated by V–t plots and the photocatalytic activity.

Acknowledgments

This study was partially supported by the Program for Element Strategy Initiative for Catalysts & Batteries (ESICB), commissioned by the Ministry of Education, Culture, Sports, Science and Technology (MEXT) of Japan, and the Precursory Research for Embryonic Science and Technology (PRESTO), supported by the Japan Science and Technology Agency (JST). Akira Yamamoto thanks the JSPS Research Fellowships for Young Scientists.

References

[1] Bosch H and Janssen F 1988 Formation and control of nitrogen oxides *Catal. Today* **2** 369

[2] Busca G, Lietti L, Ramis G and Berti F 1998 Chemical and mechanistic aspects of the selective catalytic reduction of NO_x by ammonia over oxide catalysts: a review *Appl. Catal.* B **18** 1

[3] Singoredjo L, Korver R, Kapteijn F and Moulijn J 1992 Alumina supported manganese oxides for the low-temperature selective catalytic reduction of nitric oxide with ammonia *Appl. Catal.* B **1** 297

[4] Kapteijn F, Singoredjo L, Andreini A and Moulijn J A 1994 Activity and selectivity of pure manganese oxides in the selective catalytic reduction of nitric oxide with ammonia *Appl. Catal.* B **3** 173

[5] Long R Q, Yang R T and Chang R 2002 Low temperature selective catalytic reduction (SCR) of NO with NH_3 over Fe-Mn based catalysts *Chem. Commun.* 452

[6] Li J, Chen J, Ke R, Luo C and Hao J 2007 Effects of precursors on the surface Mn species and the activities for NO reduction over MnO_x/TiO_2 catalysts *Catal. Commun.* **8** 1896

[7] Tang X, Hao J, Yi H and Li J 2007 Low-temperature SCR of NO with NH$_3$ over AC/C supported manganese-based monolithic catalysts *Catal. Today* **126** 406

[8] Li J, Chang H, Ma L, Hao J and Yang R T 2011 Low-temperature selective catalytic reduction of NO$_x$ with NH$_3$ over metal oxide and zeolite catalysts—a review *Catal. Today* **175** 147

[9] Sjoerd Kijlstra W, Biervliet M, Poels E K and Bliek A 1998 Deactivation by SO$_2$ of MnOx/Al2O3 catalysts used for the selective catalytic reduction of NO with NH$_3$ at low temperatures *Appl. Catal. B* **16** 327

[10] Yamazoe S, Okumura T, Teramura K and Tanaka T 2006 Development of the efficient TiO$_2$ photocatalyst in photoassisted selective catalytic reduction of NO with NH$_3$ *Catal. Today* **111** 266

[11] Teramura K, Tanaka T and Funabiki T 2003 Photoassisted selective catalytic reduction of NO with ammonia in the presence of oxygen over TiO$_2$ *Langmuir* **19** 1209

[12] Teramura K, Tanaka T, Yamazoe S, Arakaki K and Funabiki T 2004 Kinetic study of photo-SCR with NH$_3$ over TiO$_2$ *Appl. Catal. B* **53** 29

[13] Yamazoe S, Masutani Y, Teramura K, Hitomi Y, Shishido T and Tanaka T 2008 Promotion effect of tungsten oxide on photo-assisted selective catalytic reduction of NO with NH$_3$ over TiO$_2$ *Appl. Catal. B* **83** 123

[14] Yamamoto A, Mizuno Y, Teramura K, Shishido T and Tanaka T 2013 Effects of reaction temperature on the photocatalytic activity of photo-SCR of NO with NH$_3$ over a TiO$_2$ photocatalyst *Catal. Sci. Technol.* **3** 1771

[15] Yamamoto A, Mizuno Y, Teramura K, Hosokawa S, Shishido T and Tanaka T 2015 Visible-light-assisted selective catalytic reduction of NO with NH$_3$ on porphyrin derivative-modified TiO$_2$ photocatalysts *Catal. Sci. Technol.* **5** 556

[16] Ji Y and Luo Y 2014 First-principles study on the mechanism of photoselective catalytic reduction of NO by NH$_3$ on anatase TiO$_2$(101) surface *J. Phys. Chem. C* **118** 6359

[17] Yamazoe S, Teramura K, Hitomi Y, Shishido T and Tanaka T 2007 Visible light absorbed NH$_2$ species derived from NH$_3$ adsorbed on TiO$_2$ for photoassisted selective catalytic reduction *J. Phys. Chem. C* **111** 14189

[18] Chou Y and Ku Y 2012 Selective reduction of NO by photo-SCR with ammonia in an annular fixed-film photoreactor *Front. Environ. Sci. Eng.* **6** 149

[19] Scofield J H 1976 Hartree–Slater subshell photoionization cross-sections at 1254 and 1487 eV *J. Electron Spectrosc. Relat. Phenom.* **8** 129

[20] Chen J P and Yang R T 1993 Selective catalytic reduction of NO with NH$_3$ on SO$_4^{2-}$/TiO$_2$ superacid catalyst *J. Catal.* **139** 277

[21] Hague D C and Mayo M J 1994 Controlling crystallinity during processing of nanocrystalline titania *J. Am. Ceram. Soc.* **77** 1957

[22] Halstead W D 1970 Thermal decomposition of ammonium sulphate *J. Appl. Chem.* **20** 129

[23] Lindberg B J 1970 Substituent effects of sulfur groups: III. The influence of conjugation on ESCA spectra of sulfur substituted nitrobenzenes *Acta Chem. Scand.* **24** 3661

[24] Rodriguez J A, Jirsak T, Liu G, Hrbek J, Dvorak J and Maiti A 2001 Chemistry of NO$_2$ on oxide surfaces: formation of NO$_3$ on TiO$_2$(110) and NO$_2 \leftrightarrow$ O vacancy interactions *J. Am. Chem. Soc.* **123** 9597

[25] Haubrich J, Quiller R G, Benz L, Liu Z and Friend C M 2010 *In situ* ambient pressure studies of the chemistry of NO$_2$ and water on rutile TiO$_2$(110) *Langmuir* **26** 2445

[26] Wagner C D, Ringgs W M, Davis L E and Moulder J F 1979 *Handbook of X-Ray Photoelectron Spectroscopy* (Eden Prairie, MN: Perkin-Elmer Corp)

[27] Lu Y, Choi D-J, Nelson J, Yang O B and Parkinson B A 2006 Adsorption, desorption, and sensitization of low-index anatase and rutile surfaces by the ruthenium complex dye N$_3$ *J. Electrochem. Soc.* **153** E131

5

Engineering Dirac electrons emergent on the surface of a topological insulator

Yukinori Yoshimura[1], **Koji Kobayashi**[2], **Tomi Ohtsuki**[2] **and Ken-Ichiro Imura**[1]

[1] Department of Quantum Matter, AdSM, Hiroshima University, Higashi-Hiroshima, 739-8530, Japan
[2] Department of Physics, Sophia University, Chiyoda-ku, Tokyo, 102-8554, Japan

E-mail: imura@hiroshima-u.ac.jp

Abstract
The concept of the topological insulator (TI) has introduced a new point of view to condensed-matter physics, relating *a priori* unrelated subfields such as quantum (spin, anomalous) Hall effects, spin–orbit coupled materials, some classes of nodal superconductors, superfluid ^{3}He, etc. From a technological point of view, TIs are expected to serve as platforms for realizing dissipationless transport in a non-superconducting context. The TI exhibits a gapless surface state with a characteristic conic dispersion (a surface Dirac cone). Here, we review peculiar finite-size effects applicable to such surface states in TI nanostructures. We highlight the specific electronic properties of TI nanowires and nanoparticles, and in this context we contrast the cases of weak and strong TIs. We study the robustness of the surface and the bulk of TIs against disorder, addressing the physics of Dirac and Weyl semimetals as a new research perspective in the field.

Keywords: topological insulator, intrinsic Aharonov–Bohm effect, Dirac monopole, weak topological insulator, perfectly conducting channel

1. Introduction

Topological insulators [1–3] enable dissipationless transport in a non-superconducting context. They may allow for fixing 'neutrinos' in semiconductor chips. Known examples of the topological insulator include materials such as Bi_2Se_3 [4], Bi_2Te_3 [5], and many of their relatives [6, 7]. These are three-dimensional (3D) bulk materials under the influence of relatively strong spin–orbit coupling.

In the bulk of a sample, a topological insulator is, at least superficially, no different from a normal band insulator, but on the surface, it is quite different. A topological insulator exhibits a gapless surface state whose energy traverses the bulk energy gap. It is often said that the bulk energy gap of a topological insulator is 'inverted' [8, 9] in comparison with the normal one, though giving a precise meaning to this phrase needs further formulation of the bulk-effective Hamiltonian [10, 11]. Whether the gap is inverted or not is specified by a winding (topological) number that encodes the bulk band structure, and is in one-to-one correspondence with whether the system exhibits a gapless surface state (bulk-boundary correspondence) [12].

The surface state of a topological insulator exhibits a gapless spectrum often represented by the word 'Dirac cone', described by an (effective) two-dimensional (2D) gapless Dirac equation. Dirac electrons represented by such a Dirac equation must have a real or fictitious active spin degree of freedom. In the case of graphene, another platform for realizing such 2D Dirac electrons, this role is played by the sublattice degrees of freedom. Here, on the surface of a

(a)

(b)

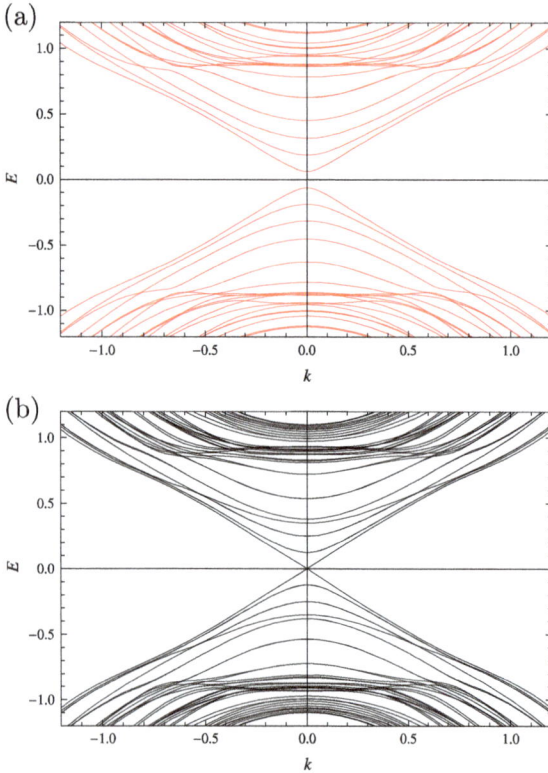

Figure 1. Typical energy spectrum of a rectangular, prism-shaped topological insulator nanowire. (a) As a consequence of the spin Berry phase, π, the spectrum of the surface state is gapped. (b) The same spectrum becomes gapless in the presence of an external flux, π, inserted along the axis of the prism to cancel the Berry phase.

topological insulator, the spin is real, and since its direction is locked to the momentum (spin-to-momentum locking), the corresponding Dirac cone is often said to be *helical*. On the generically curved surface of topological insulator samples of an arbitrary shape, the same spin also locks the local tangent to the surface (spin-to-surface locking). In topological insulator nanostructures, this second type of locking, which is effective in real space, plays an important role in determining the low-energy spectrum of the surface Dirac states. In the first half of this paper, we review such finite size effects associated with the spin-to-surface locking.

Another aspect of the 3D topological insulator is that it has two subclasses: weak and strong [13–15]. The strong topological insulator exhibits a single, or more generally, an odd number of Dirac cones on its surface, while its weak counterpart exhibits an even number of Dirac cones in the surface Brillouin zone. A single Dirac cone is robust against disorder, while an even number of Dirac cones could be more easily destroyed [16, 17]. However, being robust does not necessarily equate to being *useful* [18, 19]. Recall the differences between semiconductors and metals. Metal is always conducting, while a semiconductor is either conducting or insulating, depending on the concentration of impurities. This property of the semiconductor makes it more useful than a metal, at least for some purposes, such as functioning as a

transistor. The case of a weak topological insulator is somewhat similar; the fragility of the even number of Dirac cones makes the Dirac electrons emergent on the surface of a weak topological insulator controllable. In the second half of the paper, we argue that the controllability of the weak topological insulator surface states paves the way for constructing a dissipationless nanocircuit simply by patterning its surface by using lithography and etching.

2. Peculiar finite size effects in topological insulator surface states—part 1: the strong case

The helical Dirac electron emergent on the surface of a topological insulator is described in the $\mathbf{k} \cdot \mathbf{p}$ approximation by the following effective surface Dirac Hamiltonian,

$$H_{\text{surf}} = v_F \left(p_y \sigma_z - p_z \sigma_y \right). \tag{1}$$

Here, the surface is chosen to be normal to the x-axis, and v_F determine the aperture of the Dirac cone. The 2×2 matrix structure of equation (1) is due to the real spin degree of freedom. The form of equation (1) implies that momentum eigenstates have a spin that is oriented perpendicular to the direction of the momentum. The other way around, the spin is locked to the momentum: spin-to-momentum locking. This is the reason why the corresponding Dirac cone is said to be *helical*.

Another feature encoded in the explicit form of equation (1) is that the spin of the momentum eigenstates has no out-of-plane component. The spin is indeed locked (in-plane) to the surface. This feature, sometimes represented by the term spin-to-surface locking, also holds true on generically curved surfaces of a sample of arbitrary shape [20, 21]. Here, we will focus on cases of *cylindrical* and *spherical* shaped samples, in particular.

2.1. Intrinsic Aharonov–Bohm effect

Here, let us first focus on the *cylindrical* case by physically targeting a topological insulator nanowire. On the cylindrical surface of such a nanowire, equation (1) is modified as [22]

$$H_{\text{surf}} = v_F \begin{bmatrix} 0 & -ip_z + \frac{\hbar}{R}\left(-i\frac{\partial}{\partial\phi} + \frac{1}{2}\right) \\ ip_z + \frac{\hbar}{R}\left(-i\frac{\partial}{\partial\phi} + \frac{1}{2}\right) & 0 \end{bmatrix}, \tag{2}$$

where R is the radius of the cylinder. The additive factor, $1/2$, that appears in the two off-diagonals of equation (2) is a spin Berry phase encoding the constraint on the direction of spin as a consequence of the spin-to-surface locking. When a Dirac electron moves around the curved, cylindrical surface of the topological insulator nanowire, the spin-to-surface locking constrains the direction of the spin to follow the change of the tangential surface. As a result, when an electron completes a 2π rotation in the configuration space, its spin also performs a 2π rotation in the spin space, while the spin is naturally double-valued. This explains the origin of the additive factor, 1/2, or in other words, a Berry phase, π [19, 20, 22, 23].

The orbital part of an electron state on the surface of a cylindrical topological insulator is represented as

$$\psi(\phi, z) = e^{\frac{i}{\hbar}(L_z\phi + p_z z)}, \tag{3}$$

where $L_z = p_\phi R$ is the orbital angular momentum in the z-direction, along the axis of the cylinder. The Berry phase, π, modifies the boundary condition with respect to the polar angle, ϕ, from *periodic* to *anti-periodic*—that is,

$$e^{\frac{i}{\hbar}L_z(\phi+2\pi)} \times (-1) = e^{\frac{i}{\hbar}L_z\phi}. \tag{4}$$

This leads to the following quantization rule of the orbital angular momentum:

$$\frac{L_z}{\hbar} = \pm\frac{1}{2}, \pm\frac{3}{2}, \cdots. \tag{5}$$

In this half-integral quantization, $L_z = 0$, and therefore $p_\phi = 0$, is not allowed. On the other hand, the spectrum of the surface states takes the following Dirac form,

$$E_{\text{surf}} = \pm v_F \sqrt{p_\phi^2 + p_z^2}. \tag{6}$$

Combining equations (5) and (6), one is led to believe that the spectrum of the cylindrical topological insulator is generically *gapped*; see also a gapped spectrum shown in figure 1(a). The spectra shown in figure 1 are calculated using tight-binding implementation of the bulk 3D topological insulator of a rectangular prism shape [22].

The magnitude of this energy gap does not decrease exponentially as the size of the system increases, in sharp contrast to the case of the usual hybridization gap [24]. The energy gap, $\Delta E = v_F\hbar/R \propto R^{-1}$, decays only algebraically as a function of the size of the system. In typical topological insulators (Bi_2Se_3, Bi_2Te_3), the Fermi velocity, $v_F \simeq 4 \times 10^5 \text{m s}^{-1} \simeq c/7500$, and the gap is expected to be visible in an ARPES measurement for $R \lesssim 10^{-7}$ m at 1 K.

The spin Berry phase, π, can be interpreted in such a way that a fictitious magnetic flux is induced and pierces the cylinder. Indeed, the additive factor, 1/2, which appears in the two off-diagonals of equation (2) can be regarded as a vector potential

$$A = \frac{\Phi}{r}\hat{\phi}, \tag{7}$$

created around a flux tube of strength Φ, identical to half of a flux quantum, $\Phi_0 = h/e$. This corresponds to the Berry phase,

$$\alpha = 2\pi\frac{\Phi}{\Phi_0} = \pi. \tag{8}$$

An electron on the cylindrical surface does not touch the fictitious flux itself, since the fictitious flux tube is deep inside the cylinder. Yet quantum mechanically, the spectrum of the surface state is still influenced by this effective flux; the electron feels a vector potential, and its spectrum is affected by the vector potential, even when there is no magnetic field at any position where the electron is allowed to exist (Aharonov–

Bohm effect). Here, the effective flux is induced by a constraint on the spin of the surface state. Thus, the electron on the cylindrical surface induces an effective flux, while its quantum mechanical motion is influenced by the flux created by the electron itself; therefore, we call it the *intrinsic* Aharonov–Bohm effect. To confirm this scenario, we repeated the same numerical simulation in the presence of a real external magnetic flux designed to cancel the fictitious flux, as shown in equation (8). As expected, the spectrum of the surface state becomes gapless once the Berry phase, π, is cancelled by the external flux, as seen in figure 1(b). Related experimental results in Bi_2Se_3 nanowires have been reported in [25, 26].

2.2. Topological insulator nanoparticle as an artificial atom

Magnetic monopoles do not exist in nature. This is what we learn in courses on elementary electromagnetism, and the statement is, of course, true at the microscopic level. In matter, however, Maxwell equations are modified, or at least it is convenient to replace them with effective equations. Then, depending on the nature of the effective media, analogues of a magnetic monopole can appear. This indeed happens in the case of the spherical topological insulator.

Similarly to the cylindrical case (equation (2)), on a spherical surface of a topological insulator, equation (1) is modified to [27]

$$H_{\text{surf}} = \frac{v}{R}\begin{bmatrix} 0 & -\partial_\theta + \frac{i\partial_\phi}{\sin\theta} - \frac{1}{2}\cot\frac{\theta}{2} \\ \partial_\theta + \frac{i\partial_\phi}{\sin\theta} - \frac{1}{2}\tan\frac{\theta}{2} & 0 \end{bmatrix}. \tag{9}$$

There are corrections due to the spin Berry phase in the off diagonals of equation (9). Here, in the spherical case, they are understood as vector potentials associated with a magnetic monopole. The explicit form of the vector potential differs depending on the choice of the gauge in such a way that

$$A_{\text{I}} = \frac{g}{4\pi r}\tan\frac{\theta}{2}\hat{\phi} \tag{10}$$

when the singularity (the Dirac string) is chosen on the $-z$ axis, while

$$A_{\text{II}} = \frac{-g}{4\pi r}\cot\frac{\theta}{2}\hat{\phi} \tag{11}$$

when the Dirac string is on the $+z$ axis. Consistent with the Dirac quantization condition, equation (9) corresponds to the case of an effective magnetic monopole of strength $g = \pm 2\pi$, which is the smallest value compatible with the Dirac quantization condition. The electron on the surface of a *spherical* topological insulator behaves as if there is an effective *magnetic monopole* at the center of the sphere.

The spherical topological insulator naturally models a topological insulator nanoparticle. Taking it, therefore, as an artificial atom, let us focus on its low-lying electronic levels. The angular part of the wave function is described by an anti-periodic analogue of the spherical harmonics. A few examples are shown in figure 2, where $\alpha(\theta, \phi) = e^{im\phi}\alpha_{nm}(\theta)$

(a)   (b)

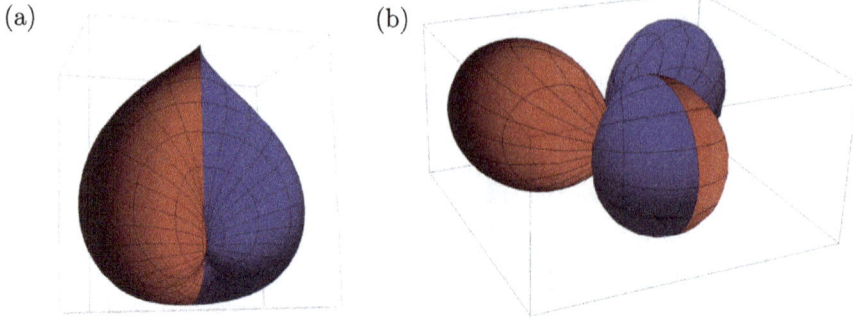

Figure 2. Topological insulator nanoparticle as an 'artificial atom'. The antiperiodic version of the low-lying (a) s-type, and (b) p-type orbitals are shown. To highlight their characters, the orbitals are painted in red (in blue) when the real part of the wave function is positive (negative).

represents the angular dependence of a surface eigenfunction, specified by quantum numbers n, m. An analogue of the s-orbital is

$$\boldsymbol{\alpha}_{0\frac{1}{2}} = \begin{bmatrix} \alpha_{0\frac{1}{2}+} \\ \alpha_{0\frac{1}{2}-} \end{bmatrix} = \frac{1}{\sqrt{4\pi}} \begin{bmatrix} \cos\dfrac{\theta}{2} \\ -\sin\dfrac{\theta}{2} \end{bmatrix}, \tag{12}$$

as seen in figure 2(a), while

$$\boldsymbol{\alpha}_{0\frac{3}{2}} = \sqrt{\frac{3}{2\pi}} \begin{bmatrix} \sin\dfrac{\theta}{2}\cos^2\dfrac{\theta}{2} \\ -\sin^2\dfrac{\theta}{2}\cos\dfrac{\theta}{2} \end{bmatrix} \tag{13}$$

can be regarded as an anti-periodic version of the p-orbital, as seen in figure 2(b). Further details on such 'monopole harmonics' [28] and the spectrum of the artificial atom are given in [27].

3. Peculiar finite size effects in topological insulator surface states—part 2: the weak case

The specificity of a 3D topological insulator is that it can be either weak or strong, as mentioned in the introduction. The strong topological insulator (STI) typically exhibits a single Dirac cone, while the weak topological insulator (WTI) exhibits an even number of Dirac cones in the surface Brillouin zone. A single Dirac cone cannot be confined (Klein tunneling), while an even number of Dirac cones can be confined. Therefore, in an STI, the single Dirac cone is extended over all the surfaces. That is actually the situation we considered in section 2. In a WTI, on the other hand, some surfaces are metallic (gapless as a consequence of an even number of Dirac cones), while others are not (no Dirac cone on such surfaces).

The WTI is characterized by three weak indices: ν_1, ν_2, ν_3 [29]. The surfaces normal to the direction specified by $\boldsymbol{\nu} = (\nu_1, \nu_2, \nu_3)$ are gapped (no Dirac cone). In this sense, the WTI is by its nature *anisotropic*, and can be regarded as a *layered material*. In the following, we demonstrate that the WTI surface states are susceptible to various even/odd

features in regard to the number of atomic layers stacked in the direction of $\boldsymbol{\nu}$.

3.1. Even/odd feature

To highlight the even/odd feature, we first consider the spectrum of a WTI sample with top, bottom, and side surfaces. The top and bottom surfaces are oriented normal to the z-axis. To realize a typical situation, we assume that the top and bottom surfaces are *gapped* surfaces (no Dirac cone). We then consider electronic states on a side surface, here chosen to be on the zx-plane. The two Dirac cones are typically located at $k_z = 0$ and at $k_z = \pi$. Low-energy electron states at or in the vicinity of the Dirac points may be represented by a plane wave:

$$\psi_1 = e^{i\left(k_x x + q_1 z\right)},$$
$$\psi_2 = e^{i\left[k_x x + (\pi + q_2)z\right]}. \tag{14}$$

Here, the crystal momentum, q_1 or q_2, of the electron is measured from the corresponding Dirac cone. Now, if we consider the presence of top and bottom surfaces, located respectively at $z = N_z$ and $z = 1$ for example, we need to confine the above electron in the region of $1 \leqslant z \leqslant N_z$. This means that we impose the boundary condition such that the wave function of the surface state, ψ, which may be expressed as a linear combination of ψ_1 and ψ_2, satisfies

$$\psi(z) = c_1\psi_1 + c_2\psi_2,$$
$$\psi(z = 0) = 0, \quad \psi(z = N_z + 1) = 0. \tag{15}$$

To cope with the boundary condition at $z = 0$, we choose the constants c_1 and c_2 such that $c_2 = -c_1 = 1$. Also, at a given energy, E, one can set q_1 and q_2 such that $q_2 = -q_1 \equiv q$. We are then left with

$$\psi(z) = e^{ik_x x}\left[e^{iqz} - (-1)^z e^{-iqz}\right], \tag{16}$$

and this must vanish at $z = N_z + 1$. Therefore, for N_z odd,

$$q = \pm\frac{\pi}{2(N_z + 1)}2n \tag{17}$$

where $n = 0, \pm1, \pm2, \cdots$ can be an arbitrary integer. Similarly, for N_z even, the vanishing of equation (16) at

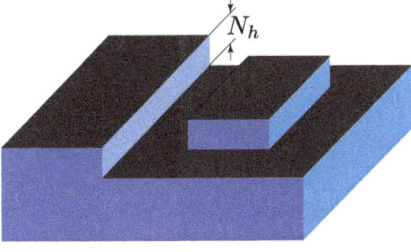

Figure 3. Patterning the surface of a weak topological insulator allows for constructing nanocircuits of one-dimensional (1D) dissipationless channels. See also [18].

$z = N_z + 1$ implies

$$q = \pm \frac{\pi}{2(N_z + 1)}(2m - 1) \qquad (18)$$

where $m = 1, 2, 3, \cdots$, and $\pm(2m - 1)$ is an arbitrary odd integer. Since the spectrum of the surface state is given as [31]

$$E = \pm v_F \sqrt{k_x^2 + \sin q^2}, \qquad (19)$$

equations (17) and (18) signify that the surface spectrum is gapless when N_z is odd, while it is gapped when N_z is even [30, 31]. Also, replacing N_z with N_h, one can equally apply equations (17) and (18) for characterizing the 1D helical modes that appear along a step formed on the surface of a WTI [18] (cf figure 3 and discussion given in the next section).

3.2. Dissipationless nanocircuit, or perfectly conducting 1D channel

We have so far argued that WTIs are susceptible to a specific type of size effect that is strongly dependent on the parity of the number of layers stacked in the direction of $\nu = (\nu_1, \nu_2, \nu_3)$. Here, we demonstrate that this even/odd feature indeed makes a WTI more useful than an STI. We consider the following Wilson–Dirac-type effective Hamiltonian on the cubic lattice,

$$H_{\text{bulk}} = v_F \sum_{\mu = x,y,z} \sin k_\mu \; \tau_x \otimes \sigma_\mu + m(\mathbf{k}) \; \tau_z \otimes 1_2,$$

$$m(\mathbf{k}) = m_0 + \sum_{\mu = x,y,z} 2m_{2\mu}(1 - \cos k_\mu), \qquad (20)$$

where two types of Pauli matrices, σ and τ, represent real and orbital spins, respectively, and 1_2 is a 2×2 identity matrix. We have chosen the parameters $m_0 = -v_F$, $m_{2x} = m_{2y} = 0.5v_F$, $m_{2z} = 0.1v_F$, and $v_F = 2$ so that a WTI with weak indices, $\nu = (0, 0, 1)$, is realized [30]. By simply patterning the surface of a WTI, as seen in figure 3, one can realize a nanocircuit of a perfectly conducting 1D channel.

The even/odd feature on the spectrum of a WTI sample with side surfaces is equally applied to a system with atomic scale steps on the gapped surfaces [18]. In figure 4, we show the spatial profile of the wave function of some low-lying states in a geometry with an atomic scale step. In figure 4(a), the height of the step N_h is 1, while in figure 4(b), the step is 2

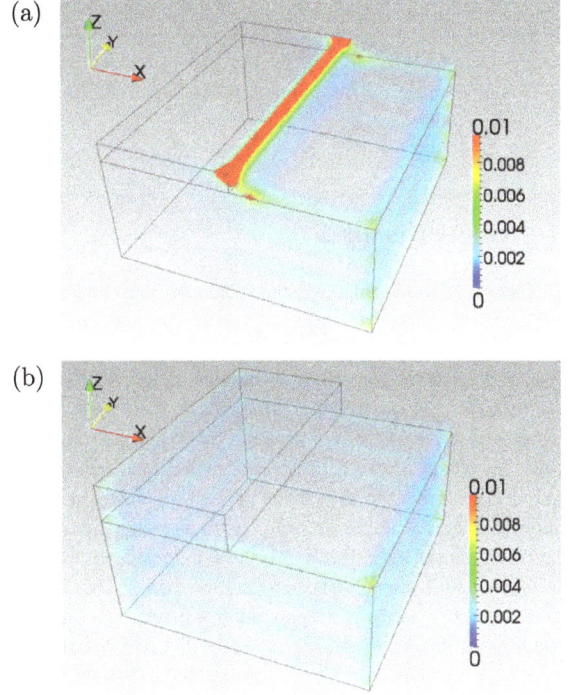

Figure 4. Spatial profile of low-lying electronic wave function along a step of height (a) one, and (b) two. In (a) a perfectly conducting channel is realized. See also [18].

atomic scale high. As shown in the two panels, the wave function has strong amplitude along the step when N_h is odd, while it has visibly no weight in the step region when N_h is even.

These 1D channels that appear when N_h is odd are also shown to be helical and robust against non-magnetic disorder [18, 32, 33]. In the supplemental material[3], we demonstrate the robustness of such an odd number of channels against disorder. Related to this, the robustness of the underlying WTI has also been studied [34, 35] (see also section 4). In any case, such a nanocircuit that appears on a patterned surface of a WTI is perfectly conducting, not only in the clean limit, but also in the presence of disorder. Therefore, a perfectly conducting 1D channel is stable, at least within the phase coherence length; to estimate the order of its magnitude, note that this is typically 200 nm for Bi_2Se_3 at 5 K [36].

Recently, a WTI was discovered experimentally in a bismuth-based layered bulk material, $Bi_{14}Rh_3I_9$, built from grapheme-like Bi-Rh sheets [37]. Other candidates for WTIs are layered semiconductors [38], superlattices of topological and normal insulators [39–41]. The superlattice has also been considered in different contexts including as a protocol for realizing a Weyl semimetal [42], and it is indeed fabricated

---

[3] Wave packet dynamics in a geometry with 1, 2, and 3 atomic scale steps on top of the sample. Initial wave packet is located at $y = 0$. Parameters are the same as in figure 4 (left) without disorder and (right) with site-diagonal disorder distributed in the range $[-W, W]$, where $W = 1.5 v_F$ (see the online supplementary material, available at stacks.iop.org/stam/16/014403/mmedia).

experimentally [43]. Still another possibility to realize a WTI-like situation is to use topological crystal insulators [44–47]. These are close relatives of the standard topological insulator, protected by crystalline symmetries instead of the time reversal symmetry.

4. Concluding remarks

In section 2, we highlighted the *enhanced* size effect characteristic to Dirac electrons emergent on the surface of topological insulators. In the classification of 3D topological insulators to weak and strong categories, this size effect was particularly applicable to the strong case. This type of size effect also protects the surface state from surface roughness [48]. In section 3, we discussed that weak topological insulators are susceptible to a different type of size effect that shows an even/odd feature with respect to the number of layers in the stacking direction. This even/odd feature makes *weak* topological insulators more *useful* than *strong* topological insulators. By simply patterning the surface of a weak topological insulator, one can achieve dissipationless transport in a non-superconducting context (i.e., nanocircuits of perfectly conducting 1D channels).

Although this paper has focused on the clean limit, the helical nature of the surface Dirac electron is responsible for its robustness against disorder. Because of the spin-to-momentum locking, backscattering is forbidden; for an incident state with momentum k, the reflected state with $-k$ state has a spin part orthogonal to that of the initial state, leading to the absence of backscattering [49]. The gapless 2D Dirac semimetal is known to possess unusual robustness against disorder [50–52]. A slightly different question is how robust the topological classification of bulk is in the clean limit against disorder. In [34], we have addressed this issue numerically and we showed that the concept of weak and strong topological insulators remains valid at finite disorder. Then, in a more recent paper [53], we extended this study to show that both distinct topological phases and a 3D Dirac semimetal that appears at their phase boundary show some robustness against disorder. A similar result has also been obtained for a Weyl semimetal [54]. To highlight such Dirac and related Weyl semimetals, especially in the context of their robustness against disorder, is a new trend in the field, both theoretically [55, 56] and experimentally [57, 58].

Acknowledgments

The authors acknowledge fruitful scientific interactions with Yositake Takane, Akihiro Tanaka, Takahiro Fukui and Igor Herbut. TO has been supported by Grants-in-Aid for Scientific Research (C) (Grant No. 23540376) and Grants-in-Aid No. 24000013.

References

[1] Moore J E 2010 *Nature* **464** 194–8
[2] Hasan M Z and Kane C L 2010 *Rev. Mod. Phys.* **82** 3045–67
[3] Qi X L and Zhang S C 2011 *Rev. Mod. Phys.* **83** 1057–110
[4] Hsieh D, Xia Y, Qian D, Wray L, Hor Y S, Cava R J and Hasan M Z 2008 *Nature* **452** 970–4
[5] Noh H J *et al* 2008 *Europhys. Lett.* **81** 57006
[6] Ando Y 2013 *J. Phys. Soc. Jpn.* **82** 102001
[7] Kuroda K *et al* 2010 *Phys. Rev. Lett.* **105** 146801
[8] Kane C L and Mele E J 2005 *Phys. Rev. Lett.* **95** 226801
[9] Bernevig B A, Hughes T L and Zhang S C 2006 *Science* **314** 1757–61
[10] Zhang H, Liu C X, Qi X L, Dai X, Fang Z and Zhang S C 2010 *Nat. Phys.* **5** 438–42
[11] Liu C X, Qi X L, Zhang H J, Dai X, Fang Z and Zhang S C 2010 *Phys. Rev.* B **82** 045122
[12] Hatsugai Y 1993 *Phys. Rev. Lett.* **71** 3697–700
[13] Moore J E and Balents L 2007 *Phys. Rev.* B **75** 121306
[14] Fu L, Kane C L and Mele E J 2007 *Phys. Rev. Lett.* **98** 106803
[15] Roy R 2009 *Phys. Rev.* B **79** 195322
[16] Mong R S K, Bardarson J H and Moore J E 2012 *Phys. Rev. Lett.* **108** 076804
[17] Liu C X, Qi X L and Zhang S C 2012 *Physica E* **44** 906–11
[18] Yoshimura Y, Matsumoto A, Takane Y and Imura K I 2013 *Phys. Rev.* B **88** 045408
[19] Imura K I, Takane Y and Tanaka A 2011 *Phys. Rev.* B **84** 035443
[20] Zhang Y, Ran Y and Vishwanath A 2009 *Phys. Rev.* B **79** 245331
[21] Takane Y and Imura K I 2013 *J. Phys. Soc. Jpn.* **82** 074712
[22] Imura K I, Takane Y and Tanaka A 2011 *Phys. Rev.* B **84** 195406
[23] Ostrovsky P M, Gornyi I V and Mirlin A D 2010 *Phys. Rev. Lett.* **105** 036803
[24] Lu H Z, Shan W Y, Yao W, Niu Q and Shen S Q 2010 *Phys. Rev.* B **81** 115407
[25] Peng H, Lai K, Kong D, Meister S, Chen Y, Qi X L, Zhang S C, Shen Z X and Cui Y 2010 *Nat. Mater.* **9** 225
[26] Dufouleur J *et al* 2013 *Phys. Rev. Lett.* **110** 186806
[27] Imura K I, Yoshimura Y, Takane Y and Fukui T 2012 *Phys. Rev.* B **86** 235119
[28] Wu T T and Yang C N 1976 *Nucl. Phys.* B **107** 365–80
[29] Fu L and Kane C L 2007 *Phys. Rev.* B **76** 045302
[30] Imura K I, Okamoto M, Yoshimura Y, Takane Y and Ohtsuki T 2012 *Phys. Rev.* B **86** 245436
[31] Arita T and Takane Y 2014 *J. Phys. Soc. Jpn.* **83** 124716
[32] Takane Y 2004 *J. Phys. Soc. Jpn.* **73** 1430–3
[33] Ringel Z, Kraus Y E and Stern A 2012 *Phys. Rev.* B **86** 045102
[34] Kobayashi K, Ohtsuki T and Imura K I 2013 *Phys. Rev. Lett.* **110** 236803
[35] Sbierski B and Brouwer P W 2014 *Phys. Rev.* B **89** 155311
[36] Matsuo S *et al* 2013 *Phys. Rev.* B **88** 155438
[37] Rasche B, Isaeva A, Ruck M, Borisenko S, Zabolotnyy V, Büchner B, Koepernik K, Ortix C, Richter M and van den Brink J 2013 *Nat. Mater.* **12** 422–5
[38] Yan B, Müchler L and Felser C 2012 *Phys. Rev. Lett.* **109** 116406
[39] Fukui T, Imura K I and Hatsugai Y 2013 *J. Phys. Soc. Jpn.* **82** 073708
[40] Yang G, Liu J, Fu L, Duan W and Liu C 2014 *Phys. Rev.* B **89** 085312
[41] Li X, Zhang F, Niu Q and Feng J 2014 *Sci. Rep.* **4** 6397
[42] Burkov A A and Balents L 2011 *Phys. Rev. Lett.* **107** 127205
[43] Nakayama K, Eto K, Tanaka Y, Sato T, Souma S, Takahashi T, Segawa K and Ando Y 2012 *Phys. Rev. Lett.* **109** 236804
[44] Fu L 2011 *Phys. Rev. Lett.* **106** 106802

[45] Tanaka Y, Ren Z, Sato T, Nakayama K, Souma S, Takahashi T, Segawa K and Ando Y 2012 *Nat. Phys.* **8** 800–3

[46] Hsieh T H, Lin H, Liu J, Duan W, Bansil A and Fu L 2012 *Nat. Commun.* **3** 982

[47] Dziawa P *et al* 2012 *Nat. Mater.* **11** 1023–7

[48] Imura K I and Takane Y 2013 *Phys. Rev.* B **87** 205409

[49] Ando T, Nakanishi T and Saito R 1998 *J. Phys. Soc. Jpn.* **67** 2857–62

[50] Nomura K, Koshino M and Ryu S 2007 *Phys. Rev. Lett.* **99** 146806

[51] Bardarson J H, Tworzydło J, Brouwer P W and Beenakker C W J 2007 *Phys. Rev. Lett.* **99** 106801

[52] Tworzydło J, Groth C W and Beenakker C W J 2008 *Phys. Rev.* B **78** 235438

[53] Kobayashi K, Ohtsuki T, Imura K I and Herbut I F 2014 *Phys. Rev. Lett.* **112** 016402

[54] Ominato Y and Koshino M 2014 *Phys. Rev.* B **89** 054202

[55] Young S M, Zaheer S, Teo J C Y, Kane C L, Mele E J and Rappe A M 2012 *Phys. Rev. Lett.* **108** 140405

[56] Nandkishore R, Huse D A and Sondhi S L 2014 *Phys. Rev.* B **89** 245110

[57] Neupane M *et al* 2014 *Nat. Commun.* **5** 3786

[58] Borisenko S, Gibson Q, Evtushinsky D, Zabolotnyy V, Büchner B and Cava R J 2014 *Phys. Rev. Lett.* **113** 027603

6

Controllable assembly of silver nanoparticles induced by femtosecond laser direct writing

Huan Wang[1], Sen Liu[1], Yong-Lai Zhang[1], Jian-Nan Wang[1], Lei Wang[1], Hong Xia[1], Qi-Dai Chen[1], Hong Ding[3] and Hong-Bo Sun[1,2]

[1] State Key Laboratory on Integrated Optoelectronics, College of Electronic Science and Engineering, Jilin University, 2699 Qianjin Street, Changchun, 130012, People's Republic of China
[2] College of Physics, Jilin University, 119 Jiefang Road, Changchun, 130023, People's Republic of China
[3] State Key Laboratory of Inorganic Synthesis and Preparative Chemistry, College of Chemistry, Jilin University, 2699 Qianjin Street, Changchun, 130012, People's Republic of China

E-mail: yonglaizhang@jlu.edu.cn and hbsun@jlu.edu.cn

Abstract

We report controllable assembly of silver nanoparticles (Ag NPs) for patterning of silver microstructures. The assembly is induced by femtosecond laser direct writing (FsLDW). A tightly focused femtosecond laser beam is capable of trapping and driving Ag NPs to form desired micropatterns with a high resolution of ~190 nm. Taking advantage of the 'direct writing' feature, three microelectrodes have been integrated with a microfluidic chip; two silver-based microdevices including a microheater and a catalytic reactor have been fabricated inside a microfluidic channel for chip functionalization. The FsLDW-induced programmable assembly of Ag NPs may open up a new way to the designable patterning of silver microstructures toward flexible fabrication and integration of functional devices.

Keywords: silver nanoparticles, femtosecond laser direct writing, patterning

1. Introduction

Recent advances in nanotechnology have resulted in rapid developments in nanomaterial synthesis methodologies [1–3]. Nanoparticles (NPs) with controllable size, shape, and polydispersity have been successfully prepared based on various material systems, ranging from inorganic carbon nanodots [4, 5] to organic nanocrystals [6, 7]. Among those zero-dimensional nanomaterials, metal nanoparticles (MNPs) that possess unique chemical/physical properties have sparked enormous research interest, revealing great potential toward a broad range of applications including electronics [8–10], sensors [11, 12], catalysis [13, 14], optics [15–20], and biomedicine [21]. Taking Ag NPs as a representative example, under certain excitation, localized surface plasmon resonance (LSPR) would occur among Ag NPs whose feature size is much smaller than the wavelength of the incident light. Consequently, Ag NPs could act as optical antennas to condense optical energy on their surface. Ag NPs have been considered a good candidate to fabricate surface-enhanced Raman scattering (SERS) substrates [22, 23]. Besides, as a non-toxic and stable inorganic material, Ag NPs have been used as fluorescent nanoclusters [24], electron acceptors [25, 26], catalysis active sites [27, 28], and a new generation of antimicrobials [29–32]. However, as compared with the refined synthetic routes for Ag NP preparation, from the practical point of view, there is a need for nanotechnologies that permit controllable assembly of these tiny NPs into desired micronanostructures for nanodevice fabrication. Nowhere is this more obvious than in some emerging fields such as plasmonic optical antennas, antibacterial, electrodes, and metasurfaces, where there is a need to organize NPs into rationally designed ensemble structures, even complex micropatterns.

This problem has motivated considerable efforts to develop improved patterning techniques toward nanodevice fabrication. For instance, electrostatic self-assembly [33–36], Langmuir–Blodgett technique [37–40], layer-by-layer assembly [41–45], and the use of solid templates [46–48] have been successfully developed for controllable Ag NP assemblies. Andrade *et al* reported a layer-by-layer assembly of Ag NPs on the surface of optical fiber tips for the development of remote SERS sensors [49]. The optical fiber tips were alternately soaked in 3-aminopropyltrimethoxysilane (APTMS) and Ag NPs several times to control the assembly. To get better control over the assembly process, solid templates could be used for patterning. Su *et al* fabricated Ag plasmonic waveguides by evaporation-induced self-assembly (EISA) of Ag NPs with the help of pillar array templates, which were fabricated beforehand through classical lithography [50]. In this regard, the patterning is limited to flat surfaces. Additionally, designable patterning of Ag NPs could also be achieved by direct printing of Ag NP inks [51–54]. Ahn *et al* fabricated Ag electrodes by moving the nozzle of Ag NP ink using a three-axis motion-controlled stage [55]. In this way, complex silver micropatterns could be directly printed. Nonetheless, the resolution is somewhat low; the narrowest line was about $2\,\mu m$. Recently, direct femtosecond laser selective NP sintering (FLSNS) has been successfully developed to fabricate silver patterns [56]. Femtosecond laser irradiation would induce the sintering of Ag NPs; and the unirradiated region could be selectively removed, forming the desired silver micropatterns. This method could be considered as a subtraction-type fabrication; loading a uniform Ag NPs film is necessary for subsequent patterning. To date, despite there existing many successful examples that realized controllable assembly of Ag NPs, continued efforts are still desired in developing novel assembly strategies that permit flexible patterning of high-resolution silver micropatterns toward nanodevice fabrication.

In this work, we report a femtosecond laser direct writing (FsLDW)-induced controllable assembly of Ag NPs for making silver micropatterns. As a designable processing technique, FsLDW enables 3D fabrication, high-resolution prototyping, mask-free patterning, and flexible fabrication on non-planar substrates, revealing great potential in microdevice fabrication [57]. Herein, a focused femtosecond laser was used to drive and assemble Ag NP colloids into desired micropatterns. Taking advantage of the 'direct writing' feature, complex silver micropatterns such as Chinese knot, microelectrodes, and letters have been successfully fabricated without the use of any templates or masks. Moreover, the laser-induced assembly of Ag NPs enables flexible integration

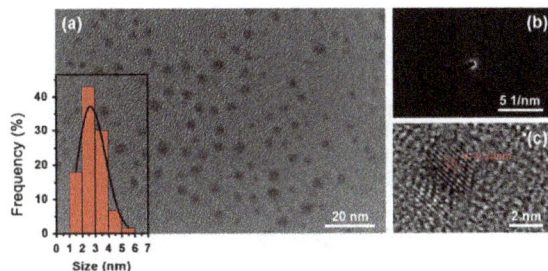

Figure 1. (a) TEM image of the as-obtained Ag NPs. The inset is the particle size distribution of the Ag NPs. (b) SAED pattern of the Ag NPs. (c) HRTEM image of a single Ag NP, (200) planes of the Ag NP with a d spacing of 0.2 nm could be observed.

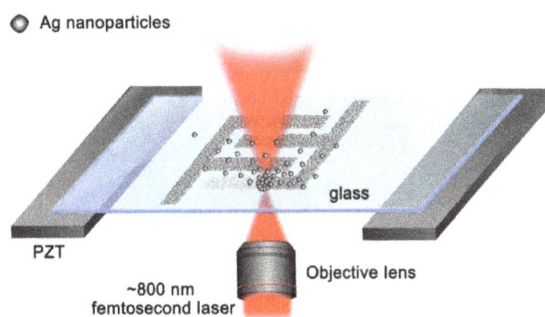

Figure 2. Schematic illustration of FsLDW-induced controllable assembly of Ag NPs. The laser beam was tightly focused by a ×100 oil immersion objective lens with a high numerical aperture NA = 1.45; the moving stage (PZT: piezoelectric transducer) was controlled by computers.

of silver microstructures with microdevices. As typical examples, we integrated a microheater inside a microfluidic chip for local heating and a catalytic reactor for H_2O_2 decomposition. The FsLDW-induced controllable assembly of Ag NPs may hold great promise for the fabrication and integration of metal NP-enabled nanodevices.

2. Experimental details

2.1. Preparation of Ag NP solution

Ag NPs were synthesized according to a reported method [58]. Typically, silver nitrate (1 mL, 0.1 mmol) and sodium malate (1 mL, 10 mmol) were added slowly into 97 mL of ultrapure water under vigorous stirring. After stirring for 10 min, sodium borohydride (1 mL, 10 mmol) was dropped into the preceding solution, followed by stirring for additional 2 min. To remove the residual salts in the solution, the obtained Ag NPs were purified by dialysis. After that, we obtained the Ag NPs aqueous solution for subsequent FsLDW.

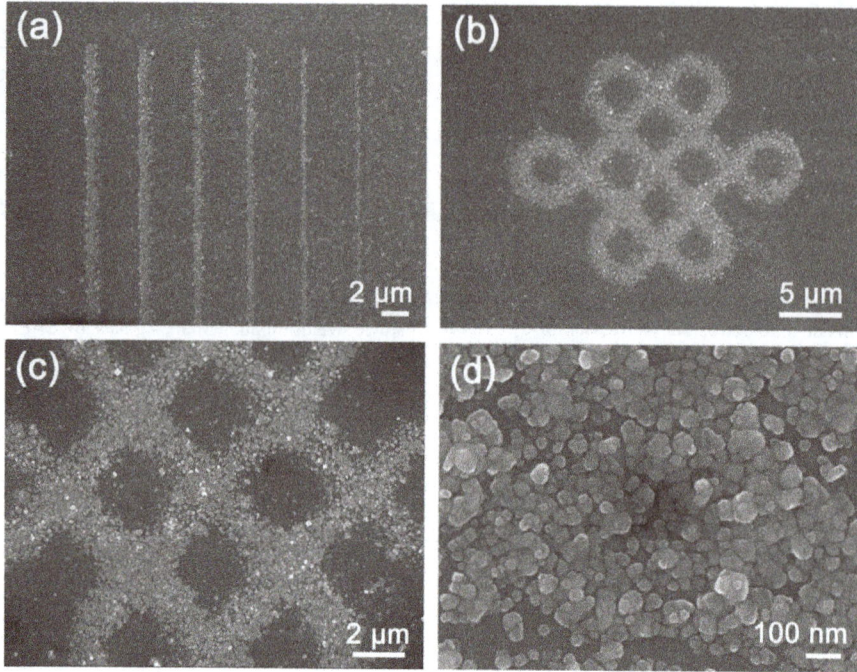

Figure 3. SEM images of silver micropatterns fabricated by FsLDW-induced controllable assembly of Ag NPs. (a) Ag microwires with different widths. From right to left, the width is 190, 290, 400, 510, 630, and 710 nm, respectively. (b) SEM image of a 'Chinese knot' pattern. (c), (d) Magnified SEM images of the 'Chinese knot' pattern.

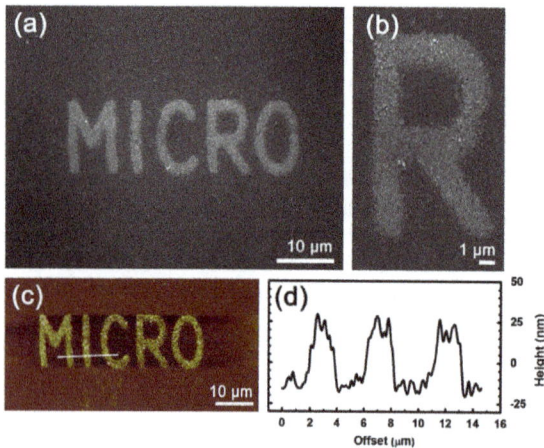

Figure 4. (a) SEM image of a silver word (MICRO) fabricated by FsLDW-induced controllable assembly of Ag NPs. (b) SEM image of the letter R. (c) AFM image of the silver word (MICRO). (d) Height profile measured along the white line shown in (c).

2.2. FsLDW fabrication and integration of silver micropatterns

In a typical FsLDW fabrication, a tightly focused femtosecond laser beam with central wavelength of 800 nm and a pulse width of 120 fs was introduced into the interface between silver NPs solution and substrates. The tightly focused femtosecond laser would drive and trap the Ag NPs within the focal spot; in this way, assembled silver

microstructures formed after sintering of Ag NPs along the traces of laser focus, which was guided by computer programs. In the integration experiment, microfluidic channels were fabricated by photopolymers (SU-8) through lithography. Gold pads for connecting Ag electrodes were deposited by vacuum thermal evaporation.

2.3. Characterization

The Ag NPs were characterized by transmission electron microscopy (TEM, JEM-2100F). Scanning electron microscopy (SEM, JSM7500), atomic force microscopy (AFM, Digital Instruments Nanoscope IIIA), and optical microscopy (OM, IBE2000) were used to observe the morphology of Ag assembly structures. Voltage applied on a microheater was offered by electrochemical workstation (CHI660E).

3. Results and discussion

In this work, malate-capped Ag NPs were used for the FsLDW fabrication. To evaluate the quality of the Ag NP suspension, the obtained Ag NPs were characterized by TEM. As shown in figure 1, the Ag NPs are uniform in size and their average diameter is 2.5 nm (inset of figure 1(a)). Figure 1(b) shows a selected area electron diffraction (SAED) pattern of the Ag NPs, indicating their crystalline structures. This was further confirmed by the high-resolution transmission electron microscopy (HRTEM) image (figure 1(c)), in which the (200)

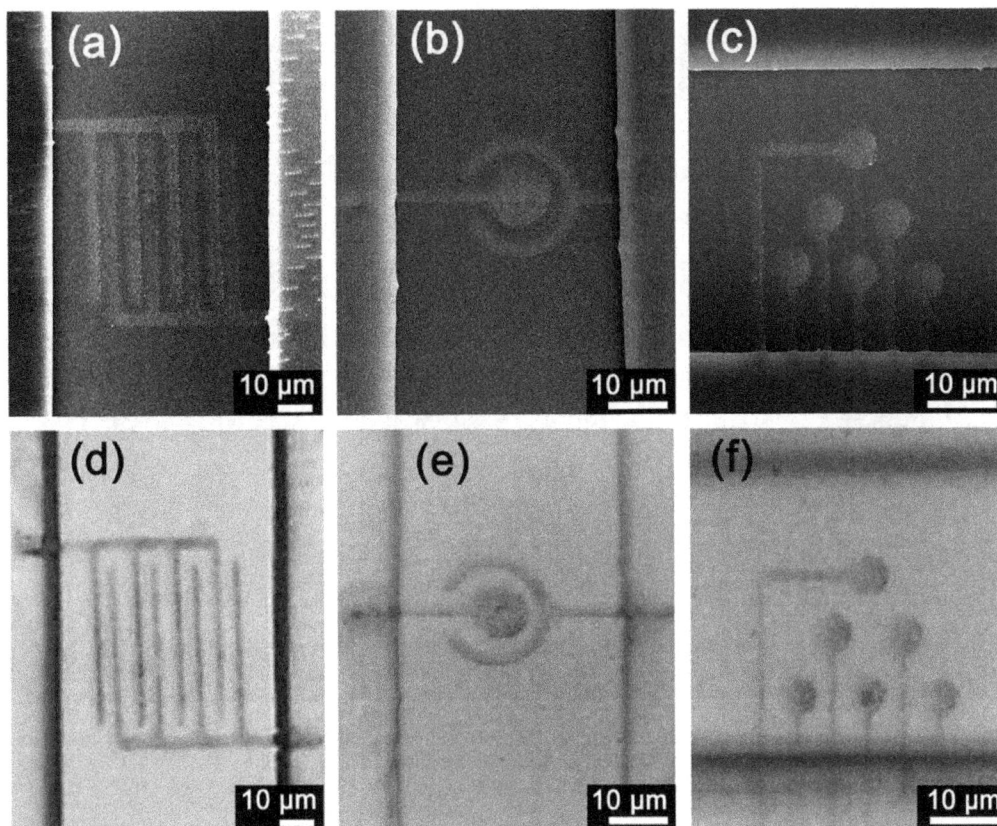

Figure 5. SEM images (a)–(c) and corresponding optical images (d)–(f) of three kinds of silver microelectrodes integrated with a microfluidic device. (a), (d) Interdigital electrode, (b), (e) two-electrode system, and (c), (f) MEA.

planes of Ag NPs with a spacing of 0.20 nm could be clearly observed.

The obtained Ag NPs have been used for FsLDW directly without the use of any photosensitive reagents. Figure 2 shows the schematic illustration of the FsLDW of Ag NPs. When a tightly focused femtosecond laser beam irradiated into the Ag NP suspension, three radiation pressures including absorption force, scattering force, and gradient force were exerted on the tiny particles. Generally, absorption force and scattering force were proportional to laser intensity, which were applied in the direction of laser beam, whereas gradient force was proportional to the gradient of laser intensity, being toward the focus [59]. Although detailed theories to calculate these forces are still under development, it is very clear that the light–matter interaction is dominated by particle sizes, refractive index, and dielectric constant of target NPs [60–62]. In this experiment, considering the metallic Rayleigh particles (diameter $d \ll$ wavelength λ) are much smaller than the wavelength of laser (here $d \approx 2.5$ nm; $\lambda \approx 800$ nm), a gradation force proportional to the polarizability of an NP and the square of optical electric field vector dominates the assembly of Ag NPs, whereas absorption force and scattering force push the Ag NPs along the laser beam direction, leading to the enrichment of Ag NPs

in the focal region. Together with the Ag NP deposition, mass transfer was performed by concentration diffusion of the Ag NPs solution. At the same time, due to the extremely high transient power density of the femtosecond laser, the aggregated Ag NPs have been sintered together. Since the near-infrared femtosecond laser beam is not absorbed by the solution, the beam penetrates deeply into the Ag NP solution with very small power loss. By scanning the laser focus along a preprogrammed trace, the desired silver micropatterns could be fabricated accordingly.

Note that the pulse width of our femtosecond laser is only 120 fs, so thermal relaxation could be effectively suppressed within a small area near the focus. Taking advantage of this feature, high-resolution silver patterns could be fabricated. Figure 3 shows SEM images of the resulting silver micropatterns. The narrowest width of a continuous silver microwire was only ~190 nm, which is much smaller than the light diffraction limit. The improved resolution could be mainly attributed to the use of an ultrafast laser, which induces nonlinear effects such as multiphoton absorption [63]. In addition to simple microwires, any desired micropatterns could be readily fabricated through FsLDW. As an example, we fabricated a 'Chinese knot' structure on a glass substrate (figure 3(b)). A magnified SEM image (figure 3(c)) showed

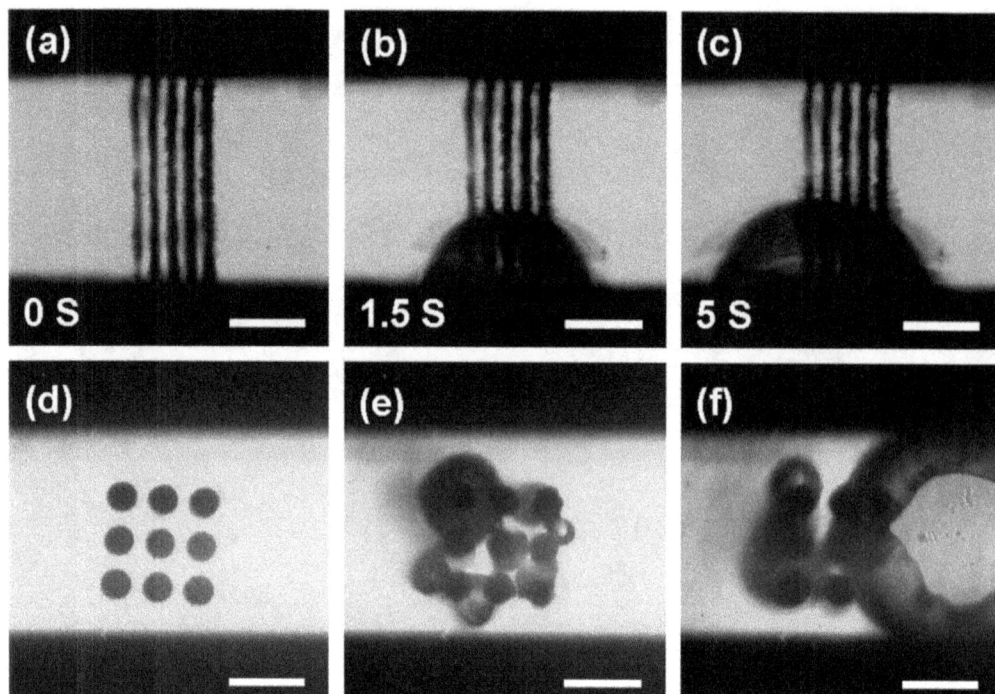

Figure 6. Silver microheater and catalytic microreactor fabricated by FsLDW-induced controllable assembly of Ag NPs. (a)–(c) Heating process of a microheater integrated within a microfluidic channel. A microbubble could be clearly observed, indicating the heating of the solvent. (d)–(f) Catalytic decomposition of H_2O_2 inside a silver microreactor. (e) Gas bubbles appeared as soon as H_2O_2 was injected into the channel; (f) 2 s later, the bubble grew bigger. The scale bar is 20 μm.

that the line width of the 'Chinese knot' was ~2 μm. By further magnifying the SEM images, we found that the silver micropattern was constructed by close-packed silver particles whose size grew to tens of nanometers. As compared with the TEM image of the pristine Ag NPs, whose size is below 5 nm, the significant increase in particle size in the silver micropattern could be attributed to the laser-induced sintering.

To illustrate its full potential in flexible patterning, a word (MICRO) was successfully fabricated through FsLDW of Ag NPs. Figures 4(a) and (b) show SEM images of the MICRO pattern and the letter 'R', respectively. The average line width of these letters is measured to be ~2 μm. To get further insight into the surface morphology of these letters, they were characterized by AFM. As shown in figure 4(c), MICRO could be clearly indentified from the image, in good agreement with the SEM image (figure 4(a)). The profile of the micropattern was also measured along a white line in figure 4(c). The micropattern is about 50 nm in thickness. Notably, the surface of the micropattern is very rough; its surface roughness (Rq) is measured to be about 11 nm. The rough surface could be attributed to the close packing of silver particles. Considering the rough surface, silver electrodes fabricated using this method may find some applications in electroanalysis.

To demonstrate the capability of integrating such silver micropatterns with given microdevices, three kinds of microelectrodes were successfully integrated within a microfluidic chip by FsLDW-induced controllable assembly of Ag NPs (figure 5). As typical examples, a pair of interdigital electrodes, and a ring-circle working-counter electrode have been patterned. Note that both the two electrodes are widely used in electroanalysis for current/impedance detection; FsLDW-induced controllable assembly of Ag NPs may find broad application in highly sensitive electroanalysis, especially for bioelectricity signal collection. In this case, microelectrode arrays (MEAs), whose sizes were similar to cells, have been fabricated, as shown in figures 5(c) and (f). Independent electrodes in MEAs could be utilized to collect potential signals transferring in organisms, or collect signals at different points from the same cells.

For practical applications, a silver microheater was fabricated by FsLDW-induced controllable assembly of Ag NPs (figure 6). Five parallel microwires were fabricated between two gold electrodes. The patterned silver structure is conductive (see supporting information figure S1), when a voltage of 2 V was supplied on the microheater, a bubble rose from the water within 1.5 s, indicating the high heating efficiency. With continuous heating, the gas bubble grew larger, indicating the increase of local temperature. Compared with general microheaters fabricated by conventional craft, FsLDW fabrication was compatible with many substrates ranging from inorganic glass and silicon to organic polymers such as polydimethylsiloxane (PDMS) and polymethyl methacrylate (PMMA). In this regard, the technique would be potentially important to multifunction integration of microfluidic chips. FsLDW-induced controllable assembly of Ag NPs enables flexible patterning of silver at any desired place. In addition to the microheater, silver microstrucutures could also be integrated with microfluidic devices for catalysis and pump applications. As shown in figures 6(d)–(f), we explored Ag-catalyzed decomposition of H_2O_2 inside a microfluidic channel; the generation of O_2 could be observed as soon as H_2O_2 was injected into the microchannel. Since the Ag catalysts could be flexibly placed at any desired position, the generated oxygen bubble could be potentially used as a pump for directional transport of target objects or to guide the flow inside a microfluidic device [64].

In conclusion, FsLDW-induced controllable assembly of Ag NPs has been successfully developed for designable patterning of silver microstructures. High-resolution silver micropatterns with the high resolution of 190 nm have been readily fabricated according to preprogrammed models. SEM and AFM images proved that silver micropatterns were constructed by close-packed Ag particles with an average size of tens of nanometers, which indicates the sintering of Ag NPs during laser irradiation. As typical models, three microelectrode patterns have been fabricated within a microfluidic device, revealing its strong capability for post-integration. Additionally, a silver microheater and a silver-catalyst array have been successfully fabricated for localized heating and catalytic decomposition of H_2O_2, respectively. As a laser-mediated assembly route, FsLDW-induced controllable assembly of Ag NPs may find broad application in the fabrication of silver-based microdevices. Especially, for microfluidic devices, FsLDW-induced controllable assembly of Ag NPs may hold great promise for flexible integration of functional devices such as SERS substrates, conductive microelectrodes, microheaters, Ag catalysts, and pumps, within a microfluidic chip, contributing to the development of multifunctional microfluidic chips.

Acknowledgment

The authors would like to acknowledge the National Basic Research Program of China (grant number: 2011CB013000; NSFC and grant numbers: 61376123, 61008014, 90923037, 6078048, 11104109.

References

[1] Sau T K and Rogach A L 2010 *Adv. Mater.* **22** 1781–804
[2] Kumar S and Nann T 2006 *Small* **2** 316–29
[3] Lim B and Xia Y 2011 *Angew. Chem. Int. Edn* **50** 76–85
[4] Bao L, Zhang Z, Tian Z, Zhang L, Liu C, Lin Y, Qi B and Pang D 2011 *Adv. Mater.* **23** 5801–6
[5] Ghosh S *et al* 2014 *Nano Lett.* **14** 5656–5661
[6] Fery-Forgues S 2013 *Nanoscale* **5** 8428–8442
[7] Jiang H *et al* 2013 *Small* **9** 990–5
[8] Lu W *et al* 2013 *Opt. Mater. Express* **3** 1660–73

[9] Luechinger N A, Athanassiou E K and Stark W J 2008 *Nanotechnology* **19** 445201
[10] Ko S H, Park I, Pan H, Grigoropoulos C P, Pisano A P, Luscombe C K and Fréchet J M J 2007 *Nano Lett.* **7** 1869–77
[11] Farcau C, Moreira H, Viallet B, Grisolia J, Ciuculescu-Pradines D, Amiens C and Ressier L 2011 *J. Phys. Chem.* C **115** 14494–9
[12] Zhang S, Wang N, Niu Y and Sun C 2005 *Sens. Actuators* B **109** 367–74
[13] Ji Q, Hill J P and Ariga K 2013 *J. Mater. Chem.* A **1** 3600–6
[14] Li Z, Liu J, Xia C and Li F 2013 *ACS Catal.* **3** 2440–8
[15] Ueno K and Misawa H 2013 *Phys. Chem. Chem. Phys.* **15** 4093–9
[16] Sun Q, Ueno K, Yu H, Kubo A, Matsuo Y and Misawa H 2013 *Light Sci. Appl.* **2** e118
[17] Lee Y, Lee S H, Park S, Park C, Lee K, Kim J and Joo J 2014 *Synthetic Met.* **187** 130–5
[18] Shrestha L K, Sathish M, Hill J P, Miyazawa K, Tsuruoka T, Sanchez-Ballester N M, Honma I, Ji Q and Ariga K 2013 *J. Mater. Chem.* C **1** 1174–81
[19] Guo C, Sun T, Feng C, Liu Q and Ren Z 2014 *Light Sci. Appl.* **3** e161
[20] Su Y, Ke Y, Cai S and Yao Q 2012 *Light Sci. Appl.* **1** e14
[21] Khan M S, Vishakante G D and Siddaramaiah H 2013 *Adv. Colloid Interface Sci.* **199–200** 44–58
[22] Rajapandiyan P and Yang J 2014 *J. Raman Spectrosc.* **45** 574–80
[23] Michaels A M, Nirmal M and Brus L E 1999 *J. Am. Chem. Soc.* **121** 9932–9
[24] Diez I, Pusa M, Kulmala S, Jiang H, Walther A, Goldmann A S, Muller A H E, Ikkala O and Ras R H A 2009 *Angew. Chem. Int. Edn* **48** 2122–5
[25] Hirakawa T and Kamat P V 2004 *Langmuir* **20** 5645–7
[26] Hirakawa T and Kamat P V 2005 *J. Am. Chem. Soc.* **127** 3928–34
[27] Liu X, Cheng H and Cui P 2014 *Appl. Surf. Sci.* **292** 695–701
[28] Merga G, Wilson R, Lynn G, Milosavljevic B H and Meisel D 2007 *J. Phys. Chem.* C **111** 12220–6
[29] Rai M, Yadav A and Gade A 2009 *Biotechnol. Adv.* **27** 76–83
[30] Eby D M, Schaeublin N M, Farrington K E, Hussain S M and Johnson G R 2009 *ACS Nano* **3** 984–94
[31] Abdullayev E, Sakakibara K, Okamoto K, Wei W, Ariga K and Lvov Y 2011 *ACS Appl. Mater. Inter.* **3** 4040–6
[32] Chernousova S and Epple M 2013 *Angew. Chem. Int. Edn* **52** 1636–53
[33] Wang Y, Chen H, Dong S and Wang E 2007 *J. Raman Spectrosc.* **38** 515–21
[34] Li X, Zhang J, Xu W, Jia H, Wang X, Yang B, Zhao B, Li B and Ozaki Y 2003 *Langmuir* **19** 4285–90
[35] Kawada S, Saeki D and Matsuyama H 2014 *Colloids Surf.* A **451** 33–7
[36] Zhou Y, Yang J, Cheng X, Zhao N, Sun L, Sun H and Li D 2012 *Carbon* **50** 4343–50
[37] Zhang Q, Lee Y, Phang I Y, Lee C K and Ling X Y 2014 *Small* **10** 2703–11
[38] Sarkar J, Pal P and Talapatra G 2005 *Chem. Phys. Lett.* **401** 400–4
[39] Ariga K, Yamauchi Y, Rydzek G, Ji Q, Yonamine Y, Wu K C-W and Hill J P 2014 *Chem. Lett.* **43** 36–68
[40] Borges J and Mano J F 2014 *Chem. Rev.* **114** 8883–942
[41] Ariga K, Yamauchi Y, Mori T and Hill J P 2013 *Adv. Mater.* **25** 6477–512
[42] Ariga K, Mori T, Ishihara S, Kawakami K and Hill J P 2014 *Chem. Mater.* **26** 519–32
[43] Liu X, Qi S, Li Y, Yang L, Cao B and Tang C Y 2013 *Water Res.* **47** 3081–92
[44] Feng Z and Yan F 2008 *Surf. Interface Anal.* **40** 1523–8

[45] Zhao S, Zhang K, An J, Sun Y and Sun C 2006 *Mater. Lett.* **60** 1215–8

[46] Henson J, Heckel J C, Dimakis E, Abell J, Bhattacharyya A, Chumanov G, Moustakas T D and Paiella R 2009 *Appl. Phys. Lett.* **95** 151109

[47] Park J *et al* 2010 *Adv. Funct. Mater.* **20** 2296–302

[48] Kim E, Baeg K, Noh Y, Kim D, Lee T, Park I and Jung G 2009 *Nanotechnology* **20** 355302

[49] Andrade G F S, Fan M and Brolo A G 2010 *Biosens. Bioelectron.* **25** 2270–5

[50] Su B, Zhang C, Chen S, Zhang X, Chen L, Wu Y, Nie Y, Kan X, Song Y and Jiang L 2014 *Adv. Mater.* **26** 2501–7

[51] Perelaer J, Gans B D and Schubert U S 2006 *Adv. Mater.* **18** 2101–4

[52] Wu J, Hsu S L, Tsai M and Hwang W 2010 *J. Phys. Chem.* C **114** 4659–62

[53] Wu J, Hsu S, Tsai M and Hwang W 2011 *J. Phys. Chem.* C **115** 10940–5

[54] Shen W, Zhang X, Huang Q, Xu Q and Song W 2014 *Nanoscale* **6** 1622–8

[55] Ahn B Y, Duoss E B, Motala M J, Guo X, Park S, Xiong Y, Yoon J, Nuzzo R G, Rogers J and Lewis J A 2009 *Science* **323** 1590–3

[56] Son Y, Yeo J, Moon H, Lim T W, Hong S, Nam K H, Yoo S, Grigoropoulos C P, Yang D and Ko S H 2011 *Adv. Mater.* **23** 3176–81

[57] Zhang Y, Chen Q, Xia H and Sun H 2010 *Nano Today* **5** 435–48

[58] Tang B, Xu S, Tao J, Wu Y and Xu W 2010 *J. Phys. Chem.* C **114** 20990–6

[59] Svoboda K and Block S M 1994 *Opt. Lett.* **19** 930

[60] Xu B, Zhang Y, Zhang R, Wang L, Xiao X, Xia H, Chen Q and Sun H 2013 *J. Mater. Chem.* C **1** 4699–704

[61] Xu B, Zhang R, Wan H, Liu X, Wang L, Ma Z, Chen Q, Xiao X, Han B and Sun H 2012 *Nanoscale* **4** 6955–8

[62] Masui K, Shoji S, Asaba K, Rodgers T C, Jin F, Duan X and Kawata S 2011 *Opt. Express* **19** 22786–96

[63] Xu B, Zhang Y, Xia H, Dong W, Ding H and Sun H 2013 *Lab. Chip* **13** 1677–90

[64] Zarzar L D, Swartzentruber B S, Harper J C, Dunphy D R, Brinker C J, Aizenberg J and Kaehr B 2012 *J. Am. Chem. Soc.* **134** 4007–10

7

Quick high-temperature hydrothermal synthesis of mesoporous materials with 3D cubic structure for the adsorption of lysozyme

Geoffrey Lawrence[1], Arun V Baskar[1], Mohammed H El-Newehy[2], Wang Soo Cha[1], Salem S Al-Deyab[2] and Ajayan Vinu[1]

[1] Australian Institute for Bioengineering and Nanotechnology, The University of Queensland, #75 Corner Cooper and College Road, Brisbane 4072, QLD, Australia
[2] Petrochemical Research Chair, Department of Chemistry, College of Science, King Saud University, Riyadh 11451, Saudi Arabia

E-mail: a.vinu@uq.edu.au

Abstract

Three-dimensional cage-like mesoporous FDU-12 materials with large tuneable pore sizes ranging from 9.9 to 15.6 nm were prepared by varying the synthesis temperature from 100 to 200 °C for the aging time of just 2 h using a tri-block copolymer F-127($EO_{106}PO_{70}EO_{106}$) as the surfactant and 1,3,5-trimethyl benzene as the swelling agent in an acidic condition. The mesoporous structure and textural features of FDU-12-HX (where H denotes the hydrothermal method and X denotes the synthesis temperature) samples were elucidated and probed using x-ray diffraction, N_2 adsorption, ^{29}Si magic angle spinning nuclear magnetic resonance, scanning electron microscopy and transmission electron microscopy. It has been demonstrated that the aging time can be significantly reduced from 72 to 2 h without affecting the structural order of the FDU-12 materials with a simple adjustment of the synthesis temperature from 100 to 200 °C. Among the materials prepared, the samples prepared at 200 °C had the highest pore volume and the largest pore diameter. Lysozyme adsorption experiments were conducted over FDU-12 samples prepared at different temperatures in order to understand their biomolecule adsorption capacity, where the FDU-12-HX samples displayed high adsorption performance of 29 μmol g^{-1} in spite of shortening the actual synthesis time from 72 to 2 h. Further, the influence of surface area, pore volume and pore diameter on the adsorption capacity of FDU-12-HX samples has been investigated and results are discussed in correlation with the textural parameters of the FDU-12-HX and other mesoporous adsorbents including SBA-15, MCM-41, KIT-5, KIT-6 and CMK-3.

Keywords: mesoporous, nanoporous, adsorption

1. Introduction

Ever since the discovery of the electron microscope, microscale and nanoscale materials have gained much prominence. Numerous investigations have been reported on controlling their properties by synthesis [1]. Nanoarchitectonics has given rise to several manipulation possibilities right from atoms and molecules to chemical modifications through controlled supra-molecular self-assembly processes to create new functional materials [2, 3]. Both inorganic and organic structures including porous nanomaterials have adopted the concept of 'nanoarchitectonics', which helped to develop new nanostructures with improved properties and their relation to the applications of the materials including drug delivery, sensing and cell growth and cell imaging [4, 5]. Mesoporous materials in particular have caught the attention of researchers from various fields and academia and are also presumed to be easy-to-make bulk nanostructures since they are produced remarkably from low cost materials through facile procedures such as mixing, heating, filtration and washing [6]. Mesoporous molecular sieves have excellent textural properties and different functions that make them vital for various applications such as catalysis, adsorption, sensing, and immobilization of biomolecules. In particular, mesoporous materials with three-dimensional (3D) cage-type structures including SBA-6, SBA-16, FDU-1, KIT-5, and FDU-12 are considered more attractive than their uni- and two-dimensional counter parts like MCM-41 and SBA-15 [7–10]. Unlike uni- and two-dimensional mesoporous molecular sieves, the shape and size of the mesopores in a 3D network structure and the connectivity between adjacent pores are considered crucial for applications involving better mass transport and diffusion [11, 12]. Mesoporous silica materials with 3D voids such as MCM-48, SBA-2, SBA-12 and FDU-1 are believed to have more advantages for catalysis and adsorption as their structures are more resistant to pore blocking and offer more adsorption sites and faster diffusion of reactant molecules. One of the important parameters of these materials is the size and shape of the pores which has a direct relation with the specific surface area and pore volume that control the catalytic or adsorption performance [13]. In addition, the spatial distribution of the necks, their sizes and their interconnection with large cavities in porous silica materials are also recognized as the key factors in adsorption studies [14].

The pore diameter of the mesoporous materials can be controlled by either adding swelling agents or using surfactants with different molecular weight or chain length [15, 16]. However, this approach ends up eventually resulting in the loss of structural ordering as it is gradually progressed by instigating pore enlargement as evidenced in the case of MCM-41 and SBA-15 [17, 18]. Zhao and his co-workers adopted a new strategy of combining salt and swelling agents to create a new mesoporous material (FDU-12) with large mesopores, cubic cage-type architecture and Fm3m symmetry [10]. FDU-12 is quite attractive not only because it has large pore entrance sizes [19] but also exhibits high specific surface area, controllable pore cage and entrance dimensions [20–26], and high thermostability [27]. Moreover, it can be easily synthesized from inexpensive reagents which are commercially available. This makes FDU-12 a unique and excellent support for the immobilization of biomolecules [10, 28], catalysis, and sensing applications [29]. Although FDU-12 has excellent structural and textural properties, the time

required for its synthesis is long compared to that of its uni- and two-dimensional counterparts such as MCM-41 and SBA-15. The synthesis time can be reduced by simply varying the synthesis temperature. This simple method has gained paramount interest among researchers as it was found to be quite simple and cost-effective. In addition, tuning the dimensions of the pores of various mesoporous siliceous materials and the interstitial voids that link them can be significantly altered with this simple approach [13].

Researchers mostly use synthesis temperatures less than $150\,^{\circ}C$ for tuning the pore size of the mesoporous materials. This is mainly due to the fact that most of the surfactants decompose at temperatures higher than $130\,^{\circ}C$. Zhao and his co-workers also realized this opportunity to tune the textural parameters of FDU-12. However, they used a long reaction time which might destroy the structure of the materials. In this work, we report on the synthesis of large pore FDU-12 materials at different temperatures between 100 and $200\,^{\circ}C$. In order to avoid the complete decomposition of the template, the synthesis time has been reduced significantly from 72 to 1 h but at the same time the synthesis temperature has been raised above $150\,^{\circ}C$. This simple approach offers a huge reduction in the synthesis time that reduces the timeframe for synthesis of FDU-12 by about 97%, without posing any threats to the structural order but at the same time offering ultra-large cage type pores. We also systematically investigate the possible influence on the structure and textural parameters of FDU-12 samples by shortening the synthesis time and the variation of temperature. The prepared materials have also been used as adsorbents for the adsorption of lysozyme. It has been found that the materials with ultra-large pores prepared at high temperature exhibit a very high adsorption capacity for the lysozyme molecule.

2. Experimental section

Pluronic F-127 triblock copolymer ($EO_{97}PO_{69}EO_{97}$, molecular mass = 12 500) as the structure directing agent and tetraethyl orthosilicate as the silica source were purchased from Sigma–Aldrich. Similarly, sodium bicarbonate, sodium hydroxide and chicken egg white lysozyme (E.C. 3.2.1.17) were purchased from Sigma–Aldrich. All the chemicals were used as received without any further purification.

2.1. Synthesis of FDU-12

The steps followed in ultra-fast synthesis of FDU-12 samples with different pore sizes are as follows: 1 g of the triblock copolymer F-127, 1 g of 1,3,5-trimethylbenzene and 1.5 g of KCl were dissolved in 60 ml of 2 M HCl and the mixture was continuously stirred in a closed PPE container at room temperature. About 4.15 g of TEOS was added to the reaction mixture and the stirring was continued at $35\,^{\circ}C$. The solution was then transferred into autoclaves and aging was done at different temperatures ranging from 100 to $200\,^{\circ}C$ for 2 h instead of the conventional duration of 72 h. The white precipitate recovered after filtration was washed with water and dried at 100 °C in a hot air oven prior to calcination. The dried sample was made into a fine powder using a mortar and pestle and then subjected to calcination at 540 °C in air to obtain mesoporous silica FDU-12-HX, where X denotes the synthesis temperature.

2.2. Characterization

Structural elucidation of the FDU-12-HX samples prepared at different synthesis temperatures was done using the powder x-ray diffraction (XRD) technique on a Rigaku diffractometer with a Cu $K\alpha$ radiation of 0.154 nm. With step time of 1 s and 2θ step size of 0.01 the diffractogram was recorded in the range from 0.3 to 10° in the 2θ region. Micromeritics ASAP 2420 surface area and porosity analyzer was used to obtain the textural properties of the FDU-12-HX samples and the measurements were done at −196 °C. Degassing was done for all the samples prior to analysis at 250 °C for 4 h. The Brunauer–Emmett–Teller (BET) specific area was obtained from the adsorption branch of the isotherm. The structural morphology of the FDU-12-HX was investigated by using a JEOL JSM 6610 HRSEM operated at an acceleration voltage of 15 kV and a working distance of 8 mm. The internal pore arrangements of FDU-12-HX samples was observed using HRTEM imaging in a JEOL JSM 2100 microscope operated at an acceleration voltage of 200 kV. Samples for HRTEM imaging were prepared by sonication with ethanol for 3 min and deposition over a copper grid. Plausible effect of synthesis temperature on the Si coordination of the FDU-12-HX framework was observed using the ^{29}Si magic angle spinning nuclear magnetic resonance (MAS NMR) on a Bruker Avance III 300 MHz NMR spectrometer with a frequency of 59.63 MHz for Si. Samples were rotated in a 4 mm zirconia rotor at a frequency of 7 kHz. With high-power decoupling, the single-pulse experiments were performed using a 4.5 μs pulse with delay time of 100 s. The decoupling scheme used was tppm15. All spectra were referenced against tetramethylsilane (TMS, 0 ppm) and about 800 scans were accumulated.

2.3. Lysozyme adsorption

Lysozyme solutions were taken in series with concentration ranging from 0.25 to 4 g l^{-1} for adsorption studies. This was achieved by dissolving different quantities of lysozyme in 25 mM buffer solution using a sodium bicarbonate buffer of pH 11. 20 mg of the adsorbent (FDU-12-HX) was suspended in 4 g of the lysozyme solution with concentration ranging from 0.25 to 4 g l^{-1} in each adsorption experiment. The mixture was consequently shaken continuously in a shaking bath with a speed of 160 shakes min^{-1} at 25 °C until equilibrium was reached at 96 h. The amount of lysozyme adsorbed was studied by subtracting the concentration in the supernatant liquid after adsorption from the amount which was initially present by using optical absorption at 281.5 nm. The instrument was calibrated before each set of measurements, using lysozyme solution buffered at the same pH as the isotherm. To avoid interference from the suspended scattering

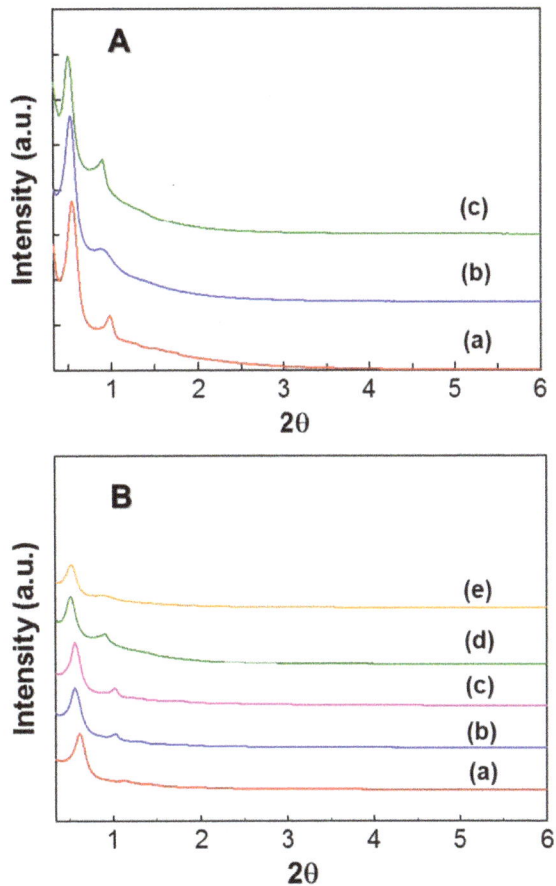

Figure 1. (A) XRD patterns of (A) FDU-12-H200 prepared at the aging durations of (a) 1 h (b) 2 h and (c) 4 h, and (B) FDU-12-H prepared at different temperatures with an aging duration of 2 h. (a) 100 °C, (b) 130 °C, (c) 150 °C, (d) 180 °C, and (e) 200 °C.

particles, centrifugation was done prior to the analysis in the UV–vis instrument. The equilibrium time required for maximum adsorption was found to be 96 h and hence it was followed for all the adsorption experiments [30].

3. Results and discussion

Since the synthesis was done at high temperatures for a shorter duration, it is important to check the mesostructure of the prepared samples in order to make sure that the structure of the samples is developed. Initially, the effect of synthesis time affecting the structural and textural properties of FDU-12 samples was studied by preparing FDU-12 samples at a synthesis temperature of 200 °C under different aging periods ranging from 1 to 4 h. Figure 1(A) shows the XRD patterns of FDU-12-H200 samples prepared at different aging periods. It can be seen that irrespective of such rapid aging duration, each of the FDU-12-H200 samples exhibited a characteristic (111) peak followed by reflections corresponding to (220),

(311), (331) and (442) lattice planes, which is the characteristic behaviour of the face centred cubic structure [7]. Among
the FDU-12-H200 samples prepared at different aging times,
the sample prepared with the aging time of 2 h seemed better
than those with the aging time of 1 and 4 h as reflected from
their respective XRD peak intensities and the well-ordered
structure formation. This was due to the fact that the synthesis
time of 1 h is not enough for the structure formation, which is
due to the poor condensation of silica species, and the
synthesis period of 4 h is too long as it destroys the structure
at the reaction temperature of 200 °C.

Figure 1(B) shows the XRD patterns of FDU-12-HX
samples prepared at different synthesis temperatures ranging
from 100 to 200 °C. Irrespective of their synthesis temperatures, all the samples except FDU-12-H100 show highly
ordered cubic structure indexed by (111), (220), (311), (331)
and (442) lattice planes as shown in figure 1(B), revealing that
the structure of the materials is highly developed within a
short period of time. FDU-12-H100 exhibits only a sharp
peak at low angle with a lower intensity as compared to that
of other samples. These results clearly reveal that the synthesis time of 2 h at 100 °C is not enough for the complete
development of well-ordered mesostructures whereas the high
temperature helps the formation of mesostructures in a shorter
duration through extensive silanol group condensation. It
should also be noted that the (111) diffraction peak of FDU-
12-HX samples is shifted to lower 2θ angle with the increase
of synthesis temperature, which clearly indicates that the unit
cell increases with increasing the temperature. The unit cell,
which is calculated from the formula $a_0 = d\sqrt{3}$, increases
from 14.5 to 17.3 nm with increasing synthesis temperature
from 100 to 200 °C (table 1). Although the aging time is
reduced by 36 fold, the 3D mesoporous silica exhibits good
structural stability revealing the $h\,k\,l$ planes of (111) (220)
and (311) even at 200 °C. This certainly shows that the
templates are quite stable even at such a high temperature
although they ought to, at temperatures beyond 150 °C. The
probable fact as seen from the literature is purely because of
the really short time, which is not enough for the complete
decomposition, or the breakdown of surfactant molecules that
still support the formation of the mesostructures. In addition,
quick condensation of the silica sources at high temperature
also helps the formation of mesostructures at relatively shorter
synthesis times [31].

The effect of the synthesis time on the textural parameters of FDU-12-HX was assessed by nitrogen adsorption–
desorption measurements. Figure 2(A) shows the N_2
adsorption–desorption isotherms of FDU-12 synthesized at
200 °C for different synthesis times. The isotherms of all the
calcined samples are of type IV isotherm with a pronounced
capillary condensation step at a higher relative pressure,
revealing the presence of large pore systems. However, the
broadness of the hysteresis loop is decreased with increasing
synthesis time. The type of hysteresis loop also changes from
H2 to H1 upon increasing the synthesis time from 1 to 4 h.
These results reveal that the shorter synthesis time favours the
cage type pores as it gives a H2 type hysteresis loop whereas
enlarged large cage type pores with narrow pore size are

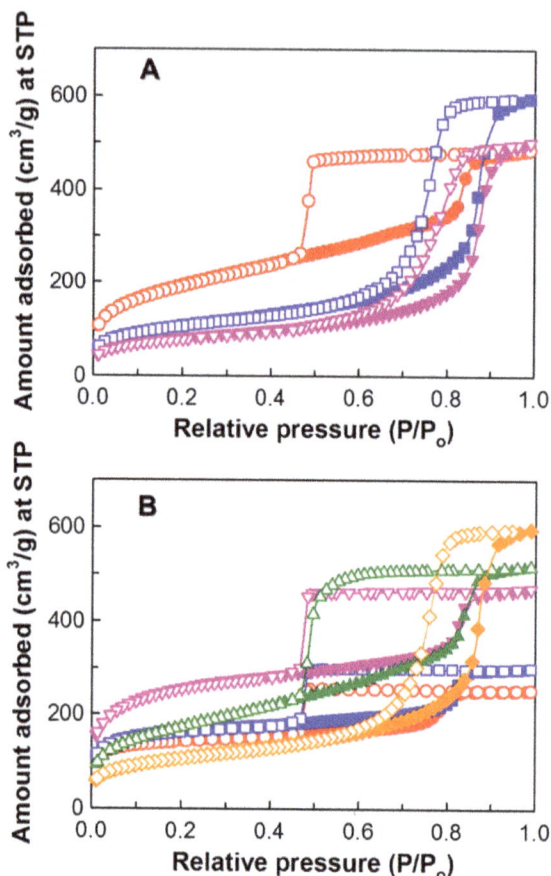

Figure 2. Nitrogen adsorption–desorption isotherms of (A) FDU-12-
H200 with different aging durations. (●) 1 h (■) 2 h, and (▼) 4 h
and (B) FDU-12-H at different temperatures with an aging duration
of 2 h. (●) 100 °C (■) 130 °C (▼) 150 °C (▲) 180 °C and (♦)
200 °C. STP stands for standard temperature and pressure.

obtained when the synthesis time was increased above 1 h. It
should also be noted that the total amount of nitrogen
adsorbed for the samples prepared with the aging time of 2 h
is higher as compared to that of other samples, indicating that
the reaction time of 2 h is the best condition to obtain samples
with a large pore volume. When the syntheses were performed at a high temperature with the aging time of 1 and 4 h,
samples with lower pore volume were obtained. It is also
interesting to note that the BET surface area decreased
from 677.3 to 290.4 $m^2\,g^{-1}$ with increasing aging time from 1
to 4 h, whereas the pore volume and pore diameters exhibited
a maximum capacity, such as 0.93 $cm^3\,g^{-1}$ and 15.6 nm,
for the sample with an aging period of 2 h (table 2). It can be
seen from the pore size distribution (figure 3(A)) that
the peaks are sharper for the sample prepared with the
shorter duration, showing better structural order of the
mesoporous network. From the nitrogen adsorption and XRD
results, it can be concluded that the aging time of 2 h at a
temperature of 200 °C is the best condition to obtain a highly

Table 1. Textural parameters of FDU-12-H prepared at different temperatures with an aging duration of 2 h (A_{BET}—specific surface area; V_p—specific pore volume, d_p—pore diameter).

Material	A_{BET} (m^2 g^{-1})	V_p (cm^3 g^{-1})	d_p (nm)	d_{111} spacing (nm)	Unit cell a_o (nm)
FDU-12-H100	547.2	0.39	9.9	14.52	25.18
FDU-12-H130	620.2	0.46	11.5	15.64	27.09
FDU-12-H150	947.5	0.73	11.8	15.69	27.18
FDU-12-H180	635.0	0.80	12.2	15.76	27.30
FDU-12-H200	387.0	0.93	15.6	17.33	30.02

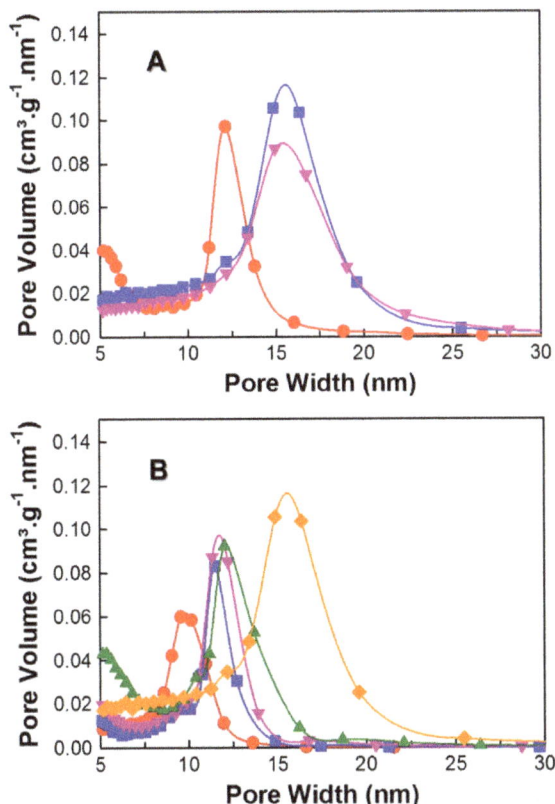

Figure 3. Pore size distribution of (A) FDU-12-H200 prepared with aging durations of (●) 1 h (■) 2 h and (▼) 4 h and (B) FDU-12-H at different temperatures with an aging duration of 2 h (●) 100 °C (■) 130 °C (▼) 150 °C (▲) 180 °C and (♦) 200 °C.

Table 2. Textural parameters of FDU-12-H200 prepared with different aging durations (a) 1 h (b) 2 h, and (c) 4 h.

Aging time (h)	A_{BET} (m^2 g^{-1})	V_p (cm^3 g^{-1})	d_p (nm)
1	677.3	0.75	12.2
2	387.0	0.93	15.6
4	290.4	0.78	15.4

ordered FDU-12 sample with larger pore volume and ordered pores.

In order to understand the role of synthesis temperature on the structure and pore diameter of FDU-12, the samples prepared at different temperatures starting from 100 to 200 °C

at a constant aging time of 2 h were analyzed by nitrogen adsorption–desorption measurements; the results are shown in figure 2(B). All the samples displayed type IV isotherms at all synthesis temperatures with broad hysteresis loops. It should be mentioned that the samples synthesized from 100 to 180 °C showed H2 type hysteresis whereas a H1 hysteresis loop was seen for the silica sample prepared at 200 °C. The shape, type and broadness of the hysteresis loop in the isotherm of the samples reveal that the cage type pores are maintained up to a synthesis temperature of 180 °C. However, when the synthesis temperature is increased to 200 °C, the pores are enlarged and connected, without loss of cage periodicity.

The presence of cage type structures was also confirmed by the XRD results. It is also interesting to note that the capillary condensation step is shifted towards higher relative pressure as the synthesis temperature is increased. When the synthesis temperature of 200 °C was used, the sample with the largest pore diameter and pore volume was obtained. However, the specific surface area of the FDU-12-H200 is much lower than that of other samples. This could be due to the presence of large pores in the sample that significantly decrease the specific surface area. The shift of the capillary condensation step towards the higher relative pressures supplements the XRD patterns, which also show enlargement of the mesopores with the increase of d spacing unit cell size with increasing synthesis temperature. The increase of the temperatures also causes the increase of the size of the entrance to these pores, which is seen by the narrowing width of the hysteresis loop for higher temperatures and is probably due to the slight decomposition of F127 into small organic molecules that act as swelling agents to enlarge the pores [31]. It is also surmised that the high temperature increases the volume fraction of the hydrophobic part of the surfactant as a result of the transformation of polyethylene oxide moieties of the surfactants into hydrophobic and the decrease of the interaction of these species with the silica species. These effects also support the enlargement of the pores in the samples prepared at high temperature. From the N$_2$ adsorption–desorption isotherms of FDU-12-HX samples, it is seen that with increasing synthesis temperature from 100 to 150 °C, the BET surface area also increased from 547.2 to 947.5 m^2 g^{-1}. However, when the temperature was increased above 150 °C, the BET surface area dropped to 635.0 and 387.0 m^2 g^{-1} for FDU-12 synthesized at 180 °C and 200 °C, respectively. However, the pore diameter and pore volume of the sample increase at higher synthesis temperature.

The pore size distribution of all the samples (figures 3(A) and (B)) reveal that the pores are highly uniform but are slightly broadened with increasing synthesis temperature. The pore volume of all the samples shows a steady increase from 0.39 to 0.93 $cm^3\,g^{-1}$ and so does the pore diameter with a gradual increase from 9.9 to 15.6 nm with increasing temperature from 100 to 200 °C. The logic behind the expansion of the pores in terms of pore cavity volume and pore entrance is that, with the rise of the synthesis temperature, the pore expansion takes place from the remnants released due to gradual decomposition of the template which enlarges the pores and the total pore volume of the samples [31]. Among all the samples prepared, FDU-12-H200 prepared with a synthesis time of 2 h shows the highest pore volume and the largest pore diameter with well-ordered porous structure. These results reveal that the FDU-12 sample with excellent structure order can be synthesized within 2 h of synthesis time by simply increasing the synthesis temperature to 200 °C. This was simply due to the high energy at high temperature that helps the quick condensation of the silanol group and the mesostructure formation within a short duration. This unique approach of high temperature synthesis with a low synthesis time would significantly reduce the total duration of the synthesis time and cost of the whole production process, which paves the way for easy commercialization.

In order to further understand the silanol group condensation at different synthesis temperatures and the quick formation of mesostructures at high synthesis temperature, the samples were analyzed by the solid state ^{29}Si MAS NMR and the data are given in figure 4(A). The spectra were deconvoluted into three well-resolved peaks of chemical shifts around −93, −103 and −112 ppm which corresponds to Q^2, Q^3, Q^4, respectively (figure 4(B)). $Q^2(Si(OSi)_2(OH)_2)$ denotes Si bound to two Si–O moieties and two OH groups whereas $Q^3(Si(OSi)_3(OH))$ represents Si bound to three Si–O moieties and one OH group. On the other hand, $Q^4(Si(OSi)_4)$ denotes Si species bound tetragonally to four other Si–O groups [31]. The percentage areas of different Si bonding in the wall structure of the samples prepared at different temperatures are given in table 3. As can be noted in table 3, when the synthesis temperature is increased, the percentage area of the Q^4 increases with the concomitant decrease of Q^3 and Q^2. This confirms the enhancement of the condensation of silanol groups to Si–O–Si when the synthesis temperature of FDU-12 is high. It should also be noted that the decrease in silanol groups in the samples prepared at high temperature offers a high hydrophobic surface which is good for the adsorption of biomolecules [31]. In addition, a high degree of silanol condensation in the wall structure within a short duration of time at high temperature particularly favours other application possibilities for FDU-12 with cage-type pores because these highly cross-linked siloxane moieties offer high hydrothermal stability due to strong Si–O–Si bonding but a lower number of Si–OH groups in the walls.

The surface morphology of the calcined FDU-12 samples prepared at different temperatures was obtained using SEM and the images are shown in figure 5. The images obtained

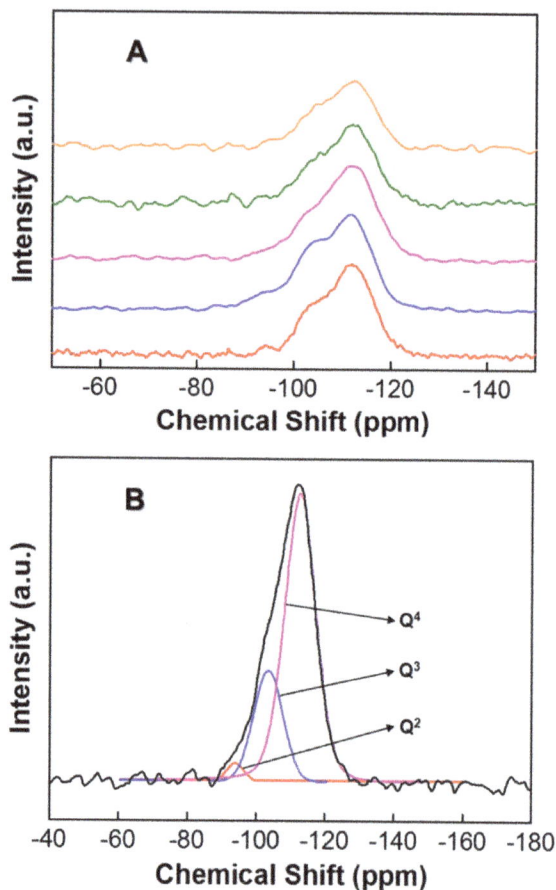

Figure 4. ^{29}Si MAS NMR spectra of (A) FDU-12-H prepared at different temperatures with an aging duration of 2 h (a) 100 °C (b) 130 °C (c) 150 °C (d) 180 °C, and (e) 200 °C, and (B) deconvoluted ^{29}Si MAS NMR spectra of FDU-12-H200.

Table 3. The percentage area of different Si species obtained from the ^{29}Si MAS NMR spectra of FDU-12-H prepared at different temperatures with an aging duration of 2 h.

Material	Q^4 (%Area)	Q^3 (%Area)	Q^2 (%Area)
FDU-12-H100	61.75	30.82	7.43
FDU-12-H130	67.1	29.46	3.44
FDU-12-H150	70.86	26.50	2.64
FDU-12-H180	73.86	23.91	2.23
FDU-12-H200	75.42	22.94	1.64

show a spherical morphology for the FDU-12 samples indicating the high morphological order of the synthesized materials. From the SEM images, the morphology of the spherical structures seems to exhibit an enlargement in the size of the particles from 1.95 μm for FDU-12-H100 to 2.35 μm for FDU-12-H200 which is in concordance with the insights derived from XRD and BET. The possible reasoning would be that at high reaction temperatures, through liquid crystal assembly, the charged surfactant molecules and the

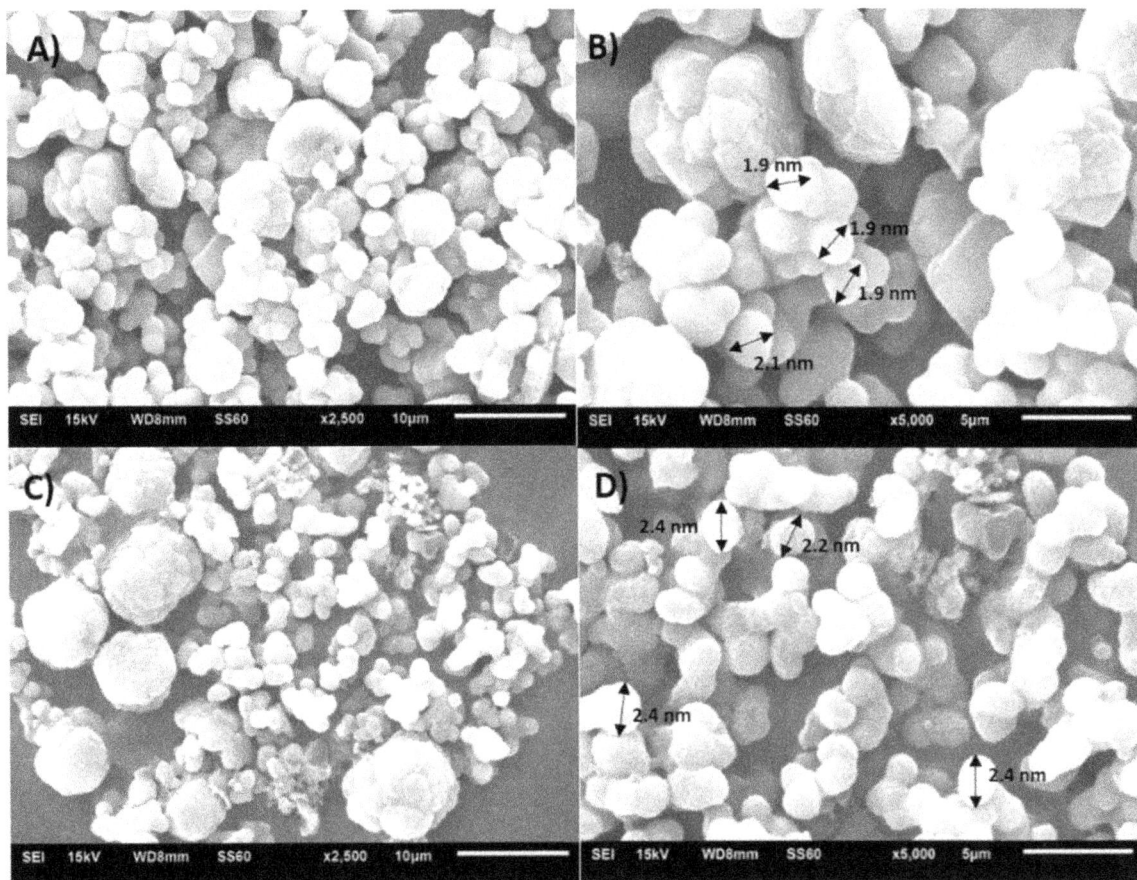

Figure 5. HRSEM images of FDU-12-H prepared at 100 °C (A), (B) and 200 °C (C), (D) for an aging duration of 2 h.

interaction between the silica species are quite notably enhanced. In addition, the enlargement of the surfactant micelles reduces the surface curvature which also results in an increase of the particle size for the sample prepared at high temperature. However, it should be noted that the particles are agglomerated and the spherical shape which is typically obtained for FDU-12 samples is not clear for these samples. This could be due to the short aging time and the high energy at high temperature that favours quick condensation with the neighbouring particles. This would significantly affect the morphological ordering of the samples [31].

The TEM images in figure 6 give a much clearer picture of the FDU-12-H200 synthesized at 200 °C with a greater magnification where we can infer the presence of well defined, uniform cubic structure with Fm3m symmetry with highly ordered array pores over large domains. In FDU-12-H200, the highly ordered pores are arranged in an orderly fashion and connected over significantly large-scale domains. Moreover, due to the short aging duration, even at high temperatures such as 200 °C, the internal porous network remains stable without any detrimental effect. A high magnification of the TEM image also confirms that the mesopores are organized in a 3D network, and the value

of the pore diameter is almost similar to that obtained using nitrogen adsorption analysis. All these results reveal that the FDU-12 materials with 3D structure and large pore and pore volumes can be prepared within a short duration by simply increasing the synthesis temperature above 180 °C.

3.1. Structural confirmation of the immobilized protein

The structural stability of the lysozyme immobilized into the pores of FDU12-200H was confirmed through Fourier transform infrared (FTIR) absorption studies. Here, a comparison is drawn between pure lysozyme protein and the lysozyme immobilized into the silica matrix. As can be seen in figure 7(a), the FTIR spectra show distinct amide I and amide II bands, which are characteristic of the structure of the enzyme depicting the confirmation of the α-helical and β-sheets of the lysozyme. The amide I of the lysozyme, which is observed near 1650 cm^{-1}, is attributed to the most sensitive part of the spectral region given by the C=O stretching of the peptide linkages. The amide II found near 1520 cm^{-1} is attributed to the NH bending vibrations of the lysozyme molecule. It is also evident from figure 7(b) that the two

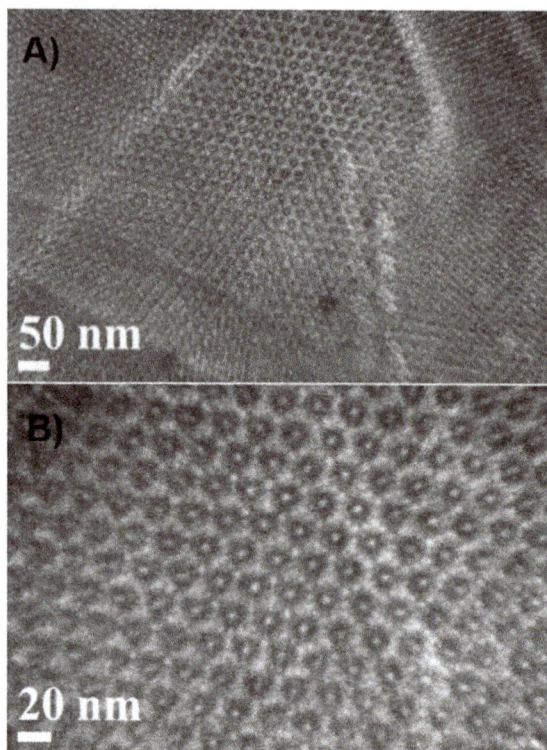

Figure 6. HRTEM images of FDU-12-H200.

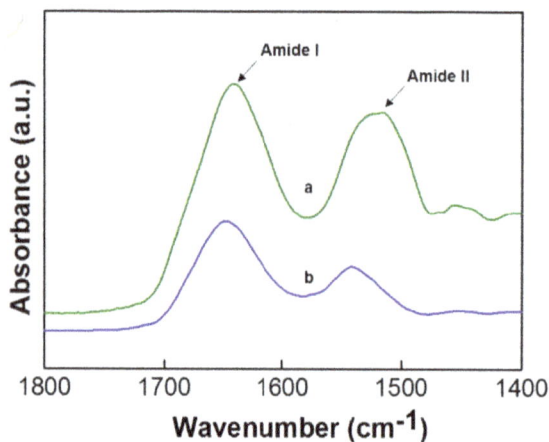

Figure 7. FTIR spectra of pure lysozyme (a) and lysozyme loaded on FDU-12-H200 (b).

amide bands are observed near 1650 and 1550 cm^{-1}. It is seen that the intensity ratio of the amide bands in both pure lysozyme and the lysozyme immobilized into the FDU-12 matrix is almost same, which reveals that the lysozyme molecules are quite stable as seen through the secondary structures such as the α-helical and β-sheets of the immobilized enzyme [32].

3.2. Lysozyme adsorption

The efficiency of FDU-12 synthesized with a short aging period was tested by performing adsorption studies using lysozyme. A schematic representation of the adsorption of lysozyme over FDU-12-HX samples is shown in figure 8. A standard optimized procedure [33] with a buffer concentration of 25 mM and a solution pH of 11 was used in the adsorption experiments. Figure 9 shows the adsorption isotherms of lysozyme adsorbed onto the FDU-12-H200 samples at the solution pH of 11 wherein the amount of lysozyme adsorbed per unit weight of the mesoporous adsorbent FDU-12 with different pore diameters is plotted against the equilibrium concentration of lysozyme solution. Optimized adsorption depends on the selection of pH which in this case is at pH 11 and is near the isoelectric point of lysozyme [30, 34]. At this pH, it is believed that the coulombic repulsive forces present in the adsorbate molecules are minimized, which significantly enhances the close packing of the protein molecules [26]. Hence, there is a maximum protein adsorption seen near the isoelectric point where the hydrophobic interaction is stronger than the electrostatic interaction, due to the neutral charge between the amido groups of lysozyme, siloxane bridges of FDU-12 and intermolecular originating from the amido groups on the lysozyme surface. The adsorption near the isoelectric point also supports the close packing of the protein molecules into the mesoporous cage by decreasing the solubility of proteins and thereby enhancing good hydration of carbonate ions under optimal ionic concentration of the buffer (25 mM). Figure 9 reveals that the lysozyme adsorption increases with an initial sharp escalation from a minimum of about $3\,\mu\text{mol}\,\text{g}^{-1}$ for FDU-12-H100 to a maximum of $29\,\mu\text{mol}\,\text{g}^{-1}$ for FDU-12-X200.

The rate and extent of adsorption is purely attributed to the textural parameters of the materials synthesized at different temperatures. It is found that the specific surface area, pore volume and pore diameters of the silica samples influence the amount of lysozyme adsorbed. More importantly, the size of the pore diameters, which spread over a broad range from 9.9 to 15.6 nm, is bound to have control over the amount of lysozyme molecules that can be immobilized into the 3D spherical pores of the FDU-12-HX materials. It can be clearly seen from figure 9 that the FDU-12-HX with the largest pore diameter and the highest pore volume offers the highest lysozyme adsorption capacity. The size of the pores not only has an influence over the diffusion of lysozyme into vacant sites in the cavities but also tends to permit close-fitting of the biomolecules possibly resulting in more layers of adsorption into the 3D cage-type mesoporous network. The 3D structure of FDU-12 with excellent textural parameters supports a high adsorption and superior mass transfer capacity of the adsorbate.

The adsorption capacity of FDU-12 prepared in this work was compared with other 1D and 3D materials. Table 4 gives a summary of the silica and carbon adsorbents used for the adsorption of lysozyme molecules. This comparison table gives brief information regarding the textural parameters, their aging duration and the amount of lysozyme adsorbed,

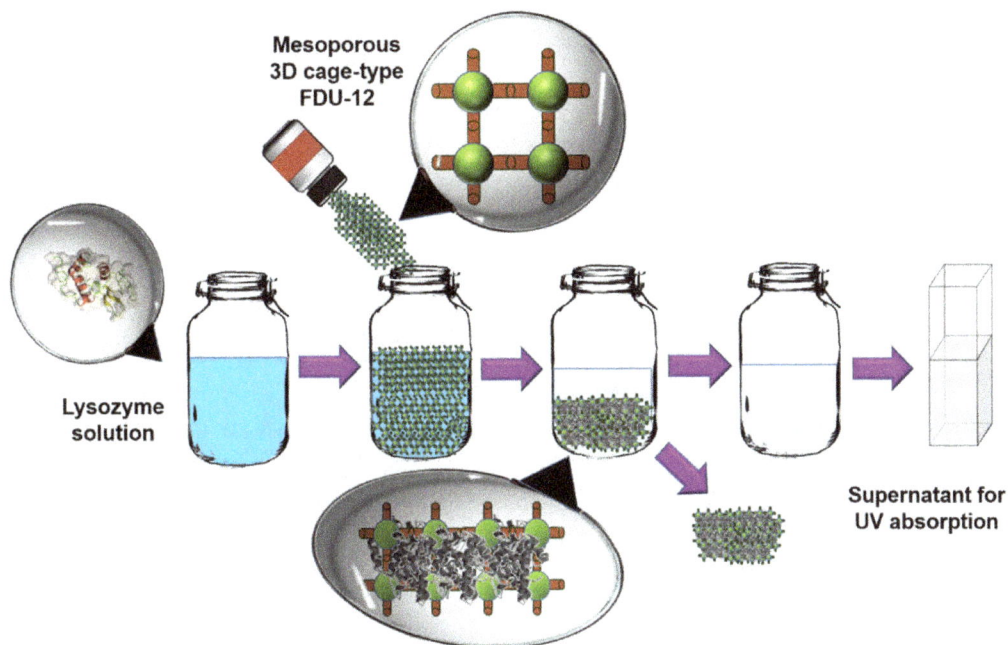

Figure 8. Schematic representation of the adsorption process using lysozyme on FDU-12 mesoporous silica.

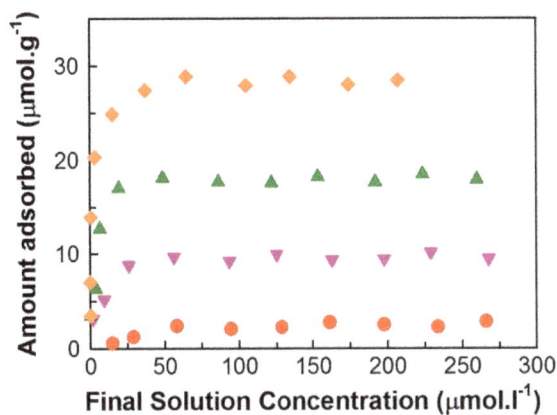

Figure 9. Adsorption of lysozyme over FDU-12-H prepared at different temperatures with an aging duration of 2 h (●) 100 °C (▼) 150 °C (▲) 180 °C and (♦) 200 °C.

which clearly shows the critical impact of the textural parameters and synthesis conditions on the amount of guest molecules adsorbed on the surface and into the porous channels of the 3D cage-type silica framework. It can be seen from table 4 that the adsorption capacity of FDU-12 is much higher than that of other samples studied, except KIT-6-150 which has the highest pore volume and 3D structure but requires a longer aging time of 24 h. These results revealed that the adsorbent with 3D cage-like network systems allow not only an easy access for lysozyme molecules to even the interior parts of the pores without any difficulties, but also easy diffusion via the pores that oriented in all different

directions [33]. Thus, it can be concluded that a well-ordered 3D structure, large pore diameter and pore volume are essential for easy access to the adsorption sites thereby accommodating large amounts of biomolecules into mesoporous spherical cages.

From the variety of mesoporous carbon and silica species studied in the literature, it can also be seen that there is a long aging duration involved in the synthesis process of silica. In the case of mesoporous carbons, their synthesis involves the silica framework which has to be synthesized first and takes more than 5 days for the preparation of the samples including the dissolution of the silica template. There are even more materials with a high surface area and pore volume available in the literature, which also require a longer synthesis time. However, with such short aging durations (2 h) as shown in our report, and yet with such high pore volume and large pore diameters, we could achieve a comparatively high amount of lysozyme adsorption into the cage-type 3D mesoporous channels of the FDU-12-HX materials. Such improvements in expediting the aging treatment could speed up processes, culminating in ultra-fast synthesis at the industrial scale and prove to be a boon towards catalysis, adsorption, energy storage, drug delivery, separation processes and other biomedical applications.

4. Conclusions

We demonstrated the high temperature synthesis of FDU-12 with well-ordered structure and large pore diameter using a non-ionic surfactant. This simple strategy allowed us not only

Table 4. Comparison of the quantity of lysozyme adsorbed over adsorbents with different structures and textural parameters.

Adsorbent	Aging duration (h)	A_{BET} (m^2 g^{-1})	V_p (cm^3 g^{-1})	d_p (nm)	Quantity of lysozyme adsorbed (μmol g^{-1})
FDU-12-H200	2	387.0	0.94	15.64	29
FDU-12-M200 [31]	2	332.8	0.98	13.08	26
FDU-12-413 [13]	72	281	0.78	12.3	6.94
MCM-41 [35]	48	1036	0.83	2.7	1.94
PMO [36]	24	908.7	1.12	7.4	13.5
SBA-15 [36]	24	885	1.24	8.89	17.5
KIT-6-150 [33]	24	555	1.53	11.3	57.2
KIT-5-150 [37]	24	470	0.75	5.7	4.99
CMK-3-150 [33]	48	1350	1.6	6.5	22.9
CKT-3A-150 [34]	24	1600	2.1	5.2	26.5

to significantly reduce the hydrothermal treatment time from 72 to 2 h, which is particularly important for practical applications, but also to prepare a material with tunable pore diameter and large pore volume. The XRD results confirmed that the well-ordered 3D mesoporous structure is still retained and the interconnected cage-type pores are formed even at 200 °C within the aging time of 2 h. The pore diameters could be tuned from 9.9 to 15.6 nm by changing the synthesis temperature from 100 to 200 °C. Among the samples studied, FDU-12-HX synthesized at 200 °C with 2 h aging displayed the highest pore volume of 0.93 cm^3 g^{-1}. We also demonstrated that the high temperature synthesis also supports the stability of the walls through perfect and quick silanol group condensation and further increases the hydrophobic nature of the sample. The prepared samples have been employed as adsorbents for the adsorption of lysozyme molecules. Among the samples studied, the sample with the largest pore diameter and the highest pore volume showed the highest adsorption capacity (29 μmol g^{-1}). We believe that this simple approach would definitely set a benchmark for much more efficient synthesis of FDU-12 and can be used for the synthesis of a variety of mesoporous materials with different structures and pore diameters within a short duration, which would create a platform for the easy commercialization of these materials and offer various application possibilities.

Acknowledgments

GL acknowledges the CMM (Centre for Microscopy and Microanalysis-AMMRF) for providing the state of art microscope facilities at the University of Queensland. AV is grateful to ARC for the award of the future fellowship and the University of Queensland for the start-up grant. The project was also financially supported by King Saud University, Vice Deanship of Scientific Research Chairs.

References

[1] Ariga K, Li M, Richards G J and Hill J P 2011 *J. Nanosci. Nanotechnol.* **11** 1

[2] Ariga K, Ji Q, Nakanishi W, Hill J P and Aono M 2015 *Mater. Horiz.* doi:10.1039/C5MH00012B

[3] Ariga K, Kawakami K, Ebara M, Kotsuchibashi Y, Ji Q and Hill J P 2014 *New J. Chem.* **38** 5149

[4] Ariga K, Ji Q, McShane M J, Lvov Y M, Vinu A and Hill J P 2012 *Chem. Mater.* **24** 728

[5] Nakanishi W, Minami K, Shrestha L K, Ji Q, Hill J P and Ariga K 2014 *Nano Today* **9** 378

[6] Ariga K, Vinu A, Yamauchi Y, Ji Q and Hill J P 2012 *Bull. Chem. Soc. Japan* **85** 1

[7] Davis M E 2002 *Nature* **417** 813

[8] Kresge C T, Leonowicz M E, Roth W J, Vartuli J C and Beck J S 1992 *Nature* **359** 710

[9] Fan J, Yu C, Lei J, Zhang Q, Li T, Tu B, Zhou W and Zhao D 2005 *J. Am. Chem. Soc.* **127** 10794

[10] Fan J, Lei J, Wang L, Yu C, Tu B and Zhao D 2003 *Chem. Commun.* 2140

[11] Gobin O C, Wan Y, Zhao D, Kleitz F and Kaliaguine S 2007 *J. Phys. Chem.* C **111** 3053

[12] Gobin O C, Huang Q, Hoang V T, Kleitz F, Eic M and Kaliaguine S 2007 *J. Phys. Chem.* C **111** 3059

[13] Fan J, Yu C, Gao F, Lei J, Tian B, Wang L, Luo Q, Tu B, Zhou W and Zhao D 2003 *Angew. Chem., Int. Edn Engl.* **42** 3146

[14] Morishige K and Yoshida K 2010 *J. Phys. Chem.* C **114** 7095

[15] Beck J S *et al* 1992 *J. Am. Chem. Soc.* **114** 10834

[16] Zhao D, Huo Q, Feng J, Chmelka B F and Stucky G D 1998 *J. Am. Chem. Soc.* **120** 6024

[17] Sayari A, Yang Y, Kruk M and Jaroniec M 1999 *J. Phys. Chem.* B **103** 3651

[18] Jana S K, Nishida R, Shindo K, Kugita Y and Namba S 2004 *Micropor. Mesopor. Mater.* **68** 133

[19] Matos J R, Kruk M, Mercuri L P, Jaroniec M, Zhao L, Kamiyama T, Terasaki O, Pinnavaia T J and Liu Y 2003 *J. Am. Chem. Soc.* **125** 821

[20] Kleitz F, Liu D, Anilkumar G M, Park I S, Solovyov L A, Shmakov A N and Ryoo R 2003 *J. Phys. Chem.* B **107** 14296

[21] Matos J R, Mercuri L P, Kruk M and Jaroniec M 2002 *Langmuir* **18** 884

[22] Kao C P, Lin H P, Chao M C, Sheu H S and Mou C Y 2001 *Stud. Surf. Sci. Catal.* **135** 1361

[23] Van D V P, Benjelloun M and Vansant E F 2002 *J. Phys. Chem.* B **106** 9027

[24] Yu C, Fan J, Tian B, Stucky G D and Zhao D 2003 *J. Phys. Chem.* B **107** 13368

[25] Antochshuk V, Kruk M and Jaroniec M 2003 *J. Phys. Chem.* B **107** 11900

[26] Kruk M, Antochshuk V, Matos J R, Mercuri L P and Jaroniec M 2002 *J. Am. Chem. Soc.* **124** 768

[27] Kruk M, Celer E B and Jaroniec M 2004 *Chem. Mater.* **16** 698

[28] Han Y J, Watson J T, Stucky G D and Butler A 2002 *J. Mol. Catal.* B **17** 1

[29] Wirnsberger G, Scott B J and Stucky G D 2001 *Chem. Commun.* 119

[30] Vinu A, Murugesan V, Tangermann O and Hartmann M 2004 *Chem. Mater.* **16** 3056

[31] Lawrence G, Anand C, Strounina E and Vinu A 2014 *Sci. Adv. Mater.* **6** 1481

[32] Vinu A, Miyahara M and Ariga K 2005 *J. Phys. Chem.* B **109** 6436

[33] Vinu A, Gokulakrishnan N, Balasubramanian V V, Alam S, Kapoor M P, Ariga K and Mori T 2008 *Chem. Eur. J.* **14** 11529

[34] Vinu A, Miyahara M, Sivamurugan V, Mori T and Ariga K 2005 *J. Mater. Chem.* **15** 5122

[35] Yang J, Stevens G W and O'Connor A J 2008 *J. Aust. Ceram. Soc.* **44** 1

[36] Qiao S Z, Djojoputro H, Hu Q and Lu G Q 2006 *Prog. Solid State Chem.* **34** 249

[37] Vinu A, Miyahara M, Hossain K Z, Takahashi M, Balasubramanian V V, Mori T and Ariga K 2007 *J. Nanosci. Nanotechnol.* **7** 828

8

Energy dispersive x-ray spectroscopy for nanostructured thin film density evaluation

Irene Prencipe, David Dellasega, Alessandro Zani[1], Daniele Rizzo and Matteo Passoni

Dipartimento di Energia, Politecnico di Milano, Milan, Italy

E-mail: irene.prencipe@polimi.it

Abstract

In this paper, we report on two fast and non-destructive methods for nanostructured film density evaluation based on a combination of energy dispersive x-ray spectroscopy for areal density measurement and scanning electron microscopy (SEM) for thickness evaluation. These techniques have been applied to films with density ranging from the density of a solid down to a few mg cm^{-3}, with different compositions and morphologies. The high resolution of an electron microprobe has been exploited to characterize non-uniform films both at the macroscopic scale and at the microscopic scale.

Keywords: density evaluation, thin film, EDS, foam, areal density

1. Introduction

In recent decades, the production and characterization of thin films with a wide range of morphologies and compositions have attracted great interest due to their applicative potential. Therefore, appropriate characterization methods are required for film properties and, in particular, for density, which is a key parameter for many applications, among which laser-plasma interaction and laser driven ion acceleration experiments have recently triggered wide interest [1, 2].

An ideal technique for thin film density measurement should be reliable in a wide density range (from the density of a solid to a few mg cm^{-3}) and for a great variety of materials and morphologies. It should allow us to evaluate the density of non-homogeneous films with a good spatial resolution. It should also be non-destructive and fast and require a simple and cheap experimental apparatus.

In general, the density of thin films can be evaluated by combining thickness and areal density measurements. The

former can be achieved, for example, through electron microscopy, by analysing cross-sectional scanning electron microscopy (SEM) images or by transmission electron microscopy (TEM), depending on the order of magnitude. Several methods can be employed to measure areal density. Commonly used nuclear-based techniques such as Rutherford backscattering spectroscopy (RBS), elastic recoil detection analysis, and nuclear reaction analysis provide accurate areal density measurements with a good spatial resolution, but they require complex experimental equipment, i.e., linear accelerators to produce MeV ion beams. Conversely, in thin film deposition facilities, a quartz crystal microbalance (QCM) is often adopted to measure the mass deposition rate on a well-defined surface and, therefore, a mean areal density value. This technique is very popular due to the simplicity of the required instrument. However, QCM only allows us to measure the average areal density of a film directly deposited on its quartz crystal surface in conditions simulating the growth configuration of the film under analysis, thus providing an indirect measurement. Moreover, this method is not reliable for very low density materials (below $\sim$30 mg cm^{-3}), as shown for carbon foams in [3].

In this context, an attractive technique satisfying most of the ideal requirements listed above is based on energy dispersive x-ray spectroscopy (EDS). The energy and intensity of characteristic x-rays produced in matter by an incident electron beam are related to the atomic number of the emitting

[1] Present affiliation: Laboratory of Molecular Biophysics, Uppsala University, SE-751 24 Uppsala, Sweden.

element and areal density of the examined layer, respectively. The penetration depth of electrons in matter is a function of the electron accelerating voltage and ranges approximately from 0.1 μm to several μm for standard electron probe beams (2–50 keV). As a consequence, a proper selection of the electron accelerating voltage allows us to characterize a surface layer of the sample under investigation, i.e., a thin coating deposited on the sample surface. In the 1960s, Sweeney et al proposed to employ EDS for the evaluation of the thickness of compact films with known density [4]. Nevertheless, this technique has never been used for density evaluation. An EDS-based method would be non-destructive and could provide local density values, allowing us to characterize non-homogeneous films. Moreover, the microanalysis equipment required for EDS is relatively simple, and it is often integrated into SEM devices, which are commonly used in material science laboratories and allow us also to achieve thickness measurements.

In this paper, we quantitatively develop and test two methods for thin film density evaluation, both based on the combined use of EDS for areal density measurement and cross-sectional SEM images for thickness assessment. The main goal of this work is to show the applicability of these methods and to study their limits. To this purpose, they have been employed to characterize compact coatings and nanostructured thin films with various compositions, a large variety of mesoscale morphologies, and density ranging from the density of a solid to a few mg cm^{-3}. Thermal evaporation and pulsed laser deposition (PLD) have been exploited for film growth, as described in section 3. To better illustrate the methods employed for areal density evaluation, we will give an account of a few EDS-based methods for thickness evaluation in section 2.1. In section 2.2, we will describe the most relevant theoretical aspects available in the literature about areal density evaluation as well as a few practical aspects concerning the experimental setup. In particular, criteria for the choice of a measurement method and for the appropriate selection of the electron accelerating voltage will be discussed. Experimental results will be illustrated and discussed in section 4.

2. EDS-based methods for thin film areal density evaluation

2.1. General background

The application of EDS to coating thickness evaluation has been widely explored since the 1960s. To this aim, a number of methods have been employed in the literature. For example, thickness evaluation was achieved by measuring the minimum accelerating voltage required to probe the whole film thickness [5] or the accelerating voltage, for which a given fraction of the x-ray intensity produced by a reference standard is emitted by the sample [6]. Here, we discuss the application of two methods proposed by Sweeney, Seebold, and Birks in 1960 [4] and by Cockett

and Davis in 1963 [7], respectively, because of their relevance to this work.

These approaches, respectively known as the *coating method* and the *substrate method*, were developed for multilayer samples composed of a known substrate and a coating with unknown thickness. In these methods, the coating layer thickness is calculated from the intensity of x-rays produced either in the sample coating or substrate by an incoming electron beam with appropriate initial energy, provided that the intensity of the x-rays produced by a bulk reference standard is known (see figure 1). The main difference between the coating method and the substrate method lies in the choice of the reference standard: the reference standard must contain an emitting element present only in the sample coating or in the substrate, respectively.

The reliability of the EDS-based methods described so far for thickness evaluation was thoroughly investigated in the literature. Thickness values measured by employing EDS-based methods were compared to values achieved using other techniques, i.e., RBS or cross-sectional SEM images. The difference between results achieved with established techniques and with EDS-based methods is generally around 15–20% [7]. Although these methods were proposed for thickness evaluation, the parameter they are directly sensitive to is the areal density, since the generation of characteristic x-rays in a layer does not depend on the material thickness t or density ρ separately, but on its areal density, $\tau = \rho t$. Therefore, this technique can be applied to density evaluation for thin films with known thickness. In the next subsection, we will illustrate theoretical aspects related to areal density evaluation and x-ray production modelling, and we discuss issues related to the experimental setup.

2.2. Areal density evaluation

The calculation of film areal density from x-ray intensity requires knowledge of the so-called probability function for x-ray production (PFXP), $\phi(\sigma)$. This function, introduced by Castaing in 1951 [8], describes the distribution in depth of the primary ionizations produced in a sample by an incoming electron. The function argument is a depth expressed in terms of areal density, and is given by $\sigma = \rho z$, where z is the depth measured in linear units.

PFXP allows us to calculate the x-ray intensity $\mathrm{d}I_1 = \phi_1(\sigma)\mathrm{d}\sigma$ emitted by an element Z_1 in a layer $\mathrm{d}\sigma$ at a depth z below the sample surface of a material with density ρ. Thus, the intensity measured for a characteristic x-ray line emitted by Z_1 in a layer with finite areal density $\Delta\tau$ is

$$I_1 = k \int_{\Delta\tau} C_1 \phi_1(\sigma) \exp\left(-\chi\sigma\right)\mathrm{d}(\sigma), \qquad (1)$$

where C_1 is the mass concentration of the element under analysis. The term $\chi = (\mu/\rho)\csc\theta$ takes into account the absorption of emitted x-rays propagating to the sample surface: μ/ρ is the mass absorption coefficient, and θ is the x-ray

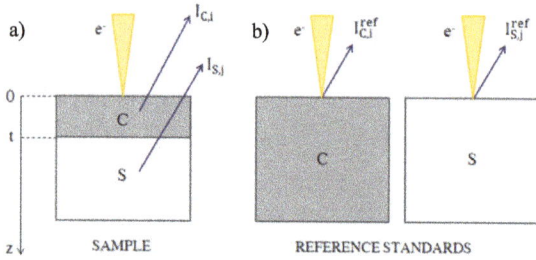

Figure 1. EDS-based film thickness measurement methods. a) X-ray emission from a coating with thickness t and from a substrate due to the incident electron beam. b) X-ray emission from reference standards for the coating method and the substrate method.

take-off angle. k is a constant given by

$$k = \frac{I^{\text{ref}}_1}{\int_0^{+\infty} C_1^{\text{ref}} \phi_1^{\text{ref}}(\sigma) \exp\left(-\chi^{\text{ref}}\sigma\right) \mathrm{d}(\sigma)}, \tag{2}$$

where I_1^{ref} is the x-ray intensity produced by Z_1 present in a reference standard with known composition. The ref superscript indicates that the concentration of Z_1, the PFXP, and the absorption term χ refer to the reference standard. X-rays intensities must be measured under the very same conditions for the sample under analysis and the reference standard, since the intensity of detected characteristic x-rays is influenced by many factors, such as the beam current, the measurement duration, and geometry.

If a model for PFXP evaluation is available, the relation between the unknown coating areal density, τ, and the x-ray intensity emitted by the sample can be calculated through equations (1) and (2) for both methods.

In the coating method, characteristic x-rays emitted by an element Z_i present in the sample coating are considered. The intensity of the selected x-ray line is calculated for both sample and reference standard from the recorded spectra by integrating the peak fitting curves. According to equations (1) and (2), the coating to reference standard intensity ratio $I_{\text{C},i}/I_{\text{C},i}^{\text{ref}}$ can be expressed in terms of film areal density through the following formula

$$\frac{I_{\text{C},i}}{I_{\text{C},i}^{\text{ref}}} = \frac{\int_0^\tau C_i \phi_{\text{C},i}(\sigma) \exp\left(-\chi_{\text{C}}\sigma\right) \mathrm{d}\sigma}{\int_0^\infty C_i^{\text{ref}} \phi_{\text{C},i}^{\text{ref}}(\sigma) \exp\left(-\chi_C^{\text{ref}}\sigma\right) \mathrm{d}\sigma}, \tag{3}$$

where the coefficients χ_{C} and χ_C^{ref} account for the absorption of x-rays in the sample coating and in the reference standard, respectively.

In the substrate method, the intensity of a characteristic x-ray line emitted by an element Z_j contained in the sample substrate is considered. The sample substrate to reference standard x-ray intensity ratio $I_{\text{S},j}/I_{\text{S},j}^{\text{ref}}$ can be expressed as a function of film areal density τ through equations (1) and (2). If x-ray absorption in the coating is taken into account, the relation between the intensity ratio and τ can be expressed as

follows

$$\frac{I_{\text{S},j}}{I_{\text{S},j}^{\text{ref}}} = \exp^{(-\chi_{\text{C}}\tau)}$$

$$\times \frac{\int_\tau^{+\infty} C_j \phi_{\text{S},j}(\sigma) \exp\left[-\chi_{\text{S}}(\sigma - \tau)\right] \mathrm{d}\sigma}{\int_0^{+\infty} C_j^{\text{ref}} \phi_{\text{S},j}^{\text{ref}}(\sigma) \exp\left(-\chi_{\text{S}}^{\text{ref}}\sigma\right) \mathrm{d}\sigma}, \tag{4}$$

where the coefficients χ_{C}, χ_{S}, and $\chi_{\text{S}}^{\text{ref}}$ introduce the effect of x-ray absorption in the sample coating and substrate and in the reference standard.

Once the x-ray intensities emitted from the sample and from an appropriate reference standard have been measured, areal density τ can be calculated by inversion of equation (3) for the coating method or (4) for the substrate method. Nevertheless, repeating this process for each set of experimental data can be time consuming. Thus, it is convenient to calculate calibration curves expressing areal density as a function of the sample to a standard intensity ratio for a given substrate–coating combination and measurement configuration.

As mentioned before, a model for PFXP evaluation is required to calculate areal density. In previous works PFXP was extrapolated from experimental data [4, 7] or from Monte Carlo simulations [9]. Later, many models were proposed for PFXP approximation as a function of experimental conditions and sample properties [10]. In the 1980s Pouchou and Pichoir proposed two of the most popular methods: the PAP (Pouchou and Pichoir) model, in which the distribution function is approximated by two smoothly joined parabolas [11], and the XPP (extended Pouchou–Pichoir) model, which is based on an exponential approximation of the PFXP and allows us to describe an experimental configuration with obliquely incident electrons [12]. In 1981, Packwood and Brown proposed the so-called modified surface-centered Gaussian (MSG) model [13], which is theoretically founded on the hypothesis that electrons move isotropically in the sample. In this model, the persistence of a directional electron propagation near the surface and the consequent deviation from a totally random walk are taken into account, introducing an exponential term which vanishes rapidly with depth. This model was employed by Stenberg and Boman [14], with slightly different parameter definitions, to assess the thickness of amorphous carbon films. In this work, the Rembach–Karduck version of the MSG model, known as the RE method [15], has been considered. The RE method generalizes the MSG model to ultrasoft x-rays emitted by low atomic number materials. The form of the distribution function employed for data analysis is

$$\phi(\sigma) = \gamma \exp\left[-\alpha^2 \sigma^2\right] \left[1 - \frac{\gamma - \phi(0)}{\gamma} e^{(-\beta\sigma)}\right]. \tag{5}$$

This equation contains four shape-parameters: α, whose inverse describes the width of the Gaussian function, is correlated with the penetration depth of incident electrons; β takes into account the deviation from a pure Gaussian function in the surface region; γ represents the amplitude of the

Gaussian function; and $\phi(0)$ is the surface ionization, the distribution value at the surface. For the definitions of these parameters in the RE model, refer to [10].

The PFXP models described so far are valid for homogeneous samples, while for multilayer samples the electron propagation and x-ray production are altered by the presence of a coating–substrate interface [16]. Thus, in principle, a modified PFXP function should be considered to take into account this effect, but, as far as we know, no analytical model is reported in the literature. As a consequence, x-ray generation distribution functions for both coating and substrate are calculated as the PFXP of a homogeneous sample with the same composition as the layer under analysis. Since the function does not show a strong dependence on the atomic number, this working assumption does not introduce a significant error if the difference between coating and substrate atomic numbers is below 5 [7]. For higher differences, the distortion due to the presence of the coating–substrate interface could introduce an error in areal density measurement.

Moreover, in general, PFXP models should consider the emission of x-rays due to fast secondary electrons (FSE) and fluorescence. In multilayer samples, these effects can be due to the composition of the layer under analysis (matrix effects) but also to the composition of the other layer. The Rehbach–Karduck model adopted in this work only takes into account the FSE matrix effect, whose contribution can be as high as 15% for low-energy x-rays if the initial electron energy is much higher than the absorption edge of the peak under analysis [17]. However, the fluorescence and secondary emission effects due to the sample multilayer structure are not considered in the model.

In addition to the theoretical formulation described so far, a few practical aspects concerning the experimental setup must be taken into account to achieve reliable areal density measurements.

A noteworthy issue is related to the method selection. The coating method and the substrate method are completely equivalent according to the theory, even though their equivalence still has to be experimentally proved. However, in a given experimental configuration one method could be more convenient or more reliable than the other for merely practical reasons. Thus, the availability of two methods is a resource which can be exploited to overcome practical difficulties related to specific experimental configurations. In a few cases, the choice of the method is determined by the specific properties of the sample's x-ray spectrum. For example, the deconvolution of overlapped x-ray peaks is a time-consuming process that could reduce the technique's reliability. Thus, the selection of non-resolved peaks should be avoided. Moreover, the choice of particular elements can present critical aspects. For instance, the extremely low energy of carbon Kα peak (277 eV) could limit the maximum detectable areal density, and the low x-ray production cross-section in the carbon reduces the signal-to-noise intensity ratio. Finally, in the case of multi-elemental coatings, the substrate method can be chosen to remove the issue related to the selection of an appropriate reference standard. In principle, it is not required for the reference standard to have the

same composition as the sample coating or substrate. The only requirement is that the emitting element must not be present in both layers. Nevertheless, adopting a standard with the same composition as the material under analysis should reduce the error due to modelling approximations. However, for multi-elemental coatings, it is usually difficult to produce a reference standard with the very same composition as the coating; thus the substrate method should be preferred.

Another issue is related to the selection of an appropriate electron accelerating voltage, which is a crucial issue for areal density measurement, since this parameter determines the methods' reliability and applicability range. A rough criterion for voltage selection can be deduced from the requirement that the electron initial energy should guarantee a significant energy loss both in the sample coating and in the substrate. Thus, electron energy must be in a range that allows us to probe the sample substrate but in which the effect of the coating on electron propagation is not negligible. Moreover, electron energy cannot be lower than the absorption edge of the emitting element. As a consequence, the lower detection limit of the technique is given by the minimum areal density required to absorb a significant fraction of electron energy for beams with initial energy slightly above the absorption edge. On the other hand, the maximum detectable areal density is lower than the electron penetration range at the maximum available accelerating voltage. In addition, the attenuation of x-rays in the sample can reduce significantly the detected spectral intensity and, as a consequence, the higher detection limit of the technique.

In section 4, the application of EDS-based areal density evaluation methods to coating density measurement will be shown for a number of experimental conditions. Due to its importance for the technique's reliability, the issue of appropriate electron accelerating voltage selection will be addressed. Moreover, particular attention will be devoted to the choice of the measurement method, and comparisons of the results achieved exploiting the two approaches will be shown whenever possible.

3. Experimental methods

Thin films exploited to study the application of EDS-based methods to density evaluation were produced by physical vapour deposition (PVD) techniques. Compact films were produced by thermal evaporation with a base pressure of 5×10^{-3} Pa. PLD [18] was employed to grow nanostructured films with density ranging from the solid density to a few mg cm^{-3}.

Reference areal density measurements were carried on for most films using QCMs provided by Inficon: reference coatings were grown on the quartz crystal resonator in the same conditions as the films under analysis.

A Zeiss Supra 40 field-emission SEM was employed to get cross-sectional images for film thickness evaluation (with an accelerating voltage of 5 kV and working distance around 4 mm) and to accelerate electrons for EDS experiments. The

Table 1. X-ray peaks considered for EDS analysis.

Element	Peak	Energy (keV)
C	Kα	0.277
Al	Kα	1.49
Si	Kα	1.74
Rh	L series	2.69
Ag	Lα	2.98
W	Lα	8.37
Au	Mα	2.12

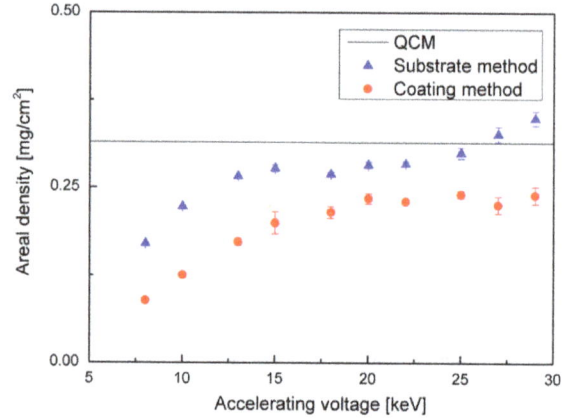

Figure 2. Areal density of a thin Ag film on Au substrate measured as a function of the electron accelerating voltage.

latter were carried out with accelerating voltage in the range of 5–30 kV, depending on the expected areal density of the film under analysis. A Si(Li) detector was employed for x-ray spectra collection. System calibration was carried out using Si Kα for accelerating voltages below 12 keV and Co Kα for electron energies above 12 keV. At least three and five spectra were recorded for each reference standard and for each sample, respectively. The acquisition times ranged between 60–120 s, depending on the signal intensity. Peaks considered for each analysed element are reported in table 1.

4. Results and discussion

In this section we discuss the results of an extensive experimental campaign intended to apply the methods described in section 2 to evaluate the density of nanostructured thin films once their thickness is known. This section is organized in three subsections corresponding to the main objectives of the experiments.

In section 4.1, we discuss the validation of the technique through a comparison with density measurements achieved using the well-established QCM technique in reliable conditions. The stability of the technique with respect to the electron accelerating voltage selection is discussed, and the effect of the atomic number difference between the sample coating and substrate is investigated. Moreover, results achieved using the substrate method and the coating method are compared to address the issue of method choice.

In section 4.2, we demonstrate the possibility of applying the technique for the evaluation of the density of nanostructured films with known thickness in a wide density and morphology range.

In 4.3, the spatial resolution of the technique is exploited to investigate the properties of non-homogeneous films at both macroscopic and microscopic scales.

4.1. Validation of the technique

The first experiment aimed at the validation of the technique was performed using compact Ag films (with expected density 10.49 g cm^{-3}) deposited through thermal evaporation on a Au-coated QCM resonator. In this way, EDS and QCM areal density measurements performed on the very same film were compared. The value achieved by the well-established QCM

technique, 0.31 mg cm^{-2}, was considered as a reference, since areal density values achieved by QCM are generally very reliable for films grown by thermal evaporation. Ag and Au were chosen as coating and substrate materials, respectively, due to the high atomic number difference between these elements. As discussed in section 2.2, this configuration is not optimal for EDS measurements and allows us to validate the technique and quantify the error in the worst-case scenario.

The coating method and the substrate method were employed to characterize the Ag film in a wide electron accelerating voltage range (8–29 kV). From the results reported in figure 2, it is evident that for both methods, an optimal range for the electron accelerating voltage exists in which the areal density values are less affected by voltage variations. For the coating method a voltage higher than 19 kV is required, while the substrate method provides stable results above 13 kV. The maximum acceptable voltage could not be determined, as it is higher than the maximum value achievable with our instruments. The results reported in figure 2 allow us to compare the accuracy of the two methods. In the optimum voltage range, the deviation from QCM-measured areal density is around 10% and 25%, respectively. Thus, in this case, the substrate method is more accurate than the coating method.

The method comparison was further investigated through experiments performed on amorphous-like tungsten films deposited by PLD on Si substrates [19]. In this case, the selection of silicon as substrate material is critical: since the Si Kα and the W Mα peaks strongly overlap, the accuracy of the substrate method could be non-optimal. In addition, the intensity of the Si Kα peak can be affected by fluorescence effects due to W Lα emission in the coating. Results are reported in figure 3. The deviation of measured values from reference density values evaluated by QCM is around 5% for the coating method, much lower than the difference observed for the substrate method, which in some cases is above 30%. Thus, only the coating method results for the W Lα peak will be discussed in section 4.2.

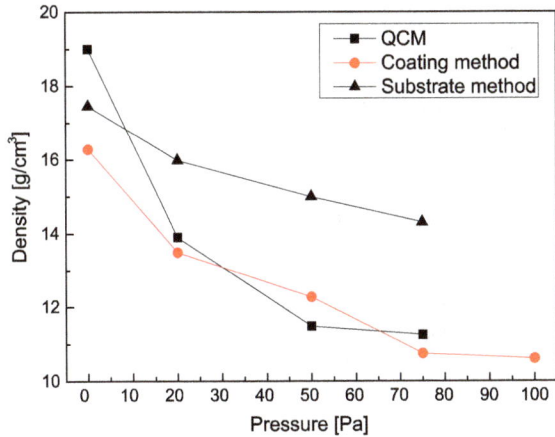

Figure 3. Density of thin tungsten films as a function of gas pressure in the deposition chamber.

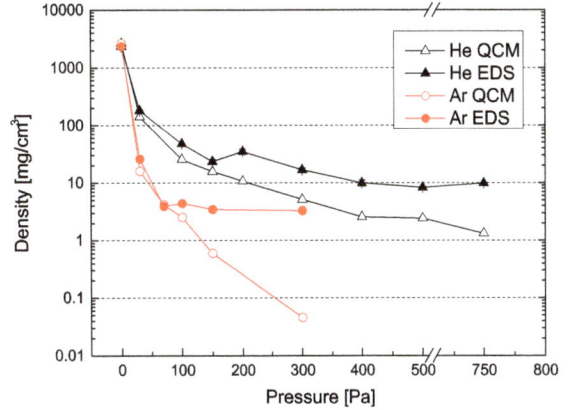

Figure 4. Density of carbon foams as a function of gas pressure in deposition chamber.

As discussed in section 2, the atomic number difference between the sample coating and substrate can produce a distortion of the PFXP affecting the measurement results. A proof-of-principle experiment was performed on commercially available Al foils with known density and thickness ($2.7 \, \mathrm{g \, cm^{-3}}$ and $750 \, \mathrm{nm} \pm 30\%$) in order to investigate the effect of the substrate atomic number on density values evaluated using the coating method. Two different experimental configurations were considered: measurements were performed on free-standing Al foils and on Al foils arranged on carbon substrates. In both cases, the foil density is systematically underestimated, but the average deviation from the nominal value is around 28% for the free-standing configuration and 20% in the presence of a carbon substrate. This difference can be interpreted, considering that the presence of a carbon substrate enhances the signal from the Al foil. Electrons emerging from the rear side of the Al foil are not necessarily lost by the system but can be scattered back to the foil by C atoms. Thus, the Al PFXP approximation is more reliable in the presence of the C substrate than in the free-standing configuration.

4.2. Density measurement of nanostructured films

In this subsection, we illustrate the application of the technique to demonstrate the possibility of evaluating the density of nanostructured films having known thickness. A variety of film morphologies are considered, resulting in films with density ranging from the solid density to a few $\mathrm{mg \, cm^{-3}}$. Results reported in this section refer to films grown by PLD. Density values achieved by QCM are considered as a reference, even though densities achieved in PLD facilities also can be affected by an error for compact films (around 5–10% for our experimental setting) due to the difficulty of placing the sensor in the very same position as the substrate.

The reliability of EDS for density measurement of nanostructured films in a wide density range was tested,

exploiting amorphous-like tungsten films on Si substrates [19].

In figure 3, density values evaluated using both the coating method and QCM are plotted as a function of the gas pressure in the deposition chamber during the film growth process. The results show a very good agreement between coating method and QCM, except for the sample deposited in vacuum. The reason for this behaviour could be the low thickness of W coatings deposited in vacuum (less than 100 nm), whose evaluation through SEM cross-section analysis might be affected by an error of around 10%.

A second experiment aimed at testing the application of the technique to nanostructured films was performed, exploiting carbon foam densities down to a few $\mathrm{mg \, cm^{-3}}$ [3]. In this case, the substrate method was selected for the reasons discussed in section 2.2. In figure 4, results for both the substrate method and QCM are shown as a function of the gas pressure in the deposition chamber. The agreement between the two methods is satisfactory only for density values above $30 \, \mathrm{mg \, cm^{-3}}$. For lower densities, values measured by QCM are unrealistically low, as QCM undergoes a sensitivity loss due to the very porous foam structure, which results in a decoupling between the film and the quartz crystal resonator [20]. On the contrary, the substrate method shows a more plausible density saturation for increasing gas pressure, which is typical of this kind of deposition process. Thus, EDS-based methods can be applied for films produced by PVD for densities down to a few $\mathrm{mg \, cm^{-3}}$, also in a density range in which QCM is not reliable.

The possibility of employing EDS for density evaluation in the case of multi-elemental coatings was tested, exploiting aluminium-doped zinc oxide nanostructured films [21]. In this case, the substrate method was chosen due to the unavailability of reference standards with the same composition as the films under analysis. In figure 5, results achieved using the substrate method and QCM are shown as a function of the target-to-substrate distance. In this case, a strong uncertainty affects density values achieved by QCM,

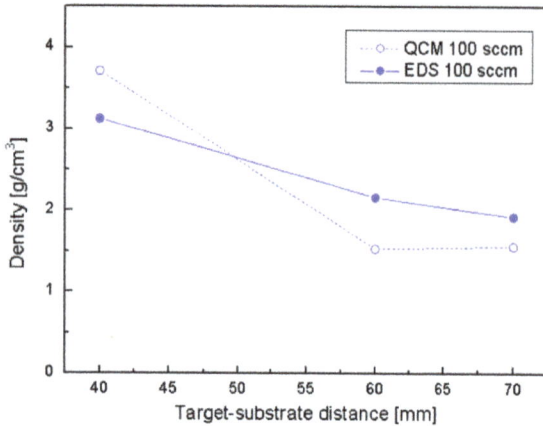

Figure 5. Density of Al-doped ZnO films as a function of the target-to-substrate distance.

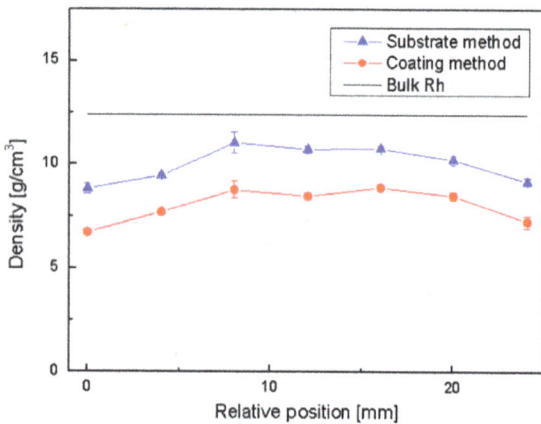

Figure 6. Density profile of a Rh nanocrystalline film.

because the deposition configurations adopted for film growth and for QCM measurements were not equivalent. However, density trends predicted by QCM are confirmed by the substrate method. Thus, as stated in [21], a decreasing trend in the film density with the target-to-substrate distance is observed.

4.3. Spatially resolved density measurements

One of the most interesting characteristics of EDS is its spatial resolution, which can be exploited for the characterization of non-uniform films at a macroscopic scale, i.e., for density profile evaluation, and at a microscopic scale, for the evaluation of the characteristic inhomogeneity length of a material.

The density profiles of Rh nanocrystalline coatings were measured along a cross-section exploiting both the coating method and the substrate method. Density was calculated from areal density, and thickness values were measured in the very same points. The results reported in figure 6 refer to a Rh film with a non-uniform thickness profile and thickness

ranging from about 70 to 135 nm. A non-uniform density profile is evident for both methods. Film density is approximately constant in the central region of the sample, while a 15% decrement is observed in the peripheral deposit region, where the film thickness is lower than 70 nm. As the film was deposited in vacuum [22], the coating density was expected to be very close to the bulk density value for Rh (12.41 g cm^{-3}). In the central region of the coated surface, film density measured by the substrate method is around 10.9 g cm^{-3}, while the coating method gives a density of about 8.7 g cm^{-3}. In both cases density is underestimated with respect to the expected value. Since the results achieved by the substrate method are closer to the expected film density, the substrate method can be considered more reliable than the coating method in this case.

The application of EDS to the analysis of mesoscale inhomogeneity was investigated, exploiting carbon films [3]. In general, EDS scans performed on a wide film area result in a relatively low areal density standard deviation, since measured areal density values are averaged on a large surface. As the sampled region is reduced, standard deviation increases if the film presents inhomogeneities with a length scale comparable with the diameter of the sampled region. Thus, the inhomogeneity length scale can be estimated as the sampled area for which areal density standard deviation suddenly starts increasing. This approach was developed to introduce a quantitative criterion to compare films with qualitatively similar mesoscale structures. To this aim, we analysed carbon films produced by PLD with different inhomogeneity length scales: a compact coating produced in vacuum and two foams produced using argon as buffer gas with pressures around 30 Pa and 300 Pa. These films have different mesoscale morphologies (see figure 7) and, as a consequence, different densities (2.5 g cm^{-3}, 26 mg cm^{-3}, and 4 mg cm^{-3}, respectively). Results are illustrated in figure 8. The sampled surface area ranges from 10 to 10^4 μm^2. The areal density standard deviation for the compact coating is stable even for high magnifications. For carbon foam layers produced in argon at 30 and 300 Pa, a sudden increase is observed as the sampled area decreases below 65 μm^2 and 100 μm^2, respectively. Thus, the inhomogeneity length scales can be estimated as 8 μm and 10 μm. These values confirm the morphological difference evident from SEM images and provide a quantitative criterion to compare the inhomogeneity length scale of films with similar morphology.

4.4. Discussion

The experimental results reported in this paper allow us to draw general conclusions regarding the application of EDS to the evaluation of nanostructured thin film density and provide useful indications regarding its practical applications for density measurements.

The first observation is related to the choice of the EDS-based method. Although the technique accuracy is strongly dependent on the specific experimental configuration and, in particular, on the substrate–coating combination, in general the substrate method can be considered more reliable than the

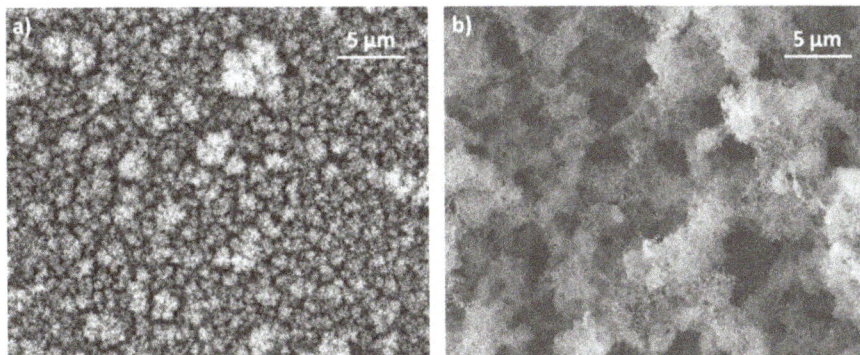

Figure 7. SEM images of carbon films deposited in Ar at 30 Pa (a) and 300 Pa (b).

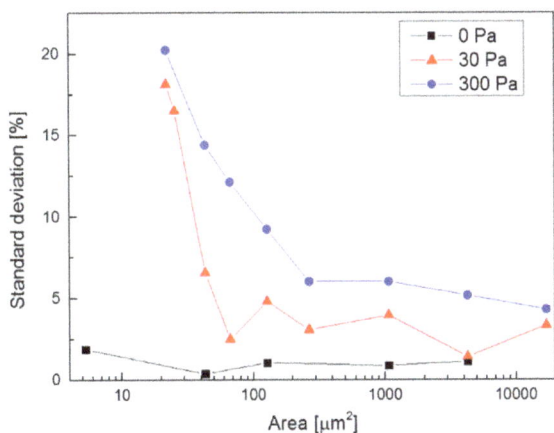

Figure 8. Standard deviation of areal density measurements for carbon films deposited in vacuum and in argon with 30 Pa and 300 Pa.

coating method. For the substrate method, the error with respect to values measured by QCM is around 10–15%, but it can reach values up to 30% if the substrate–coating combination is particularly unfavourable, for example, because of peak overlapping. Nevertheless, in a few cases (i.e., W films on Si substrates) the coating method allows us to achieve very reliable measurements with an extremely low deviation from nominal density values.

The strong dependence of the technique accuracy on the substrate–coating combination has been considered in subsection 4.1. Apparently, this factor is the main error source in density measurements performed by EDS. In principle, the precision could be enhanced by developing an appropriate model for PFXP in multilayer samples, taking into account also fluorescence and secondary emission effects. However, the accuracy of both the coating method and the substrate method can be enhanced by adopting suitable conditions, namely selecting substrates with an atomic number similar to the coating atomic number. Moreover, the effect of the characteristic properties of the model used to evaluate the PFXP—in this case the MSG model—on the accuracy of the two methods should be considered.

Also, the electron accelerating voltage selection plays an important role in areal density evaluation. This issue has been extensively discussed in section 2. Moreover, the stability of the technique with respect to accelerating voltage variations has been studied from 5–30 keV for Ag films deposited on Au substrates. An empirical method to verify that the selected voltage is included in the stability voltage range consists in checking the measure and repeating it with slightly higher and lower acceleration voltages for a test film. If the measured density does not change, the accelerating voltage falls in the optimum range.

5. Conclusions

In conclusion, we discussed the quantitative development and test of two methods based on EDS and cross-sectional SEM images for thin film density measurement. We demonstrated the applicability of these methods to a number of different experimental conditions: thin films with various compositions, different coating–substrate combinations, various mesoscale morphologies, and with densities in an extremely wide range (few $mg\,cm^{-3}$–$20\,g\,cm^{-3}$). Moreover, a novel application of EDS to the analysis of coating inhomogeneity at the macroscopic and microscopic scales has been shown.

Although the results can be affected by an error up to 30% in a few unfavourable configurations, in general this technique guarantees a reliable, fast, simple, and cheap measurement process to evaluate the density of nanostructured thin films in a wide range of morphologies and compositions, exploiting the common integrated EDS–SEM equipment present in most material science laboratories.

Acknowledgments

We acknowledge A Uccello, P Gondoni, and A Pezzoli for technical support in the production of nanostructured films by PLD and A Maffini and P Mazzolini for fruitful discussions.

References

[1] Willingale L *et al* 2006 *Phys. Rev. Lett.* **96** 245002

[2] Passoni M, Zani A, Sgattoni A, Dellasega D, Macchi A, Prencipe I, Floquet V, Martin P, Liseykina T V and Ceccotti T 2014 *Plasma Phys. Control. Fusion* **56** 045001

[3] Zani A, Dellasega D, Russo V and Passoni M 2013 *Carbon* **56** 358

[4] Sweeney W E, Seebold R E and Birks L S 1960 *J. Appl. Phys.* **31** 1061

[5] Shumacher B W and Mitra S S 1962 *Electronics Reliability and Microminiaturization* **1** 321

[6] Bentzon M D, Nielsen P S and Eskildsen S S 1993 *Diam. Relat. Mater.* **2** 893

[7] Cockett G H and Davis C D 1963 *Brit. J. Appl. Phys.* **14** 813

[8] Castaing R 1951 *PhD thesis* University of Paris

[9] Bishop H E and Poole D M 1973 *J. Phys. D: Appl. Phys.* **6** 1142

[10] Lavrent'ev Yu G, Korolyuk V N and Usova L V 2004 *J. Anal. Chem.* **59** 600

[11] Pouchou J L and Pichoir F 1984 *Rech. Aerospatiale* **3** 350

[12] Pouchou J L and Pichoir F 1986 *J. Microsc. Spectrosc. Electron* **11** 229

[13] Packwood R H and Brown J D 1981 *X-ray Spectrom.* **10** 138

[14] Stenberg G and Boman M 1996 *Diam. Relat. Mater.* **5** 1444

[15] Rehbach W and Karduck P 1988 *Microbeam Analysis* (San Francisco, CA: San Francisco Press) 285 p

[16] Pouchou J L 1993 *Anal. Chim. Acta* **283** 81

[17] Karduck P and Rehbach W 1988 *Microbeam Analysis* (San Francisco, CA: San Francisco Press) 277 p

[18] Dellasega D, Facibeni A, di Fonzo F, Russo V, Conti C, Ducati C, Casari C S, Li Bassi A and Bottani C E 2009 *Appl. Surf. Sci.* **255** 5248

[19] Dellasega D, Merlo G, Conti C, Bottani C E and Passoni M 2012 *J. Appl. Phys.* **112** 084328

[20] Kanazawa K K and Gordon J G 1985 *Anal. Chem.* **57** 1170

[21] Gondoni P, Russo V, Bottani C E, Li Bassi A and Casari C S 2013 *J. Vis. Exp.* **72** 50297

[22] Uccello A, Dellasega D, Perissinotto S, Lecis N and Passoni M 2013 *J. Nucl. Mater.* **432** 261

9

Recent progress in advanced optical materials based on gadolinium aluminate garnet (Gd$_3$Al$_5$O$_{12}$)

Ji-Guang Li and Yoshio Sakka

Advanced Materials Processing Unit, National Institute for Materials Science, 1-2-1 Sengen, Tsukuba, Ibaraki 305-0047, Japan

E-mail: li.jiguang@nims.go.jp

Abstract

This review article summarizes the recent achievements in stabilization of the metastable lattice of gadolinium aluminate garnet (Gd$_3$Al$_5$O$_{12}$, GAG) and the related developments of advanced optical materials, including down-conversion phosphors, up-conversion phosphors, transparent ceramics, and single crystals. Whenever possible, the materials are compared with their better known YAG and LuAG counterparts to demonstrate the merits of the GAG host. It is shown that novel emission features and significantly improved luminescence can be attained for a number of phosphor systems with the more covalent GAG lattice and the efficient energy transfer from Gd^{3+} to the activator. Ce^{3+} doped GAG-based single crystals and transparent ceramics are also shown to simultaneously possess the advantages of high theoretical density, fast scintillation decay, and high light yields, and hold great potential as scintillators for a wide range of applications. The unresolved issues are also pointed out.

Keywords: gadolinium aluminate garnet, lattice stabilization, down-/up-conversion phosphors, single crystal, transparent ceramic, scintillator, energy transfer

1. Introduction

Rare-earth aluminate garnets, having a general formula of Ln$_3$Al$_5$O$_{12}$ (LnAG, Ln: lanthanide and Y), are an important family of multi-functional ceramic materials. The compounds crystallize in a bcc structure (space group: Ia$\bar{3}$d) with 160 (80) atoms in the cubic (primitive) cell, where the Ln occupies the 24c sites (D$_2$ point symmetry, CN = 8; CN: coordination number) and the oxygen atoms take the 96h sites. The Al atoms have two positions to reside on: the 16a sites with an octahedral point symmetry (C$_{3i}$, 40%; CN = 6) and the 24d sites with a tetragonal point symmetry (S$_4$, 60%; CN = 4) [1].

The garnet structure can be viewed as a framework built up via corner sharing of the Al–O polyhedra, with the Ln residing in dodecahedral interstices [1]. A schematic diagram of the crystal structure is shown in figure 1.

YAG might be the best-known garnet compound owing to its excellent chemical stability, high creep resistance, optical isotropy, and particularly the ability to accept substantial trivalent Ln^{3+} for diverse optical functionalities. In the bulk form, the YAG:Nd single crystal is one of the most widely used solid laser materials since its discovery in the 1960s [2]. Transparent YAG:Nd ceramics that are equal to or superior to single crystals in optical transmittance and laser performance have also been successfully developed since the 1990s via advanced powder processing and sintering technologies [3, 4]. YAG:Ho^{3+} and YAG:Er^{3+} are important infrared (IR) laser materials for medical surgery, since their fluorescence lines ($\sim$2 μm for Ho^{3+} and $\sim$3 μm for Er^{3+})

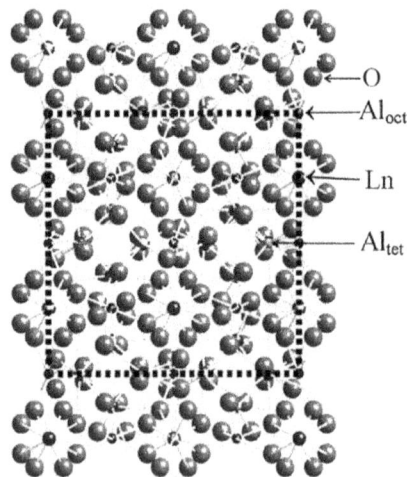

Figure 1. A schematic illustration of the crystal structure of LnAG, where Al_{oct} and Al_{tet} represent the Al atoms taking octahedral and tetrahedral lattice sites, respectively. Adapted with permission from [1], copyright 1999 by the American Physical Society.

match well with the water absorptions of the human body [5]. Transparent YAG:Ce^{3+} ceramic is nowadays being developed as an encapsulant for white-light emitting diodes (LEDs) to replace the widely used but readily degradable resin sealant [6]. Since the parity-law allowed $4f^05d^1 \rightarrow 4f^15d^0$ transition of Ce^{3+} has a very short fluorescence lifetime of $\sim$10–100 ns and the intrinsic quenching temperature of Ce^{3+} in YAG is very high ($>$700 K) [7], YAG:Ce^{3+} transparent ceramic has been considered as a scintillator [8], but does not seem to have a satisfactory stopping power for the incident radiations (x-, α- or γ-rays) owing to the relatively low theoretical density of YAG ($\sim$4.55 g cm^{-3}) and the small atomic weight of Y ($\sim$89). In the powder form, rare-earth (Eu^{3+}, Tb^{3+}, Ce^{3+} etc) activated YAG is being widely studied and used as phosphors for fluorescent lamps, field emission displays (FEDs), and white LEDs.

The occurrence and thermal stability of compounds in the Ln$_2$O$_3$–Al$_2$O$_3$ binary system heavily depend on the ionic radius of Ln^{3+}, conforming to lanthanide contraction. Earlier studies by Mizuno et al [9–13] on the phase diagram of Ln$_2$O$_3$–Al$_2$O$_3$ found the two intermediate compounds of Ln$_2$O$_3$·11Al$_2$O$_3$ (β-Al$_2$O$_3$ type) and LnAlO$_3$ (commonly referred to as aluminate perovskite or LnAP, an orthorhombic modification of ideal perovskite) for the large ions of La^{3+}–Nd^{3+}, the two compounds of LnAP and monoclinic Ln$_4$Al$_2$O$_9$ (commonly called LnAM) for the intermediately sized Sm^{3+}–Gd^{3+}, and the three compounds of LnAM, LnAP, and LnAG for the small Ln^{3+} ions of Tb^{3+}–Er^{3+}. Though GdAG (further abridged as gadolinium aluminate garnet (GAG) hereafter) was not identified in the work of Mizuno et al, it was successfully synthesized via flux growth by Van Uitert et al [14] and Manabe and Egashira [15] for potential optical applications. Later on, Shishido et al [16] found via annealing Gd$_2$O$_3$·5/3Al$_2$O$_3$ amorphous glass that GAG is metastable and would completely decompose to α-Al$_2$O$_3$ and GdAP

(Gd$_3$Al$_5$O$_{12}$ $\rightarrow$ Al$_2$O$_3$ + 3GdAlO$_3$) upon prolonged annealing at 1500 °C. A recent work by Li et al [17] further showed that the stoichiometric GAG synthesized via low temperature combustion starts to decompose at $\sim$1300 °C. All these studies point to the fact that thermodynamically stable LnAG only exists for the Ln^{3+} smaller than Gd^{3+} and Gd^{3+} is the boundary for a LnAG to be formed. This is understandable from the crystal structure shown in figure 1. That is, the dodecahedral interstice has a certain geometric shape and dimension, and thus a size limit exists for Ln^{3+} to enter the space without disintegrating the Al–O framework. Mainly due to its structural metastability, GAG has been much less explored than YAG for its properties and applications, though its specific heat and thermal expansion coefficient were experimentally determined by Chaudhury et al [18]. Compared with YAG, however, GAG may hold a number of merits for optical applications: (1) the intrinsic $^8S_{7/2} \rightarrow {}^6I_J$ transition of Gd^{3+} (usually centered at $\sim$275 nm) can be utilized as a new excitation source for some types of rare-earth activators, and enhanced luminescence may also be attained via an efficient energy transfer from Gd^{3+} to the activator [19–22], (2) the GAG lattice is more covalent than YAG due to the lower electronegativity of Gd^{3+} ($\chi = 1.20$) than Y^{3+} ($\chi = 1.22$), which may produce new emission features and result in improved emission intensity, and (3) GAG has a significantly higher theoretical density (5.97 g cm^{-3}) than YAG (4.55 g cm^{-3}) and the atomic weight of Gd (157, close to the 175 of Lu) is much higher than Y, and thus GAG is more desirable for scintillation applications. Similar to the growth of single crystals and sintering of transparent ceramics, a reasonably high processing temperature is usually needed to produce high quality phosphors through crystal perfection. In this context, lattice stabilization becomes a prerequisite for any practical application of GAG in advanced optical materials. This review article summarizes the recent achievements in GAG, including lattice stabilization via doping and its application in down-/up-conversion (UC) phosphors and transparent ceramic/single crystal scintillators.

2. Lattice stabilization of GAG by modifying the Gd/Al sites

There are two primary ways to stabilize the garnet lattice of GAG, as can be perceived from the crystal structure shown in figure 1, with the first one partially replacing the Al sites with suitably larger trivalent ions to enlarge the dodecahedral interstices via forming Gd$_3$(Al$_{1-x}$M$_x$)$_5$O$_{12}$ solid solution and the second one being partially replacing Gd^{3+} with a smaller Ln^{3+} to form (Gd$_{1-x}$Ln$_x$)$_3$Al$_5$O$_{12}$. Ga^{3+} is the main choice in the former case, and Gd$_3$Ga$_5$O$_{12}$ (GGG), known as a thermodynamically stable garnet host for phosphors and solid lasers [2], is an extreme example. The effectiveness of Ga^{3+} doping was experimentally demonstrated by Chiang et al [23], who found that phase-pure garnet can be crystallized from chemically precipitated precursors at $\sim$1400 °C in the presence of 10 at% of Ga^{3+} and the crystallization temperature decreases to 1300 °C with 20 at% of Ga^{3+} addition. Without

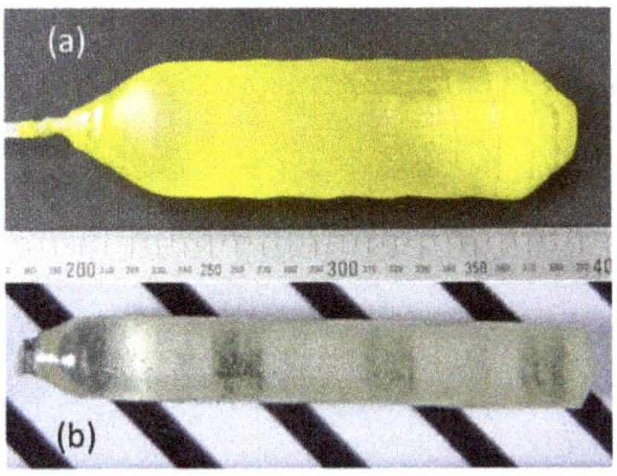

Figure 2. Appearance of the $Gd_3(Al_2Ga_3)O_{12}$ single crystals doped with 1 at% of Ce^{3+} (a) and 1 at% of Pr^{3+} (b). Part (a) reproduced with permission from [24] and part (b) reproduced with permission from [25], copyright 2012 by Elsevier.

Ga^{3+} doping, only a phase mixture of LnAP, LnAG and amorphous alumina was formed. By applying the same stabilization strategy, Kamada *et al* were able to grow two-inch-diameter $Gd_3(Al_2Ga_3)O_{12}$:Ce^{3+} single crystals by the Czochralski (Cz) method using [100] oriented seeds [24] and $Gd_3(Ga,Al)_5O_{12}$:Pr^{3+} single crystals by a micro-pulling down (μ-PD) technique [25] (figure 2). Though Ga^{3+} was thought to exclusively replace Al^{3+} in these studies, atomistic modeling using the static lattice computational approach and pairwise (Buckingham) interatomic potentials by Maglia *et al* [26] revealed that Ga^{3+}, though it prefers to take the octahedral Al^{3+} site, can also be inserted into the dodecahedral position of Gd^{3+} with the generation of anti-site defects owing to its relatively large ionic radius. In addition, suppressing activator oxidation (such as Pr^{3+}, Ce^{3+}, and Tb^{3+}) and Ga^{3+} reduction should be made at the same time to avoid lattice defects and deterioration of optical performance.

Lu^{3+} (0.0977 nm for CN = 8) is the tiniest Ln^{3+} ion and would thus be the most effective to stabilize GAG via replacing the Gd^{3+} site to form a $(Gd_{1-x}Lu_x)$AG solid

solution. With coprecipitated carbonate precursors, Li *et al* [17] thoroughly studied the effects of Lu content on phase evolution and also properties of the resultant $(Gd_{1-x}Lu_x)$AG ($x = 0$–0.5). It was shown that the garnet phase generally crystallizes via LnAM and LnAP intermediates, as is commonly observed for YAG, but the crystallization temperature substantially decreases towards a higher Lu content. With $x = 0.3$–0.5, phase-pure garnet can even be crystallized at a temperature as low as 1000 °C (figure 3(a)), revealing the significant effectiveness of Lu^{3+} doping. Again, only a phase mixture of LnAG, LnAP and amorphous Al_2O_3 was produced in the absence of Lu^{3+} (figure 3(b)).

A simultaneous advantage of Lu doping is that it improves the already high theoretical density of GAG (5.97 g cm^{-3}). The $(Gd_{0.5}Lu_{0.5})$AG solid solution, for example, reaches the high value of ~6.44 g cm^{-3} (figure 4(a)), being close to that of the heavy LuAG (6.73 g cm^{-3}). Since Gd is commercially much cheaper than Lu, the $(Gd_{1-x}Lu_x)$AG solid solutions may replace LuAG to be used as cost effective and high density scintillation materials. Assayed from UV/vis absorption, the $(Gd_{1-x}Lu_x)$AG solid solutions were found to have increasing optical bandgaps of ~5.87, 5.97, 6.07, 6.17, 6.27, and 6.37 eV with increasing x from 0 to 0.5 (0.1 interval, figure 4(b)) [17], and the bandgap of $(Gd_{0.5}Lu_{0.5})$AG has been close to that (~6.40 eV) of a YAG single crystal [27]. The results may also imply that the luminescence property of a (Gd, Lu)AG based phosphor can be finely tuned by varying the Lu content and the onset of optical transmittance of a transparent (Gd, Lu)AG bulk (single crystal or transparent ceramic) would shift towards a shorter wavelength with increasing Lu incorporation.

Tb^{3+} is the largest single Ln^{3+} for a stable LnAG to be formed, and thus its size can be taken as a reference for a lattice stabilization study of GAG. The minimum amount of Lu^{3+} (~17 at%) calculated from the ionic size of Tb^{3+} (0.1040 nm for CN = 8), however, is significantly larger than the ~10 at% found in practice (figure 3(b)). This indicates that stable garnet solid solutions exist if the average ionic size of

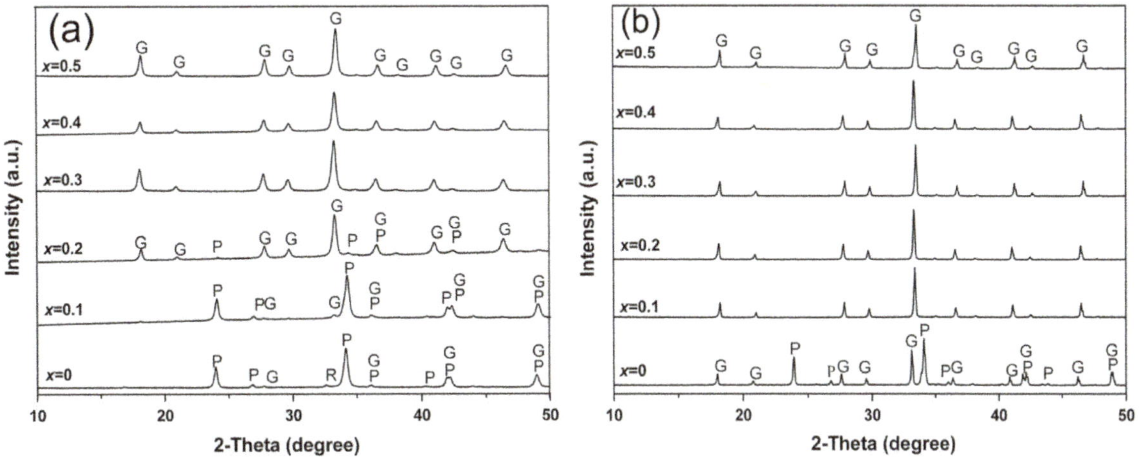

Figure 3. The effects of Lu content (x value) on phase evolution of $(Gd_{1-x}Lu_x)$AG solid solution. Parts (a) and (b) are for calcination temperatures of 1000 and 1500 °C, respectively. Reproduced with permission from [17], copyright 2012 by the American Ceramic Society.

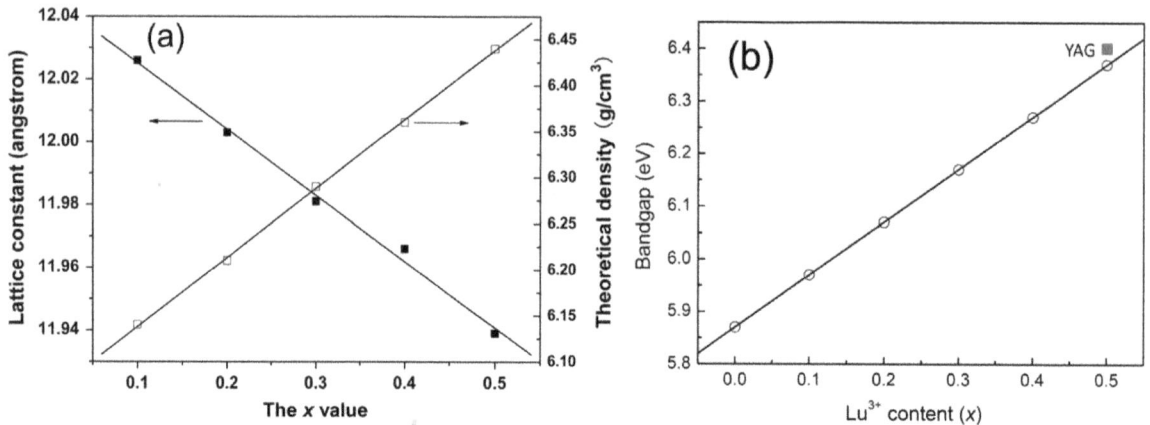

Figure 4. Lattice constant and theoretical density (a) and bandgap (b) of the $(Gd_{1-x}Lu_x)AG$ solid solution, as a function of the Lu content. Part (a) reproduced with permission from [17], copyright 2012 by the American Ceramic Society.

$(Ln_1,Ln_2)^{3+}$ pair lies in between those of Gd^{3+} (0.1053 nm for CN = 8) and Tb^{3+}, in agreement with the fact that TbAG [28–30] and even $(Gd_{0.9}Lu_{0.1})AG$ [31, 32] can be further doped with larger Eu^{3+} (0.1066 nm, CN = 8) and Ce^{3+} (0.1143 nm, CN = 8) for luminescence. Taking the average ionic size of $(Gd_{0.9}Lu_{0.1})^{3+}$ (~0.1045 nm) as a standard, Li *et al* [33] analyzed the minimum amounts of various small Ln^{3+} that are needed for GAG stabilization, and the x value was predicted to be ~0.5 for Tb^{3+}, 0.3 for Dy^{3+} (0.1027 nm), 0.22 for Y^{3+} (0.1019 nm), 0.2 for Ho^{3+} (0.1015 nm), 0.15 for Er^{3+} (0.1004 nm), 0.13 for Tm^{3+} (0.0994 nm), and 0.11 for Yb^{3+} (0.0985 nm). Practical powder synthesis indeed shows that $(Gd_{1-x}Ln_x)AG$ garnet can be obtained in high phase purity with incorporation of the calculated amount of dopant in each case (figure 5). The results may thus lay a base for flexible materials design by properly combining different types of stabilizers to achieve diverse optical functionalities. The characteristic emission of Ln^{3+} in $(Gd_{0.5}Ln_{0.5})AG$ was observed by the authors for Tb (green), Dy (similarly strong blue and yellow), Ho (green), and Tm (blue), despite the high Ln^{3+} concentration (figure 6).

3. Down-conversion (DC) phosphors based on GAG

DC phosphors are generally referred to those that absorb high-energy photons and re-emit them at longer wavelengths in the visible range. There are many host lattices for DC luminescence, and the most extensively employed ones may include single-/multi-cation oxides, oxysulfides, phosphates, vanadates, borates, and molybdates/tungstates. Different hosts are used in practice to meet different application needs. Though almost all of the optically active Ln^{3+} can be used as the activator for DC luminescence, the four ions of Ce^{3+}, Eu^{3+}, Tb^{3+}, and Dy^{3+} are the most efficient since their energy gaps between the lowest emission state and ground state are sufficiently wide to avoid significant non-radiative cross relaxations. The emission behaviors of these activators are generally governed by their site symmetry in the host lattice,

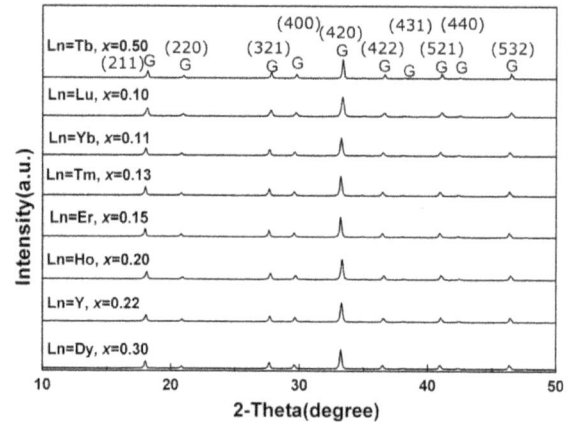

Figure 5. Powder XRD patterns of the phase pure $(Gd_{1-x}Ln_x)AG$ garnets obtained by doping GAG with the calculated amount (x value) of different Ln^{3+}. Reproduced with permission from [33], copyright 2013 by Trans Tech Publications.

lattice covalency, lattice defects/distortion, and particle size/shape (surface effects).

3.1. Eu^{3+} doping for red luminescence

Eu^{3+} is well known for its orange red/red emission arising from the $^5D_0 \rightarrow {}^7F_{1,2}$ electronic transitions. The Eu^{3+} ions doped in LnAG are assumed to replace the Ln^{3+} sites and thus inherit a D_2 point symmetry, which is only slightly perturbed from the highly symmetric D_{2h} point group [34]. For this, the emission of YAG:Eu and LuAG:Eu is dominated by the parity-law allowed $^5D_0 \rightarrow {}^7F_1$ magnetic dipole transition at ~590 nm rather than the forced $^5D_0 \rightarrow {}^7F_2$ electric dipole transition at ~610 nm as observed from the well-known Y_2O_3:Eu red phosphor. A $[(Gd_{1-x}Lu_x)_{1-y}Eu_y]AG$ solid solution has recently been developed as efficient red phosphor with Lu^{3+} as the lattice stabilizer, and the effects of various factors on optical properties were thoroughly investigated [31]. Taking $[(Gd_{0.7}Lu_{0.3})_{1-y}Eu_y]AG$ for example, the

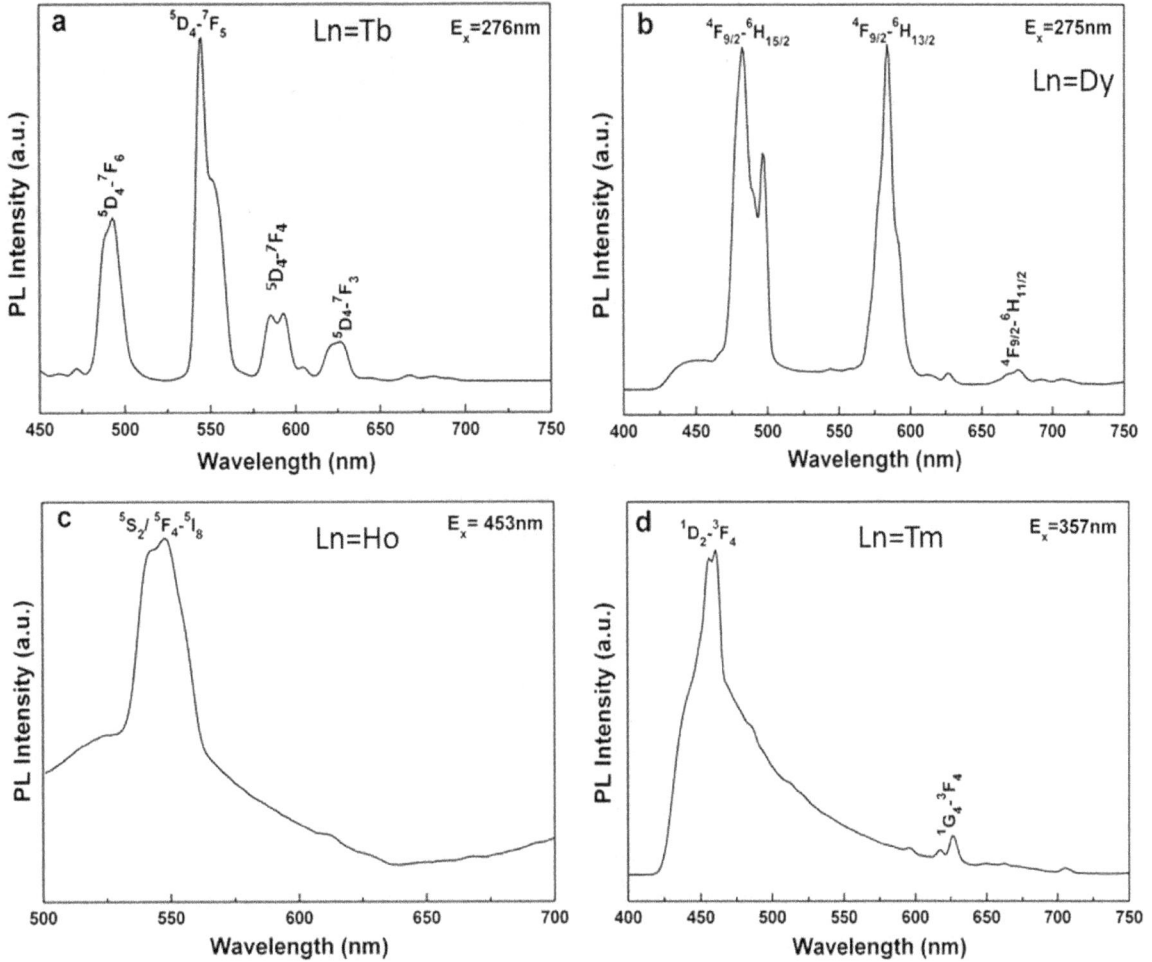

Figure 6. Emission spectra for the $(Gd_{0.5}Ln_{0.5})AG$ solid solution, with the excitation wavelength (E_x) and the origin of luminescence indicated. Reproduced with permission from [33], copyright 2013 by Trans Tech Publications.

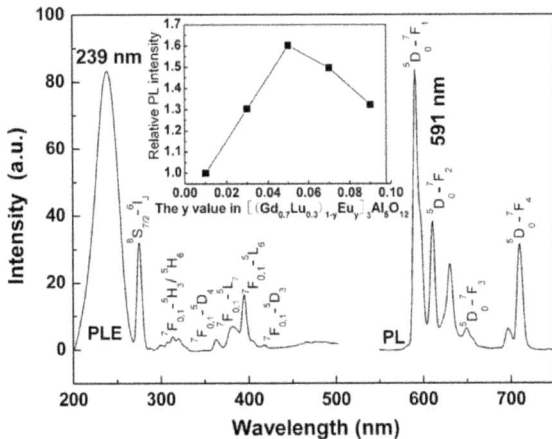

Figure 7. Excitation and emission behaviors of the $[(Gd_{0.7}Lu_{0.3})_{1-y}Eu_y]AG$ red phosphors. Reproduced with permission from [31], copyright 2012 by the National Institute for Materials Science.

material was shown to be efficiently excitable with the charge transfer band (CTB) at ~239 nm to produce a sharp orange–red emission at 591 nm (figure 7), with CIE chromaticity coordinates of (0.62, 0.38) and a full width at half maximum of only ~6 nm for the emission peak. The optimal Eu^{3+} content was experimentally determined to be ~5 at% ($y = 0.05$), and the quenching mainly resulted from exchange interactions, possibly via phonon assisted three Eu^{3+} ion nonresonant interactions. Greatly improved emission intensity and quantum yield, shortened fluorescence lifetime, and increased asymmetry factor of luminescence (the I_{591}/I_{610} intensity ratio) were observed along with increasing synthesis temperature up to 1500 °C [31], primarily owing to lattice perfection, defect elimination, and particle growth. The $[(Gd_{0.7}Lu_{0.3})_{0.95}Eu_{0.05}]AG$ phosphor synthesized at 1500 °C has internal/external quantum efficiencies (%) of 83.2/56.1, an asymmetry factor of ~2.85, and a fluorescence lifetime of ~4.1 ms for the 591 nm emission [31]. The lifetime is close to that (4.66 ms) reported for YAG:Eu [35] but is significantly longer than those (generally 0.5–2.5 ms) of the well-known red phosphors of Y_2O_3:Eu [36, 37], $(Gd_{1-x}Ln_x)_2O_3$:Eu (Ln = Y, Lu) [22, 38], and La_2O_2S:Eu [39], since the Eu^{3+} activator takes the highly symmetric D_2 lattice site in garnet. Increasing Lu incorporation up to $x = 0.5$ would lower excitation/emission and also blue-shift the CTB due to gradually decreased covalency of the host lattice ($\chi = 1.27$ for Lu^{3+}), for which a minimized Lu content was recommended as long as the garnet lattice can be effectively stabilized [31]. Similar phenomena were observed in the development of $(Gd_{1-x}Ln_x)_2O_3$:Eu red phosphors (Ln = Y, Lu) [38]. Compared with YAG:Eu, the GAG-based phosphor (figure 7) has an additional excitation band arising from the $^8S_{7/2} \rightarrow {}^6I_J$ Gd^{3+} transition at ~275 nm (significantly stronger than the strongest $^7F_{0,1} \rightarrow {}^5L_6$ intra-$4f^6$ transition of Eu^{3+} at ~395 nm), suggesting substantial energy migration from Gd^{3+} to Eu^{3+}. The advantage of GAG over YAG as a host lattice was demonstrated in another study by the authors [40]. For example, the internal quantum yield (~76%) of $[(Gd_{0.9}Lu_{0.1})_{0.95}Eu_{0.05}]AG$ is appreciably higher than that (~71%) of $(Y_{0.95}Eu_{0.05})AG$ at the same temperature of phosphor synthesis. This is primarily owing to the higher lattice covalency of the former, which allows improved excitation absorption and higher probability of electronic transitions. Though $[(Gd_{1-x}Lu_x)_{0.95}Eu_{0.05}]AG$ has had a sufficiently high theoretical density and emission intensity, its fluorescence lifetime is too long for scintillation. Shortening the lifetime to below ~1.0 ms via codoping (such as Pr^{3+}) seems necessary for it to compete with the commercialized $(Y,Gd)_2O_3$:Eu scintillator [41].

The carbonate coprecipitation technique, with ammonium bicarbonate as the precipitant, has been able to produce low-aggregation garnet powders, but the primary particles are not separable from each other and are not in a spherical shape (figure 8(a)). Current advances in high-resolution displays not only need finer phosphor particles to improve the resolution by decreasing pixel size but also prefer a spherical particle shape to build a uniform luminescence screen and to improve the brightness of luminescence by minimizing the light scattering on particle surfaces. Urea-based homogeneous precipitation (UBHP) frequently finds success in synthesizing well-dispersed spherical particles of uniform size (monospheres) for single-/multi-cation oxides of the lanthanides [19, 42–48], but failed for YAG in most of the previous studies [49, 50] owing to substantially different solution chemistries of the constituent Y^{3+} and Al^{3+} ions. YAG:Ce phosphor microspheres are thus alternatively made via crystallizing the glassy microbeads quenched from melt droplets produced with laser heating [51]. Xu et al [52] identified that, with nitrate as the rare-earth source in UBHP, the aluminum source plays an essential role in the formation of precursor microspheres for YAG:Nd. They found that ammonium aluminum sulfate (alum) is indispensable and the optimal alum/$Al(NO_3)_3$ molar ratio is 1/1. Mechanistic study further revealed that microspheres of the Al component are formed at the early stage of precipitation, followed by Y^{3+} precipitation as basic carbonate. Annealing the sulfate-containing precursor at 1100 °C produced YAG:Nd microspheres that can be

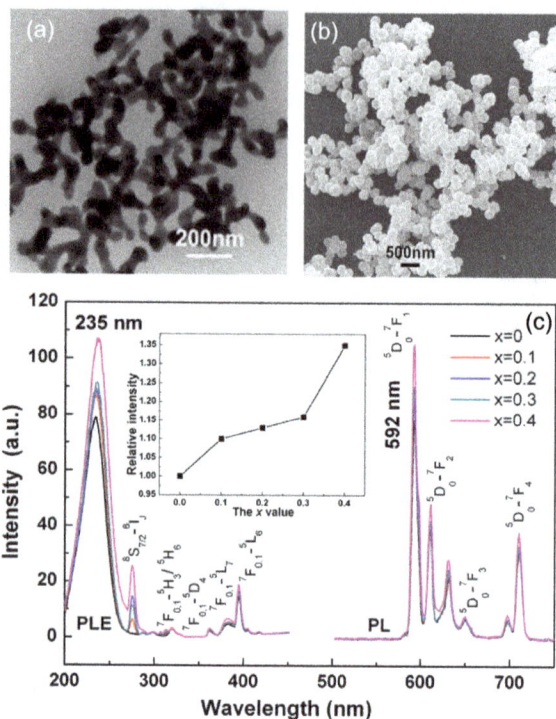

Figure 8. Typical TEM (a) and SEM (b) morphologies of the $[(Lu_{1-x}Gd_x)_{0.95}Eu_{0.05}]AG$ red phosphor particles obtained via carbonate coprecipitation and urea-based homogeneous precipitation, respectively. Part (c) shows that the emission intensity of the phosphor spheres improves with increasing Gd content. Parts (a) and (b) reproduced from [31], copyright 2012 by the National Institute for Materials Science.

densified to a translucent state via vacuum sintering of the dry-compacted green body at 1650 °C for 3.5 h. Such a strategy proved similarly successful for $[(Lu_{1-x}Gd_x)_{0.95}Eu_{0.05}]$ AG red phosphor microspheres ($x \leqslant 0.4$, figure 8(b)) [53], though Gd^{3+} and Lu^{3+} are different from Y^{3+} in solution chemistry owing to lanthanide contraction. Again, the best results were obtained with alum/$Al(NO_3)_3 = 1:1$ molar ratio. When $Al(NO_3)_3$ is the sole Al source, only a gelatinous precursor that would aggregate into a glasslike hard mass upon drying was produced, implying that the sulfate anions from alum have significantly modified the solution chemistry of cations, particularly that of the significantly smaller Al^{3+}, and have taken part in precipitation. When the alum/$Al(NO_3)_3$ ratio is over 1, the primary spheres tend to glue together to form porous precipitates as observed for YAG [50], suggesting that superfluous SO_4^{2-} may serve as a flocculant. The diameter of $[(Lu_{1-x}Gd_x)_{0.95}Eu_{0.05}]AG$ microspheres can be finely tuned from ~500 to 150 nm by increasing the urea/(Al+Ln) molar ratio from 20 to 100, showing the flexibility of the UBHP technique. A photoluminescence study [53] found successively stronger $^5D_0 \rightarrow {}^7F_1$ emission (591 nm) with increasing x (the Gd content), owing to increased lattice covalency (figure 8(c)), and gradually weaker emission at a decreasing particle size owing to surface effects [47, 54].

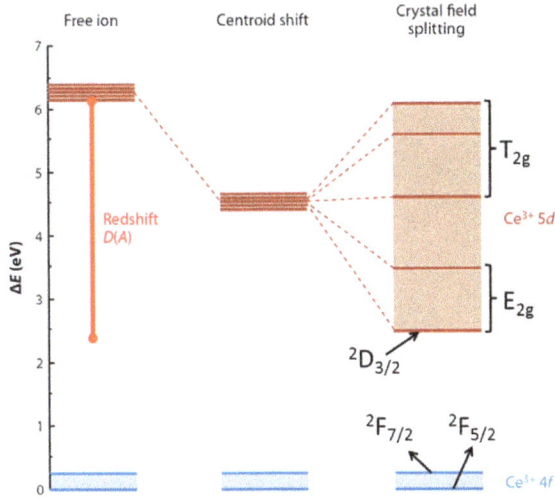

Figure 9. A schematic energy diagram showing the effects of host lattice A on centroid shift and crystal field splitting of the Ce^{3+} 5d energy level. Such effects shorten the energy difference between the 5d excited state and 4f ground state, known as red-shift D(A). Reproduced with permission from [55], copyright 2013 by Annual Reviews Inc.

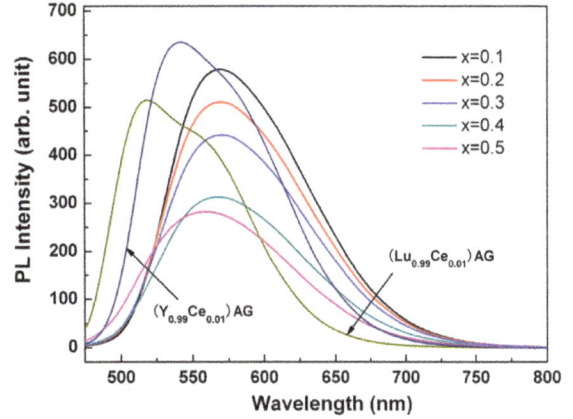

Figure 10. Emission spectra for the $[(Gd_{1-x}Lu_x)_{0.99}Ce_{0.01}]AG$, $(Lu_{0.99}Ce_{0.01})AG$, and $(Y_{0.99}Ce_{0.01})AG$ yellow phosphors. Reproduced with permission from [32], copyright 2013 by the National Institute for Materials Science.

3.2. Ce^{3+} doping for yellow luminescence

The yellow emission of Ce^{3+} arises from the $4f^0 5d^1 \rightarrow 4f^1 5d^0$ ($^2F_{5/2}$ and $^2F_{7/2}$ ground states) inter-configurational electronic transition. As the exposed 5d electron readily interacts with the surrounding anion ligands, Ce^{3+} emission is strongly influenced by centroid shifting and crystal field splitting of the 5d energy level (figure 9) [55]. YAG:Ce^{3+} has been the most prominent and widely used yellow phosphor in LED lighting, since it can be efficiently excited with (Ga, In)N blue LED chips (~450 nm) and exhibits a high quantum yield of ~90% for its ~540 nm emission [7]. One shortcoming is that, for warm-white lighting, YAG:Ce^{3+} has low color rendering and high correlated color temperature due to its lack of a sufficient red portion in the emission spectrum. To overcome this, partially replacing the Y^{3+} sites with less electronegative La^{3+} or Gd^{3+} [6, 56] and more recently doping YAG with Si_3N_4 to form the oxynitride solid solution of $Y_{2.925}Ce_{0.075}Al_{5-x}Si_xO_{12-x}N_x$ (x < 0.4 for phase pure garnet) [57] were shown to be able to push down the lowest crystal-splitting component ($^2D_{3/2}$) of the $5d^1$ excited level to yield red-shifted emissions. Red-shifting can also be achieved by partially substituting Al^{3+} with Mg^{2+}–Si^{4+} pairs on the octahedral and tetrahedral sites, respectively, to enhance lattice covalency [58, 59]. Alternatively, red-shifted Ce^{3+} emission can be directly attained with a more covalent host lattice, such as TbAG and GAG-based garnets. With single crystal films, Zorenko et al [29] found that TbAG:Ce exhibits broad band emission peaking at 550 nm under 470 nm excitation [$4f^1(^2F_{5/2}) \rightarrow 5d^1(E_{2g})$ Ce^{3+} transition], with light yields of ~62–71% depending on the Ce^{3+} content. An efficient $Tb^{3+} \rightarrow Ce^{3+}$ energy transfer was identified through directly exciting the Tb^{3+} ions in the host lattice at 325 nm. Chiang et al [23] found that the emission wavelength of Ga^{3+}-stabilized

$(Gd_{0.97}Ce_{0.03})_3(Al_{1-x}Ga_x)_5O_{12}$ yellow phosphors gradually shortens from ~565 to 552 nm ($\lambda_{ex} = 470$ nm) with increasing Ga^{3+} substitution from x = 0.1 to 0.3, yet substantially longer than the ~540 nm emission of YAG:Ce, and the shortening was ascribed to the higher electronegativity of Ga^{3+} ($\chi = 1.81$) than Al^{3+} ($\chi = 1.61$). Li et al [32] studied in detail the synthesis and optical properties of $[(Gd_{1-x}Lu_x)_{1-y}Ce_y]AG$ yellow phosphors. It was found that 1 at% (y = 0.01) of much larger Ce^{3+} (0.1143 nm) can be doped into the garnet lattice in the presence of 10 at% of Lu (x = 0.1) and more Ce^{3+} needs more Lu^{3+}. The optimal Ce^{3+} concentration was experimentally determined to be ~1 at%, and luminescence quenching mainly resulted from exchange interactions. Intensity ratio (I_b/I_a) of the 460 nm [$4f^1(^2F_{5/2}) \rightarrow 5d^1(E_{2g})$] to 340 nm [$4f^1(^2F_{5/2}) \rightarrow 5d^1(T_{2g})$] excitations was observed to significantly increase from ~4.8 at y = 0.01 to ~9.0 at y = 0.02 and then to ~12 at y = 0.03, due to successively stronger non-radiative absorptions. Energy transfer from Gd^{3+} to Ce^{3+} was identified from the appearance of $^8S_{7/2} \rightarrow {}^6I_J$ Gd^{3+} transition at ~275 nm. Figure 10 compares the emission spectra of $[(Gd_{1-x}Lu_x)_{0.99}Ce_{0.01}]AG$ ($\lambda_{ex} = 455$ nm), $(Y_{0.99}Ce_{0.01})AG$ ($\lambda_{ex} = 454$ nm), and $(Lu_{0.99}Ce_{0.01})AG$ ($\lambda_{ex} = 448$ nm), from which it is seen that the emission covers the broad range of ~475–750 nm in each case and the peak wavelength of $[(Gd_{1-x}Lu_x)_{0.99}Ce_{0.01}]AG$ red-shifts relative to those of YAG:Ce and LuAG:Ce even at the high Lu content of 50 at% (x = 0.5). Increasing Lu incorporation steadily shortens the emission wavelength due to decreased lattice covalency by the high electronegativity of Lu^{3+} ($\chi = 1.27$) and monotonically lowers the emission intensity possibly owing to lattice distortion and defect introduction. The best luminescent $[(Gd_{0.9}Lu_{0.1})_{0.99}Ce_{0.01}]AG$ has an integrated emission intensity ~97% of $(Y_{0.99}Ce_{0.01})AG$ and ~128% of $(Lu_{0.99}Ce_{0.01})AG$ at the same temperature of powder synthesis. The excellent emission, high theoretical density, and relatively low cost of $[(Gd_{0.9}Lu_{0.1})_{0.99}Ce_{0.01}]AG$ may allow it to compete with YAG:Ce and particularly LuAG:Ce for scintillation applications. CIE chromaticity coordinates (figure 11) of the three phosphors are

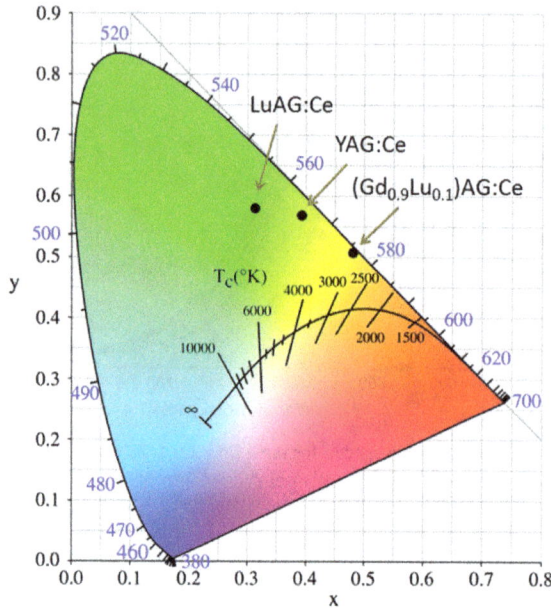

Figure 11. Emission color coordinates for the (Gd, Lu)AG:Ce, YAG: Ce, and LuAG:Ce yellow phosphors.

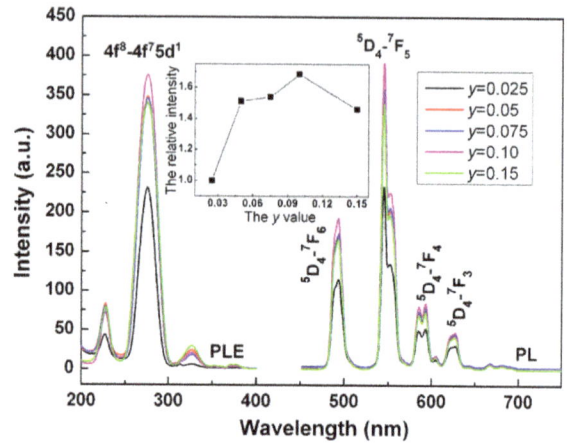

Figure 12. Excitation (λ_{em} = 545 nm) and emission (λ_{ex} = 276 nm) spectra for the [(Gd$_{0.8}$Lu$_{0.2}$)$_{1-y}$Tb$_y$]AG green phosphors.

around (0.48, 0.51), (0.39, 0.57), and (0.31, 0.58), corresponding to color temperatures of ~3044, 4612 and 6010 K, respectively. The chromaticity data again confirm that the GAG-based yellow phosphor has a stronger red component in its emission and is more desirable for warm-white LED lighting.

3.3. Tb^{3+} doping for green luminescence

When a Tb^{3+} activator is excited with light of sufficient energy, such as UV light, its 4f^8 electrons would be raised to the higher 4f^{7}5d^1 level and then fed to the ^{5}D$_{3,4}$ excited states, from which fluorescence is produced by transitions to the ^{7}F$_J$ (J = 1–6) ground states. Though the excited 5d electron is exposed, Tb^{3+} transitions involve only a redistribution of electrons within the inner 4f sub-shell [60], and thus similar emissions are usually observed from different types of host lattices, with the ^{5}D$_4 \rightarrow {}^7$F$_5$ green emission at ~545 nm being dominant.

YAG:Tb green phosphors are widely studied for applications in cathode ray tubes and flat panel displays such as FED and electroluminescent display, since it is thermally stable and resists saturation under high-current excitations [61, 62]. The GAG-based green phosphor of (Gd, Lu)AG:Tb could be an alternative choice for these purposes and improved performance might also be expected from the possible Gd$^{3+} \rightarrow$ Tb^{3+} energy transfer. The effects of Tb^{3+} content on photoluminescence of [(Gd$_{0.8}$Lu$_{0.2}$)$_{1-x}$Tb$_x$]AG were studied in figure 12, from which the quenching concentration was determined to be ~10 at% (x = 0.1), almost identical to that of YAG:Tb [63], and luminescence quenching was suggested to occur via exchange interactions [53].

The excitation spectrum consists of three 4f$^8 \rightarrow$ 4f^{7}5d^1 transition bands at ~227 nm (E_3^{1-2} level, spin allowed), 276 nm (E_2^{1-3} level, spin allowed), and 323 nm (E_1 level, spin forbidden), with the 276 nm one being dominant as widely observed [29]. It should be noted that the ^{8}S$_{7/2} \rightarrow {}^6$I$_J$ Gd^{3+} transition (~275 nm) well overlaps the 276 nm excitation, suggesting the likelihood of Gd$^{3+} \rightarrow$ Tb^{3+} energy transfer, since the ^{6}I$_J$ state of Gd^{3+} lies higher than the ^{5}D$_{3,4}$ emission states of Tb^{3+} in the energy diagram of excited states for Ln^{3+} [64–66]. The emission spectrum obtained under 276 nm excitation has four groups of bands at ~490 (blue), 545 (green, the strongest), 589 (yellow), and 623 nm (red), corresponding to the ^{5}D$_4 \rightarrow {}^7$F$_{6,5,4,3}$ transitions, respectively. Emission from the higher ^{5}D$_3$ excited level, usually in the 450–490 nm region, is hardly observable, which can be explained by cross-relaxation via resonance between the excited and ground states of two Tb^{3+} ions, that is, populating the ^{5}D$_4$ level by quenching the ^{5}D$_3$ level via Tb^{3+} (^{5}D$_3$) + Tb^{3+} (^{7}F$_0$) $\rightarrow$ Tb^{3+} (^{5}D$_4$) + Tb^{3+} (^{7}F$_6$) [67]. Emission from the ^{5}D$_3$ state was experimentally found for YAG doped with 1 at% [61] but not with 5 at% of Tb^{3+} [62], and 1 at% is generally accepted as the up-limit for the ^{5}D$_3$ emission to appear in many hosts. It is noteworthy that the charge transfer state (CTS) of the host lattice also determines the occurrence of ^{5}D$_3$ emission [68]. In La$_2$O$_2$S:Tb^{3+}, for example, it is completely quenched even at very low Tb^{3+} concentrations, not owing to cross relaxation but by thermal excitation of the ^{5}D$_3$ electrons into CTS since the two states have similar energies [68]. Comparative studies showed that the [(Gd$_{1-x}$Lu$_x$)$_{0.9}$Tb$_{0.1}$]AG phosphors with x = 0.1 and 0.2 have emission intensities close to (Y$_{0.9}$Tb$_{0.1}$)AG and (Lu$_{0.9}$Tb$_{0.1}$) AG, though the latter two have better crystallinity owing to their ease of crystallization, and have fluorescence lifetimes of ~3.31 ms (3.18 ms for YAG:Tb) and CIE color coordinates of (0.35, 0.57) [53]. Electron-beam excited luminescence of (Gd, Lu)AG:Tb^{3+} is yet needed to study for the aforesaid applications.

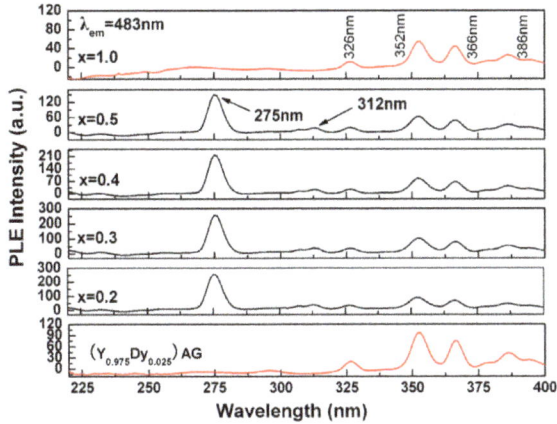

Figure 13. Excitation spectra for the [(Gd$_{1-x}$Lu$_x$)$_{0.975}$Dy$_{0.025}$]AG, (Lu$_{0.975}$Dy$_{0.025}$)AG, and (Y$_{0.975}$Dy$_{0.025}$)AG white phosphors (λ_{em} = 483 nm). Reproduced with permission from [74], copyright 2013 by the Royal Society of Chemistry.

3.4. Dy^{3+} doping for white luminescence

The primary interest in using Dy^{3+} as an activator is that it simultaneously emits blue (~483 nm, ^{4}F$_{9/2} \rightarrow$ ^{6}H$_{15/2}$ transition) and yellow (~584 nm, ^{4}F$_{9/2} \rightarrow$ ^{6}H$_{13/2}$ transition) lights, which are needed to develop white light in LEDs and optical display systems [69]. Dy^{3+}-containing compounds are also used as thermographic phosphors to measure surface temperature by applying a thin coating of the phosphor to the substrate [70]. The luminescence behavior of Dy^{3+} is governed by parity law, and in the cubic lattice of Ln$_2$O$_3$ sesquioxide the emission spectrum is dominated by the yellow band at ~584 nm [71, 72]. This is because the blue (parity allowed) and yellow (parity forbidden) emissions come from the Dy^{3+} ions taking symmetric and non-symmetric (or low symmetric) lattice sites, respectively, while in this type of oxide the centrosymmetric S$_6$ site has a much lower occupancy (25%) than the non-centrosymmetric C$_2$ site (75%) [73]. Relative intensity of the blue emission can be improved in

YAG lattice owing to higher site symmetry, but the overall emission intensity is rather limited since within the 4f^9 configuration of Dy^{3+} the excited electrons have high probabilities of non-radiative cross relaxation owing to the relatively limited energy gap between the excited and ground states and also the abundant energy multiplets for both the states [64, 66]. (Gd,Lu) AG was recently demonstrated to be significantly superior to YAG as the host for Dy^{3+} emission [74]. With the more covalent lattice and particularly via an efficient Gd$^{3+} \rightarrow$ Dy^{3+} energy transfer, greatly enhanced blue and yellow emissions were simultaneously attained. The optimal Dy^{3+} concentration was found to be ~2.5 at%, close to the ~2.0 at% reported for YAG [69], above which luminescence quenching occurs via dipole-dipole interactions. The excitation behaviors of [(Gd$_{1-x}$Lu$_x$)$_{0.975}$Dy$_{0.025}$]AG, (Y$_{0.975}$Dy$_{0.025}$)AG, and (Lu$_{0.975}$Dy$_{0.025}$)AG (x = 1.0) are compared in figure 13, where the intra-4f^9 excitations of Dy^{3+} are similarly found at ~326, 352, 366, and 386 nm for the ^{6}H$_{15/2}$ to ^{6}P$_{3/2}$, ^{4}I$_{11/2}$+^{4}M$_{15/2}$+^{6}P$_{7/2}$, ^{4}P$_{3/2}$+^{6}P$_{3/2,5/2}$, and ^{4}I$_{13/2}$+^{4}F$_{7/2}$+^{4}K$_{17/2}$+^{4}M$_{19/2,21/2}$ transitions, respectively. It is also seen that the main excitation at 352 nm is generally stronger for [(Gd$_{1-x}$Lu$_x$)$_{0.975}$Dy$_{0.025}$]AG than (Y$_{0.975}$Dy$_{0.025}$)AG and particularly (Lu$_{0.975}$Dy$_{0.025}$)AG owing to the lower electronegativity of the (Gd$_{1-x}$Lu$_x$)$^{3+}$ pair. A significant difference is that [(Gd$_{1-x}$Lu$_x$)$_{0.975}$Dy$_{0.025}$]AG has an additional excitation band at 275 nm, being the strongest in the whole excitation spectrum, that corresponds to the ^{8}S$_{7/2} \rightarrow$ ^{6}I$_J$ Gd^{3+} transition, indicating the happening of efficient Gd$^{3+} \rightarrow$ Dy^{3+} energy transfer. The ^{8}S$_{7/2}$ $\rightarrow$ ^{6}P$_J$ Gd^{3+} transition appears at ~312 nm. Figure 14 compares luminescence spectra of the three types of phosphors, from which it is seen that neither the Lu content nor excitation wavelength (275 or 352 nm) brings about appreciable change to the peak position. Emission intensity of the Gd-containing phosphor under 275 nm excitation is roughly two times that under 352 nm excitation, implying that the energy transfer is of high efficiency. Exciting the most luminescent [(Gd$_{0.8}$Lu$_{0.2}$)$_{0.975}$Dy$_{0.025}$]AG phosphor under 275 nm produced an emission intensity roughly six and three times those of (Lu$_{0.975}$Dy$_{0.025}$)AG and (Y$_{0.975}$Dy$_{0.025}$)AG under 352 nm excitation, respectively

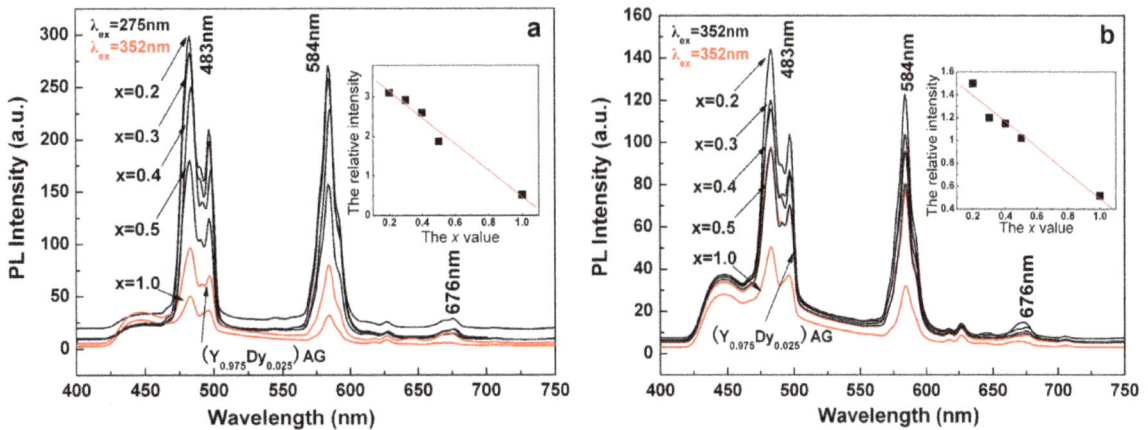

Figure 14. Emission spectra for the [(Gd$_{1-x}$Lu$_x$)$_{0.975}$Dy$_{0.025}$]AG, (Lu$_{0.975}$Dy$_{0.025}$)AG, and (Y$_{0.975}$Dy$_{0.025}$)AG white phosphors, taken under excitations with the ^{8}S$_{7/2} \rightarrow$ ^{6}I$_J$ Gd^{3+} transition at 275 nm (part (a)) and the intra-4f^9 Dy^{3+} transition at 352 nm (part (b)). Reproduced with permission from [74], copyright 2013 by the Royal Society of Chemistry.

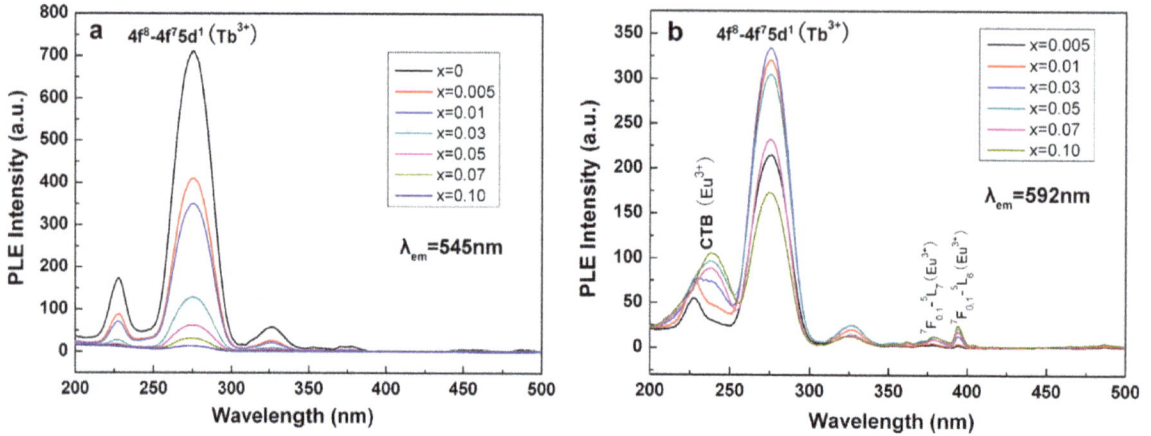

Figure 15. Excitation spectra for the $[(Gd_{0.8}Lu_{0.2})_{0.9-x}Tb_{0.1}Eu_x]AG$ phosphors, taken by monitoring the green Tb^{3+} emission at 545 nm (a) and the red Eu^{3+} emission at 592 nm (b).

(figure 14(a)). Even under identical excitation at 352 nm, the emission intensity of $[(Gd_{0.8}Lu_{0.2})_{0.975}Dy_{0.025}]AG$ is about 3.1 and 1.5 times those of $(Lu_{0.975}Dy_{0.025})AG$ and $(Y_{0.975}Dy_{0.025})$ AG, respectively (figure 14(b)). Furthermore, the $[(Gd_{1-x}Lu_x)_{0.975}Dy_{0.025}]AG$ phosphor has color coordinates of (0.33, 0.35), very close to the ideal white point of (0.33, 0.33) in the CIE chromaticity diagram, with a color temperature of ~5609 K [74].

3.5. Eu^{3+}/Tb^{3+} codoping for color tunable luminescence

Energy transfer between two types of activators is widely utilized in the phosphor field to tune the emission color, to produce a specific color that cannot be attained with one single type of activator, and to enhance the desired emission. The Ce^{3+}/Tb^{3+} and Tb^{3+}/Eu^{3+} combinations are among the most frequently adopted activator pairs. In the former, the direction of energy transfer largely depends on the $5d^1$ energy level of Ce^{3+}, which is as aforesaid readily subjected to centroid shift and crystal field splitting [55]. For example, $Ce^{3+} \rightarrow Tb^{3+}$ energy transfer is found in $CePO_4$:Tb [75] while $Tb^{3+} \rightarrow Ce^{3+}$ in TbAG:Ce [28, 29]. Dorenbos [76] determined that crystal field splitting of the Ce^{3+} 5d level is affected by coordination geometry, and tends to decrease following the order: octahedral > cubic > dodecahedral > tricapped trigonal prisms and cuboctahedral. Only $Tb^{3+} \rightarrow Eu^{3+}$ transfer can be observed for the Tb^{3+}/Eu^{3+} pair, since the $^5D_{3,4}$ excited states of Tb^{3+} lie higher than the $^5D_{0,1}$ emission states of Eu^{3+} and both the ions have relatively fixed energy levels for the 4f electrons [64–66]. The $Tb^{3+} \rightarrow Eu^{3+}$ energy transfer is of high efficiency (can be ~90%, for example), because of significant overlapping of the emission spectrum of Tb^{3+} with the excitation spectrum of Eu^{3+} [77, 78]. With such an energy transfer, occurring via electric multipole interactions [78], the emission color of Tb^{3+}/Eu^{3+} codoped Y_2O_3 can be finely tuned between red and green by varying the atomic ratio of the two activators [78]. Energy transfer and emission control were recently studied for the GAG-based phosphor of $[(Gd_{0.8}Lu_{0.2})_{0.9-x}Tb_{0.1}Eu_x]AG$ [53], where the Eu content was

varied from $x = 0$ to 0.1. The excitation spectra taken for the Tb^{3+} green emission at ~545 nm and the Eu^{3+} red emission at ~592 nm are shown in figure 15. For Tb^{3+} emission (figure 15(a)), only the characteristic excitation bands of Tb^{3+} are resolved, with the inter-configurational $4f^8 \rightarrow 4f^75d^1$ transition at ~276 nm being dominant as found for (Gd,Lu) AG:Tb^{3+}. Intensity of the excitation significantly decreases with increasing Eu^{3+} addition and finally becomes negligible at $x = 0.1$, primarily owing to $Tb^{3+} \rightarrow Eu^{3+}$ energy transfer and also concentration quenching at high total contents of the two activators. The excitation spectra taken for Eu^{3+} emission are, however, dominated by Tb^{3+} transitions, and only very weak CTB and intra-$4f^6$ transitions originated from Eu^{3+} are found (figure 15(b)). This indicates that, in the codoped system, exciting Tb^{3+} is the only efficient way to produce Eu^{3+} luminescence through energy transfer. Intensity of the 276 nm excitation reaches its maximum at $x = 0.03$, followed by a steady decrease at higher Eu contents owing to concentration quenching. Figure 16(a) analyzes intensities of the 592 nm Eu^{3+} and 545 nm Tb^{3+} emissions ($\lambda_{ex} = 276$ nm), where the strongest emission is normalized to 10 for both the activators. It is seen that the Tb^{3+} emission is monotonically weakened at a higher Eu content while the Eu^{3+} emission gradually gains intensity up to $x = 0.03$ and then decreases, following tendency found from the excitation spectra. The I_{592}/I_{545} intensity ratio steadily increases with increasing Eu^{3+} incorporation, which may suggest a persistent energy transfer from Tb^{3+} to Eu^{3+} or the quenching of Eu^{3+} emission is less than that of Tb^{3+}. The CIE color coordinates shown in figure 16(b) indicate that the emission can be well tuned from green to orange red via yellow (figure 17). Further analysis indicated that energy transfer may have occurred via electric dipole-quadrupole interactions [53]. It should be noted that the energy process is more complicated for (Gd, Lu)AG than Y_2O_3 owing to the presence of optically active Gd^{3+}. Since the $^8S_{7/2} \rightarrow ^6I_J$ Gd^{3+} transition well overlaps the $4f^8 \rightarrow 4f^75d^1$ Tb^{3+} transition at ~276 nm, multichannel energy transfer is highly possible, including $Gd^{3+} \rightarrow Tb^{3+} \rightarrow Eu^{3+}$, $Tb^{3+} \rightarrow Eu^{3+}$

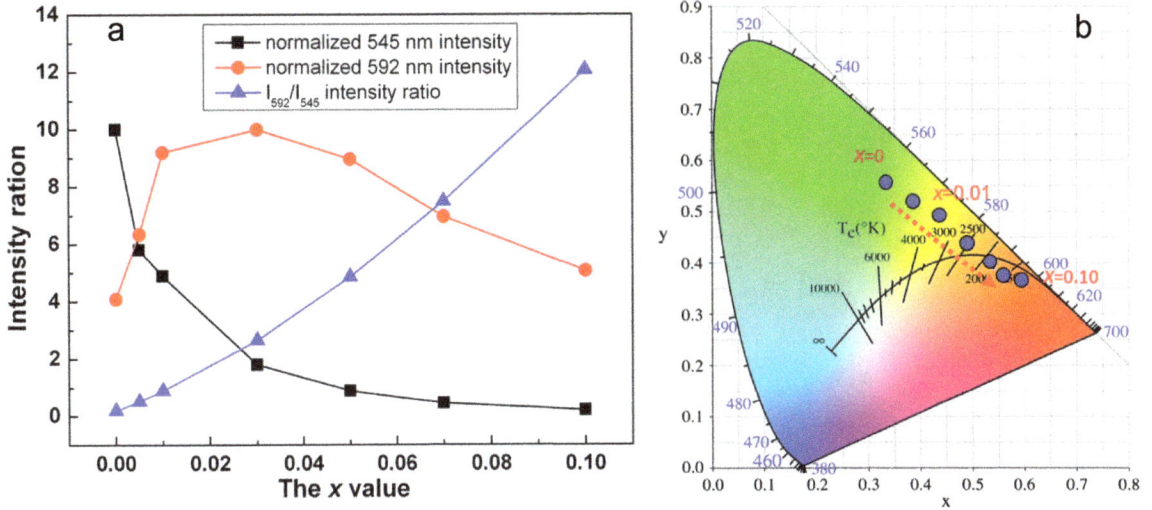

Figure 16. Relative intensities (a) of the 545 nm Tb^{3+} and 592 nm Eu^{3+} emissions and color coordinates (b) of the Tb^{3+}/Eu^{3+} co-activated $[(Gd_{0.8}Lu_{0.2})_{0.9-x}Tb_{0.1}Eu_x]AG$ phosphors.

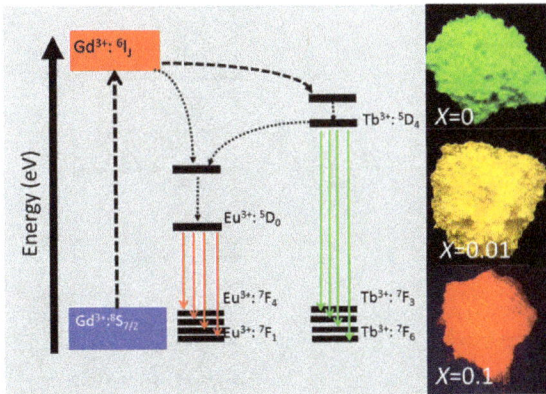

Figure 17. A scheme showing possible pathways of energy transfer (left) in the $[(Gd_{0.8}Lu_{0.2})_{0.9-x}Tb_{0.1}Eu_x]AG$ phosphor and digital pictures (right) showing color-tunable emission through the energy transfer (excitation: 275 nm).

and $Gd^{3+} \rightarrow Eu^{3+}$ (figure 17), though further studies are needed to clarify the exact routes. The excitation behavior of Eu^{3+} and the significantly lowered Tb^{3+} while improved Eu^{3+} emissions up to $x = 0.03$, however, unambiguously reveal the presence of $Tb^{3+} \rightarrow Eu^{3+}$ transfer path.

4. UC phosphors based on GAG

UC luminescence is an anti-Stokes process in which a longer wavelength radiation, usually near infrared (NIR) or IR light, is converted to a shorter wavelength such as UV or visible light via a two-photon or multi-photon mechanism [79]. The materials are drawing wide interest since they can be used as biological labels for medical diagnosis and

therapy, in photovoltaic cells to efficiently harvest solar energy, and also in laser and anti-counterfeit fields [79]. The activators used for UC are those that usually do not exhibit efficient DC luminescence, including Pr^{3+}, Sm^{3+}, Ho^{3+}, Er^{3+}, and Tm^{3+}. Though when properly doped in a host the above activators themselves have been able to produce UC emission, the efficiency is usually rather limited owing to their unsatisfactory NIR/IR excitations. Yb^{3+} is thus widely employed as a codopant to improve NIR absorption (at ~980 nm, the $^2F_{7/2} \rightarrow {}^2F_{5/2}$ Yb^{3+} transition) and to sensitize UC emission via nonradiative energy transfer from Yb^{3+} to the activator. The most preferred and widely used hosts for UC luminescence are fluorides owing to their low phonon energy, though other material types, such as Y_2O_3 transparent ceramics [80, 81] and Gd_2O_2S powder [82], were also explored. YAG was suggested to possess a large ground-state Stark splitting and has a quasi three-level energy structure, which may enable a broad and intense absorption of Yb^{3+} in it [83]. The energy transferred from Yb^{3+} may effectively populate the upper level of Tm^{3+} in YAG [83]. UC performances were recently studied for stabilized GAG with the compositions of $[(Gd_{1-x}Lu_x)_{0.948}Yb_{0.05}Ln_{0.002}]AG$ ($x = 0.1$–0.5, Ln = Er, Ho, and Tm) [53, 84, 85]. Despite the dilute Yb and Ln concentrations, strong UC luminescence was clearly observed in each case upon laser exciting Yb^{3+} at 978 nm, as shown in figure 18 together with the mechanism of UC. The UC luminescence presents as an intense blue band at ~487 nm ($^1G_4 \rightarrow {}^3H_6$ transition) and a weaker red one at ~650 nm ($^1G_4 \rightarrow {}^3F_4$) for Tm^{3+}, as a fairly strong green band at 543 nm ($^5F_4, {}^5S_2 \rightarrow {}^5I_8$) and a strong red band at 668 nm ($^5F_5 \rightarrow {}^5I_8$) for Ho^{3+}, and as three bands at 525 nm (green, $^2H_{11/2} \rightarrow {}^4I_{15/2}$), 556 nm (yellow, $^4S_{3/2} \rightarrow {}^4I_{15/2}$) and 655 nm (red, $^4F_{9/2} \rightarrow {}^4I_{15/2}$) for Er^{3+}. CIE chromaticity coordinates of the UC luminescence were found to be around (0.14, 0.19) for Tm^{3+}, (0.38, 0.58) for Ho^{3+}, and

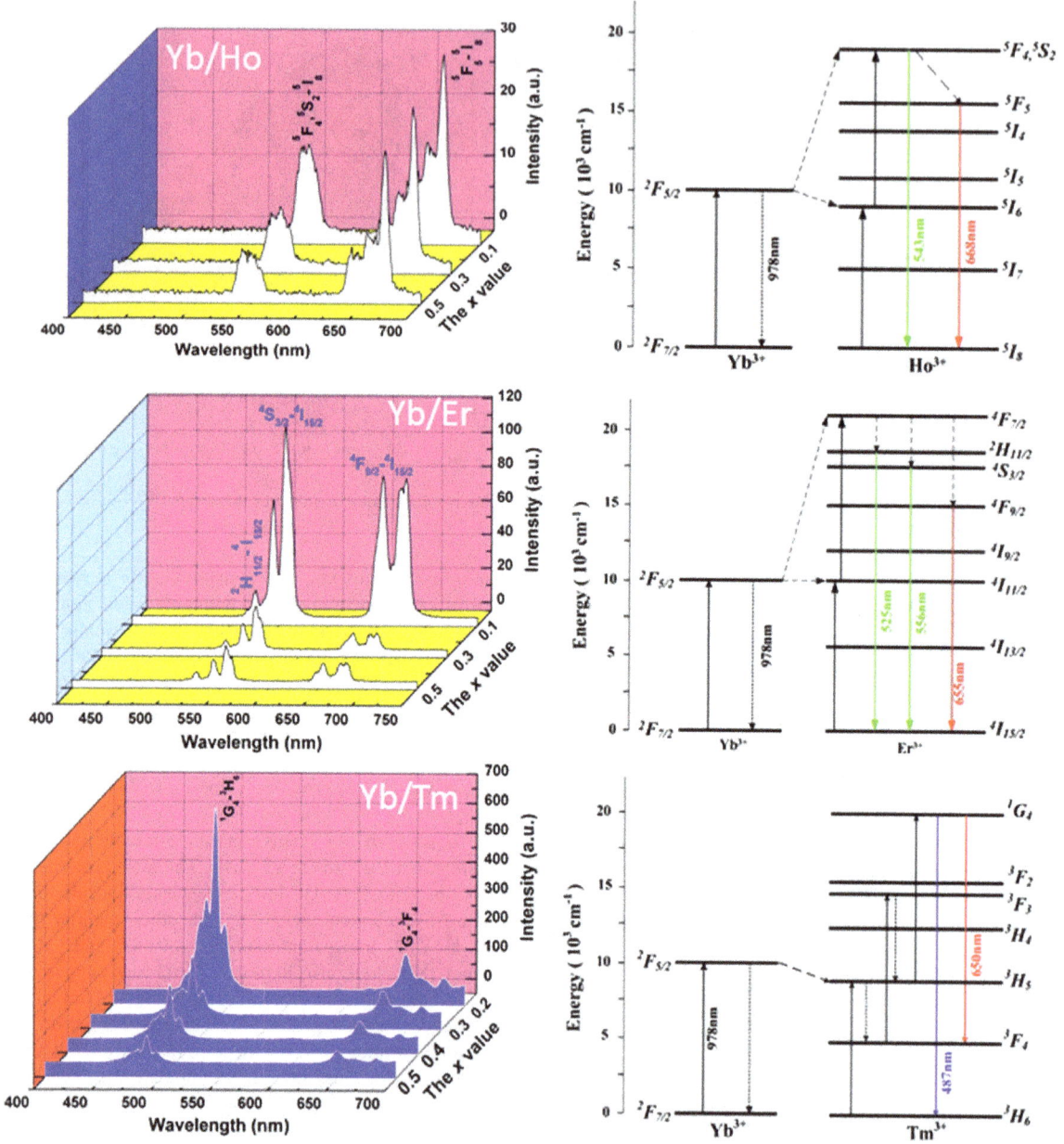

Figure 18. Up-conversion (UC) luminescence spectra of the Yb/Ho, Yb/Er, and Yb/Tm codoped $(Gd_{1-x}Lu_x)AG$ solid solutions. Mechanisms of the UC processes are presented in the right-hand schemes. Reproduced with permission from [85], copyright 2014 by Trans Tech Publications, and [84], copyright 2014 by Elsevier.

(0.34, 0.64) for Er^{3+}, corresponding to blue, greenish yellow, and green colors, respectively (figure 19). Lowered luminescence with increasing Lu incorporation and enhanced emission with increasing temperature of phosphor synthesis were found for the above UC systems. Analyzing the emission intensity against excitation power indicated that the UC luminescence may have occurred via a three-photon process for Tm^{3+} and a two-photon process for both Ho^{3+} and Er^{3+}, which are schematically shown in the right part of figure 18 [53, 84, 85].

5. Single-crystal and transparent-ceramic scintillators based on GAG

A scintillator is essentially a luminescent material that absorbs high-energy photons and then emits visible light, for which efficient absorption of the excitation source is a fundamental requirement [41, 86]. Since the relation among absorption coefficient (η_{abs}), theoretical density (ρ), and effective atomic number (Z_{eff}) can be expressed as $\eta_{abs} = \rho Z_{eff}^4$ [87], high theoretical density (generally >6 g cm^{-3}) and particularly high

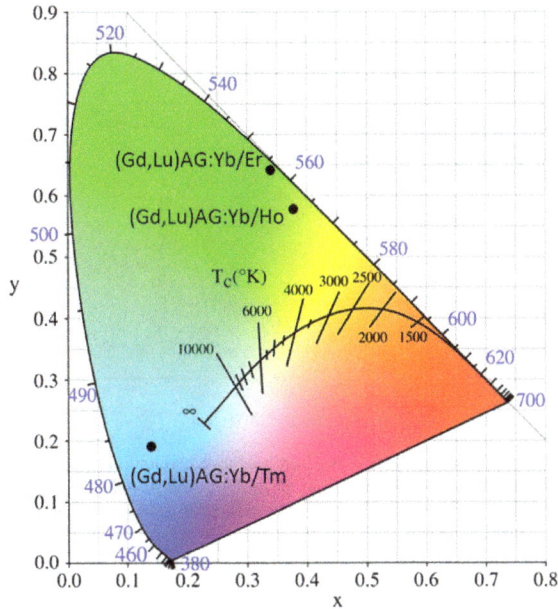

Figure 19. Color coordinates for the UC luminescence of (Gd, Lu) AG:Yb/Ln (Ln = Er, Ho, and Tm).

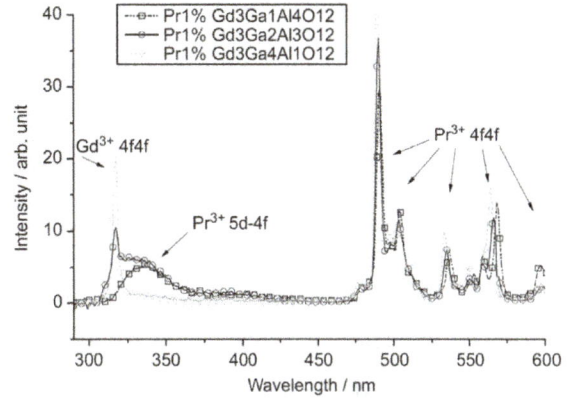

Figure 20. Radioluminescence spectra of the $Gd_3(Al_{5-x}Ga_x)$ O_{12}:1 at%Pr single crystals under γ-ray excitations. Reproduced with permission from [25], copyright 2012 by Elsevier.

effective atomic number are thus needed for an excellent scintillator, though other characteristics such as high light yield and fast response (10–100 ns, dominant decay generally <3 μs to avoid signal pile-up with standard shaping electronics) are also essential [86]. Scintillators combined with photodetectors are widely used in various medical imaging technologies, such as x-ray computed tomography, positron emission tomography (PET), and single-photon emission computed tomography (SPECT), and also in high energy and nuclear physics. The most common scintillators up to date are CsI:Tl, CdWO$_4$, and Bi$_4$Ge$_3$O$_{12}$ (BGO) single crystals, but they have their respective shortcomings such as hygroscopicity, poor machinability, insufficient light output, and slow blinking [86]. For these reasons, Ce-doped silicates, such as Gd$_2$SiO$_5$ (GSO), Lu$_2$SiO$_5$ (LSO), (Lu$_{1-x}$Y$_x$)$_2$SiO$_5$ (LYSO), and LaBr$_3$ are being developed as alternatives [88–94].

GAG-based single crystal scintillators are mostly studied by Kamada *et al* [24, 25, 95–97] through crystal growth by the μ-PD and CZ techniques. The CZ-grown Gd$_3$(Al$_2$Ga$_3$) O$_{12}$:1 at%Ce single crystal, where Ga^{3+} is a lattice stabilizer, was reported to have a high light yield of 46 000 photons/ MeV for the Ce^{3+} emission (31 000 photons/MeV for LYSO: Ce), an energy resolution of 4.9% at 662 keV, a primary decay time of 88 ns (91%), and a high theoretical density of 6.63 g cm^{-3} [24]. The crystals are thus regarded as promising scintillators for PET, SPECT, and gamma camera. The μ-PD grown Gd$_3$(Al$_{5-x}$Ga$_x$)O$_{12}$:1 at%Pr single crystals (x=1–5) were found to exhibit the 5d → 4f and intra-4f^2 transition emissions of Pr^{3+} in the 300–400 and 480–600 nm (dominant) regions, respectively, together with the intra-4f^7 Gd^{3+} emission at 310 nm (figure 20) [25]. A higher Ga^{3+} content would

suppress the 5d → 4f Pr^{3+} emission while enhance the f–f transitions of both Gd^{3+} and Pr^{3+}. A low light output of ~4500 photons/MeV, only about 1/5 of the CZ-grown LuAG: Pr standard, and a relatively long primary decay time of 214 ns (98.8%) were reported for the Gd$_3$(Al$_2$Ga$_3$)O$_{12}$:1 at% Pr single crystal [25]. The poor light output was suggested to be associated with an energy transfer from the 5d state of Pr^{3+} to the 4f state of Gd^{3+} and non-radiative relaxations from the 5d to 4f states of Pr^{3+}, with the former being dominant, as schematically shown in figure 21 [98]. Pr^{3+} was thus suggested not to be a proper activator for Gd^{3+}-containing scintillators. By simultaneously modifying GAG with Lu^{3+} for Gd^{3+} and Ga^{3+} for Al^{3+}, (Gd$_2$Lu$_1$)(Al$_{5-x}$Ga$_x$)O$_{12}$:Ce single crystals (x=1–5; Ce^{3+} content: 0.2, 1.0, and 3 at%) were grown by the μ-PD technique and thoroughly investigated for their luminescence properties [95]. It was shown that decay of the 5d → 4f Ce^{3+} emission at ~520 nm accelerates with increasing Ga or Ce concentration, and the best composition of (Gd$_2$Lu$_1$)(Al$_2$Ga$_3$)O$_{12}$:1 at%Ce has a light yield of ~22 000 photons/MeV, about 70% of LYSO:Ce (31 000 photons/ MeV), a theoretical density of 6.88 g cm^{-3}, and a decay time of 76.5 ns (83%) and 282 ns (17%). With the combinatorial approach, Kamada *et al* made comprehensive composition optimization for 0.2 at% Ce^{3+} doped (Gd$_{3-x}$Lu$_x$)(Al$_{5-y}$Ga$_y$) O$_{12}$ [96] and (Gd$_{3-x}$Y$_x$)(Al$_{5-y}$Ga$_y$)O$_{12}$ [97] single crystals. The light output of Ce^{3+} in the best hosts of (Gd$_2$Y$_1$)(Al$_2$Ga$_3$) O$_{12}$ and Gd$_3$(Al$_2$Ga$_3$)O$_{12}$ reached ~42 000–44 000 photons/ MeV, being ~150% of the value of LYSO:Ce and 730% of that of BGO (5700 photons/MeV), with the scintillation decay time dominated by 50–80 ns. The energy resolution at 662 kV was determined to be 8.3% [96], which, though inferior to the 6.7% for high-quality CZ-grown LuAG:Ce, is comparable to the 8.7% measured for LYSO:Ce. Bandgap engineering was pointed out to be crucial in developing high quality scintillators, as also suggested by previous studies. For example, the 5d → 4f luminescence of Ce^{3+} is quenched in the high density garnets of Lu$_3$Ga$_5$O$_{12}$ (7.4 g cm^{-3}) and Gd$_3$Ga$_5$O$_{12}$ (7.04 g cm^{-3}) owing to positioning of the Ce^{3+} 5d states in the conduction band of the host [99], and the performance of

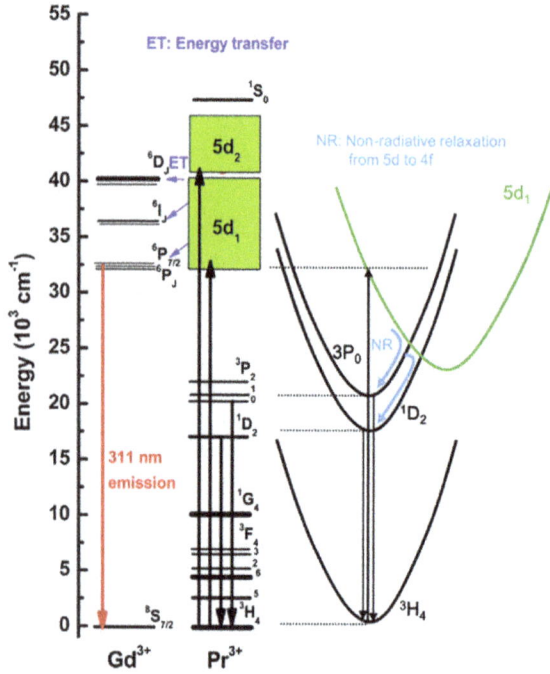

Figure 21. Energy diagram for the Gd^{3+} and Pr^{3+} centers in $(Gd_xLu_{3-x})(Ga_3Al_2)O_{12}$ ($x < 0.2$), with the energy transfer channel from Pr^{3+} to Gd^{3+} indicated. The right-hand scheme depicts non-radiative relaxation from the lowest $5d^1$ to low-lying $^3P_0$ and $^1D_2$ levels of Pr^{3+}. Reproduced with permission from [98], copyright 2013 by Elsevier B V.

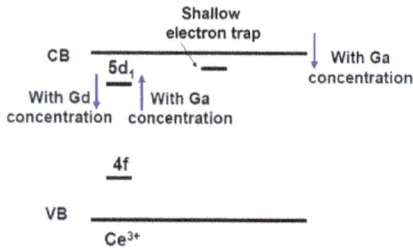

Figure 22. Energy level scheme showing bandgap and $5d^1$ level engineering for the $(Gd, Lu)_3(Ga, Al)_5O_{12}$:Ce scintillation crystals. The Ga component helps to lower the conduction band (CB) to bury the shallow traps while the Gd component pushes away the $5d^1$ level of Ce^{3+} from the bottom of CB to avoid Ce^{3+} ionization. Reproduced with permission from [96], copyright 2011 by the American Chemical Society.

LuAG:Ce is strongly degraded by shallow electron traps (Lu_{Al} anti-site defects) via delaying energy transfer to Ce^{3+} and giving rise to quite slow components in the scintillation response [100, 101]. A balanced Ga and Gd admixture may eliminate the trapping effects by burying the shallow traps in the bottom of the conduction band and at the same time avoid ionization of the Ce^{3+} activators by separating the 5d excited level from the conduction band, as schematically shown in figure 22 [96, 97].

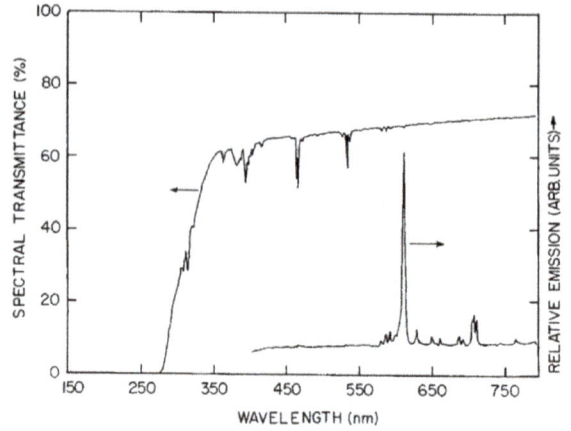

Figure 23. Transmittance and luminescence (excitation: 254 nm UV light) spectra of a 1.5 mm thick $(Y_{0.67}Gd_{0.30}Eu_{0.03})_2O_3$ ceramic scintillator. Reproduced with permission from [86], copyright 1997 by Annual Reviews Inc.

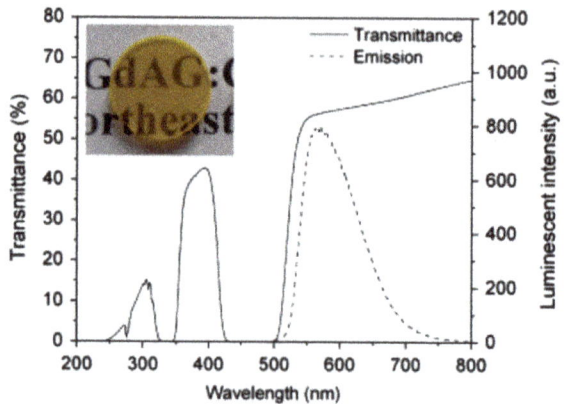

Figure 24. In-line transmittance and luminescence (excitation: 340 nm UV light) spectra of a 1.3 mm thick $(Y_{1.48}Gd_{1.5}Ce_{0.02})AG$ scintillation ceramic. Reproduced with permission from [104], copyright 2010 by the American Ceramic Society.

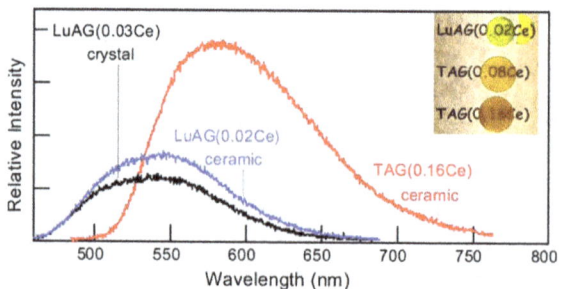

Figure 25. Beta-excited radioluminescence spectra and appearances of the Ce^{3+} doped TAG and LuAG ceramics with scatter mean free path >1 cm. Reproduced with permission from [105], copyright 2009 by IEEE.

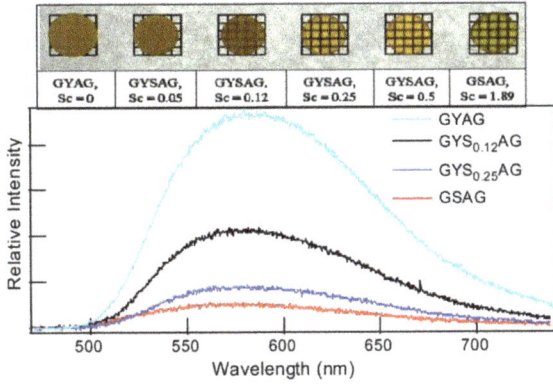

Figure 26. Beta-excited radioluminescence spectra and appearances of the Ce^{3+} doped gadolinium-based garnet ceramics. Reproduced with permission from [105], copyright 2009 by IEEE.

Aside from single crystals, transparent ceramics are under active development for scintillation applications, since the sintering technique generally has advantages in product size and fabrication cost. The $(Y,Gd)_2O_3$:Eu transparent ceramic (inline transmittance ~73% at the 610 nm emission of Eu^{3+} or ~90% of the theoretical value, figure 23), fabricated via pressureless sintering, pressureless sintering plus hot isostatic pressing, and vacuum hot pressing, has been the first commercialized polycrystalline scintillator used in medical x-ray detectors [41, 86]. The $(Y_{0.67}Gd_{0.30}Eu_{0.03})_2O_3$ composition (5.92 g cm^{-3}) was reported to have a relative light output 2.5 times higher and ~30% lower than those of $CdWO_4$ and CsI:Tl single crystals, respectively [41]. Li et al [102] fabricated transparent $(Y_{0.3}Gd_{0.67}Eu_{0.03})_2O_3$ ceramic with a higher Gd content, an inline transmittance of ~68% at the 610 nm

Eu^{3+} emission and a higher theoretical density of 6.87 g cm^{-3}, by vacuum sintering at 1670 °C for 2 h of the oxide powders calcined from coprecipitated carbonate precursors. The main problem associated with Eu^{3+} emission is the long fluorescence decay, which is usually ~1 ms [41, 86]. Yanagida et al [8, 103] made transparent YAG:Ce and (Gd,Y)AG:Ce ceramics for possible scintillation applications (the Gd content and optical transmittance not specified), and the dominant decay time was found to be ~90 ns under γ-ray excitations. Light yield of the latter was reported to be 30% of the former, without specifying the reason, and the energy resolution was found to be ~8% at 662 keV when coupled with an avalanche photodiode [103]. The stopping power of (Gd, Y)AG:Ce was reported to be five times as high as that of YAG:Ce [103]. Li et al [104] studied sintering of $(Y_{1-x}Gd_x)$AG:Ce ($x = 0.01$, 0.5, 0.75, and 1; Ce content: 2 at%) transparent ceramics with commercially available nano-sized powders. It was shown that the GAG composition ($x = 1.0$) cracks owing to the stress (volume change) arising from thermal decomposition while all the other compositions can be densified to transparency by vacuum sintering at 1600–1700 °C for 5 h. The $(Y_{1.48}Gd_{1.5}Ce_{0.02})$AG ceramic, with an inline transmittance of ~65% at 800 nm, exhibits Ce^{3+} emission at ~570 nm under 340 nm UV excitation, which is red-shifted ~40 nm relative to YAG:Ce (~530 nm) owing to the high content of less electronegative Gd^{3+} (figure 24). Cherepy et al [105] studied radioluminescence properties of translucent $(Tb_{3-x}Ce_x)$AG ceramics (figure 25), and much higher light yields (80 000 photons/MeV) than LuAG:Ce (~30 000 photons/MeV) were found under β-ray excitations. The decay, however, is slow owing to energy migration on the Tb^{3+} sites, and the principal decay time reaches <1 μs only at the high Ce^{3+} content of $x = 0.12$. By applying vacuum sintering plus hot isostatic

Table 1. A summary of the scintillation properties of the promising single-crystal/ceramic scintillators discussed in this work.

Host lattice	Ce content (at%)	Material form	Density (g cm^{-3})	Excitation source	Light yield (photons/MeV)	Resolution at 662 kV (%)	Primary decay (ns)	Reference
LuAG	1.0	Crystal	6.73	γ-ray	7500	66.2 (88%)	—	[95]
LuAG	3.0	Crystal	6.73	γ-ray	28 000	8.6	—	[105]
				β-ray	30 000			
LuAG	2.0	Ceramic	6.73	γ-ray	20 000	11.4	—	[105]
				β-ray	30 000			
YAG	1.0	Crystal	4.55	γ-ray	22 700	—	83.1 (71%)	[97]
YAG	—	Ceramic	4.55	γ-ray	30 000	7.3	90	[103, 105]
				β-ray	40 000			
TbAG	2-16	Ceramic	—	γ-ray	19 500	10.6	4500–600	[105]
				β-ray	80 000			
$(Gd_{1.5}Y_{1.5})Al_5O_{12}$	3.0	Ceramic	—	γ-ray	16 500	11.2	100–200	[105]
				β-ray	>80 000			
$Gd_3(Sc_{1.89}Al_{3.11})O_{12}$	3.0	Ceramic	—	γ-ray	7500	10.8	100–200	[105]
				β-ray	20 000			
$Gd_3(Al_2Ga_3)O_{12}$	1.0	Crystal	6.63	γ-ray	46 000	4.9	88 (91%)	[24]
$Gd_3(Al_2Ga_3)O_{12}$	0.2	Crystal	6.63	γ-ray	42 217	8.3	52.8 (73%)	[96]
$(Gd_2Lu_1)(Al_2Ga_3)O_{12}$	1.0	Crystal	6.88	γ-ray	22 000	11.2	76.5 (83%)	[95]
$(Gd_2Y_1)(Al_2Ga_3)O_{12}$	1.0	Crystal	—	γ-ray	44 000	8.2	56.9 (66%)	[97]

pressing, Cherepy *et al* [105] also fabricated a series of GAG-based scintillation ceramics (Ce^{3+} content: 3 at%), including $(Gd_{1.5}Y_{1.5})_3Al_5O_{12}$ (GYAG), $(Gd_{1.5}Y_{1.5})_3(Al_{5-x}Sc_x)O_{12}$ (GYSAG), and $Gd_3(Al_3Sc_2)O_{12}$ (GSAG), using Y^{3+}/Sc^{3+} as dopant for lattice stabilization (figure 26). GYAG:Ce was reported to have a very high light yield of ~100 000 photons/MeV under β-ray excitation, due to an efficient energy transfer from Gd^{3+}, and have decay time in the 100–200 ns range. Transparency, however, was not achieved for this composition due to the presence of a small amount of $GdAlO_3$ secondary phase. In contrast, GSAG and GYSAG only produced the moderate light yields of ~20 000 photons/MeV, with the primary decay time <200 ns, but formed phase-pure garnet with excellent transparency for the former and acceptable transparency for the latter ($x \geqslant 0.12$, transmittance data not shown). In general, the Ce^{3+} emission in Gd-containing scintillators simultaneously has fast primary decay (<200 ns) and high light yield, as compared from table 1 for the typical garnet compounds discussed in this work, showing substantial advantages of the GAG host lattice.

6. Summary and outlook

Lattice stabilization of GAG ($Gd_3Al_5O_{12}$, GAG) and the related development of advanced optical materials are summarized in this article, including down-/up-conversion phosphors, transparent ceramics, and single crystals. It is shown that novel emission features and significantly improved luminescence can be achieved for a number of phosphor systems with the more covalent GAG lattice and the efficient energy transfer from Gd^{3+} to the activator. GAG-based single crystals and transparent ceramics with Ce^{3+} as the activator are shown to have fast scintillation decay and high light yields, and thus hold great potential as scintillators for a wide range of applications. Anti-site defects commonly exist in aluminate garnets [106–108], and their occurrence and energy level have profound influences on various aspects of emission through interacting with the excited electrons. These issues, however, have rarely been tackled for the GAG-based garnets, either when Lu^{3+} or Ga^{3+} is used as lattice stabilizer, and thus need to clarify in future studies for a better control of optical properties. In addition, GAG-based transparent ceramics have been much less developed than their YAG and LuAG counterparts, and need significantly more studies on powder processing and sintering technologies to improve their overall transmittance and other optical performances.

Acknowledgment

Experimental contribution to GAG-based phosphors by Dr Jinkai Li, through his PhD thesis work under the supervision of Dr Ji-Guang Li, is acknowledged. Sincere thanks are due to the editorial board of STAM for inviting this review.

References

[1] Xu Y-N and Ching W Y 1999 *Phys. Rev. B* **59** 10530
[2] Geusic G E, Narcos H M and Van Uitert L G 1964 *Appl. Phys. Lett.* **4** 182
[3] Ikesue A, Konoshita T, Kamata K and Yoshida K 1995 *J. Am. Ceram. Soc.* **78** 1033
[4] Ikesue A and Aung Y L 2006 *J. Am. Ceram. Soc.* **89** 1936
[5] Fornasiero L, Mix E, Peters V, Petermann K and Huber G 2000 *Ceram. Int.* **26** 589
[6] Nishiura S, Tanabe S, Fujioka K and Fujimoto Y 2011 *IOP Conf. Ser.: Mater. Sci. Eng.* **18** 102005
[7] Bachmann V, Ronda C and Meijerink A 2009 *Chem. Mater.* **21** 2077
[8] Yanagida T *et al* 2005 *IEEE Trans. Nucl. Sci.* **52** 1836
[9] Mizuno M, Yamada T and Noguchi T 1977 *Yogyo-Kyokai-Shi* **85** 30
[10] Mizuno M, Yamada T and Noguchi T 1977 *Yogyo-Kyokai-Shi* **85** 90
[11] Mizuno M, Yamada T and Noguchi T 1977 *Yogyo-Kyokai-Shi* **85** 374
[12] Mizuno M, Yamada T and Noguchi T 1977 *Yogyo-Kyokai-Shi* **85** 543
[13] Mizuno M, Yamada T and Noguchi T 1978 *Yogyo-Kyokai-Shi* **86** 360
[14] Van Uitert L G, Grodkiewicz W H and Dearborn E F 1965 *J. Am. Ceram. Soc.* **48** 105
[15] Manabe T and Egashira K 1971 *Mater. Res. Bull.* **6** 1167
[16] Shishido T, Okamura K and Yajima S 1978 *J. Am. Ceram. Soc.* **61** 373
[17] Li J K, Li J-G, Zhang Z J, Wu X L, Liu S H, Li X D, Sun X D and Sakka Y 2012 *J. Am. Ceram. Soc.* **95** 931
[18] Chaudhury S, Parida S C, Pillai K T and Mudher K D S 2007 *J. Solid State Chem.* **180** 2393
[19] Li J-G, Li X D, Sun X D and Ishigaki T 2008 *J. Phys. Chem. C* **112** 11707
[20] Li Y H and Hong G Y 2007 *J. Lumin.* **124** 297
[21] Wu X L, Li J-G, Zhu Q, Li J K, Ma R Z, Sasaki T, Li X D, Sun X D and Sakka Y 2012 *Dalton Trans.* **41** 1854
[22] Wu X L, Li J-G, Ping D-H, Li J K, Zhu Q, Li X D, Sun X D and Sakka Y 2013 *J. Alloys Compd.* **559** 188
[23] Chiang C-C, Tsai M-S and Hon M-H 2007 *J. Electrochem. Soc.* **154** J326
[24] Kamada K, Yanagida T, Endo T, Tsutumi K, Usuki Y, Nikl M, Fujimoto Y, Fukabori A and Yoshikawa A 2012 *J. Cryst. Growth* **352** 88
[25] Kamada K, Yanagida T, Pejchal J, Nikl M, Endo T, Tsutumi K, Usuki Y, Fujimoto Y, Fukabori A and Yoshikawa A 2012 *J. Cryst. Growth* **352** 84
[26] Maglia F, Buscaglia V, Gennari S, Ghigna P, Dapiaggi M, Speghini A and Bettinelli M 2006 *J. Phys. Chem. B* **110** 6561
[27] Tomiki T, Isa Y, Kadekawa Y, Ganaha Y, Toyokawa N, Miyazato T, Miyazato M, Kohatsu T, Shimabukuro H and Tamashiro J 1996 *J. Phys. Soc. Japan* **65** 1106
[28] Zorenko Y, Gorbenko V, Voznyak T, Batentschuk M, Osvet A and Winnacker A 2008 *J. Lumin.* **128** 652
[29] Zorenko Y, Gorbenko V, Voznyak T, Zorenko T, Kuklinski B, Turos-Matysyak R and Grinberg M 2009 *Opt. Spectrosc.* **106** 365
[30] Bi J, Li J-G, Guan W M, Chen J L and Sun XD 2014 *Key Eng. Mater.* 1028
[31] Li J K, Li J-G, Zhang Z J, Wu X L, Liu S H, Li X D, Sun X D and Sakka Y 2012 *Sci. Technol. Adv. Mater.* **13** 035007
[32] Li J K, Li J-G, Liu S H, Li X D, Sun X D and Sakka Y 2013 *Sci. Technol. Adv. Mater.* **14** 054201
[33] Li J K, Li J-G, Wu X L, Liu S H, Li X D and Sun X D 2013 *Key Eng. Mater.* **544** 245

[34] Ravichandran D, Roy R, Chakhovskoi A G, Hunt C E, White W B and Erdei S 1997 *J. Lumin.* **71** 291
[35] Pereira P F S, Caiut J M A, Ribeiro S J L, Messaddeq Y, CiuffiK J, Rocha L A, Molina E F and Nassar E J 2007 *J. Lumin.* **126** 378
[36] Zhu Q, Li J-G, Zhi C, Ma R, Sasaki T, Xu J X, Liu C H, Li X D, Sun X D and Sakka Y 2011 *J. Mater. Chem.* **21** 6903
[37] Zhu Q, Li J-G, Li X D, Sun X D, Yang Q, Zhu M Y and Sakka Y 2014 *Sci. Technol. Adv. Mater.* **15** 014203
[38] Lu B, Li J-G and Sakka Y 2013 *Sci. Technol. Adv. Mater.* **14** 064202
[39] Wang X J, Li J-G, Zhu Q, Li X D, Sun X D and Sakka Y 2014 *Sci. Technol. Adv. Mater.* **15** 014204
[40] Li J K, Li J-G, Li J, Liu S H, Li X D, Sun X D and Sakka Y 2013 *J. Solid State Chem.* **206** 104
[41] Greskovich C D, Cusano D, Hoffman D and Riedner R J 1992 *Am. Ceram. Soc. Bull.* **71** 1120
[42] Matijević E and Hsu W P 1987 *J. Colloid Interface Sci.* **118** 506
[43] Hsu W P, Rönnquist L and Matijević E 1988 *Langmuir* **4** 31
[44] Aiken B, Hsu W P and Matijević E 1988 *J. Am. Ceram. Soc.* **71** 845
[45] Matijević E. 1993 *Chem. Mater.* **5** 412
[46] Li J-G, Li X D, Sun X D, Ikegami T and Ishigaki T 2008 *Chem. Mater.* **20** 2274
[47] Zhu Q, Li J-G, Li X D, Sun X D and Sakka Y 2011 *Sci. Technol. Adv. Mater.* **12** 055001
[48] Li J-G, Zhu Q, Li X D, Sun X D and Sakka Y 2011 *Acta Mater.* **59** 3688
[49] Sordelet D J, Akinc M, Panchula M L, Han Y and Han M H 1994 *J. Eur. Ceram. Soc.* **14** 123
[50] Matsushita N, Tsuchiya N, Nakatsuka K and Yanagitani T 1999 *J. Am. Ceram. Soc.* **82** 1977
[51] Hui Y, Sun X D, Chen J L, Li X D, Huo D, Liu S H, Zhu Q, Zhang M and Li J-G 2014 *IEEE Trans. Nucl. Sci.* **61** 362
[52] Xu X J, Sun X D, Liu H, Li J-G, Li X D, Huo D and Liu S H 2012 *J. Am. Ceram. Soc.* **95** 3821
[53] Li J K 2013 Crystal structure stabilization of gadolinium aluminate garnet (Gd₃Al₅O₁₂) and development of new phosphors *PhD Thesis* Northeastern University
[54] Zhu Q, Li J-G, Ma R Z, Sasaki T, Yang X J, Li X D, Sun X D and Sakka Y 2012 *J. Solid State Chem.* **192** 229
[55] George N C, Denault K A and Seshadri R 2013 *Annu. Rev. Mater. Res.* **43** 481
[56] Shao Q, Li H, Gong Y, Jiang J, Liang C and He J 2010 *J. Alloys Compound.* **498** 199
[57] Wang X J, Zhou G H, Zhang H L, Li H L, Zhang Z J and Sun Z 2012 *J. Alloys Compd.* **519** 149
[58] Katelnikovas A, Bettentrup H, Uhlich D, Sakirzanovas S, üstel T J and Kareiva A 2009 *J. Lumin.* **129** 1356
[59] Maniquiz M C, Jung K Y and Jeong S M 2010 *J. Electrochem. Soc.* **157** H1135
[60] Hipolito M G, Ocampo A C, Fregoso O A, Martinez E, Mendoza J G and Falcony C 2004 *Phys. Status Solidi* a **201** 72
[61] Choe J Y, Ravichandran D, Blomquist S M, Morton D C, Kirchner K W, Ervin M H and Lee U 2001 *Appl. Phys. Lett.* **78** 3800
[62] Hakuta Y, Seino K, Ura H, Adschiri T, Takizawa H and Arai K 1999 *J. Mater. Chem.* **9** 2671
[63] Guo K, Huang M-L, Chen H-H, Yang X-X and Zhao J-T 2012 *J. Non-Cryst. Solids* **358** 88
[64] Dieke G H and Crosswhite H M 1963 *Appl. Opt.* **2** 675
[65] Wegh R T, Meijerink A, Lamminmaki R J and Holsa J 2000 *J. Lumin.* **87–89** 1002
[66] Peijzel P S, Meijerink A, Wegh R T, Reid M F and Burdick G W 2005 *J. Solid State Chem.* **178** 448
[67] Song H W and Wang J W 2006 *J. Lumin.* **118** 220
[68] Ratinen H 1972 *Phys. Solid State* A **12** 447
[69] Klimczak M, Malinowski M, Sarnecki J and Piramidowicz R 2009 *J. Lumin.* **129** 1869
[70] Heyes A L, Seefeldt S and Feist J P 2006 *Opt. Laser Technol.* **38** 257
[71] Wang H, Yang J, Zhang C M and Lin J 2009 *J. Solid State Chem.* **182** 2716
[72] Yang J, Li C, Quan Z, Zhang C, Yang P, Li Y, Yu C and Lin J 2008 *J. Phys. Chem.* C **112** 12777
[73] Zhu Q, Li J-G, Li X D and Sun X D 2009 *Acta Mater.* **57** 5975
[74] Li J K, Li J-G, Liu S H, Li X D, Sun X D and Sakka Y 2013 *J. Mater. Chem.* C **1** 7614
[75] Di W H, Shirahata N, Zeng H B and Sakka Y 2010 *Nanotechnology* **21** 365501
[76] Dorenbos P 2002 *J. Alloys Compound.* **341** 156
[77] Di W, Wang X, Zhu P and Chen B 2007 *J. Solid State Chem.* **180** 467
[78] Wu X L, Li J-G, Li J K, Zhu Q, Li X D, Sun X D and Sakka Y 2013 *Sci. Technol. Adv. Mater.* **14** 015006
[79] Azuel F 2004 *Chem. Rev.* **104** 139
[80] Hinklin T R, Rand S C and Laine R M 2008 *Adv. Mater.* **20** 1270
[81] Hou X R, Zhou S M, Jia T T, Lin H and Teng H 2011 *Phys.: Condens. Matter* **406** 3931
[82] Martin-Rodriguez R, Fischer S, Ivaturi A, Froehlich B, Kramer K W, Goldschmidt J C, Richards B S and Meijerink A 2013 *Chem. Mater.* **25** 1912
[83] Xu X D, Wu F, Xu W W, Zong Y H, Wang X D, Zhao Z W, Zhou G Q and Xu J 2008 *J. Alloys Compound.* **462** 347
[84] Li J K, Li J-G, Li J, Liu S H, Li X D, Sun X D and Sakka Y 2014 *J. Alloys Compound.* **582** 623
[85] Li J K, Li J-G, Liu S H, Li X D, Sun X D and Sakka Y 2014 *Key Eng. Mater.* **602–603** 1034
[86] Greskovich C and Duclos S 1997 *Annu. Rev. Mater. Sci.* **27** 69
[87] Zorenko Y, Gorbenko V, Konstantkevych I, Grinev B and Globus M 2002 *Nucl. Instrum. Methods Phys. Res.* A **486** 309
[88] Nikl M 2006 *Meas. Sci. Technol.* **17** R37
[89] Suzuki H, Tombrello T A, Melcher C L and Schweitzer J S 1922 *Nucl. Instrum. Methods Phys. Res.* A **320** 263
[90] Cooke D W, McClellan K J, Bennett B L, Roper J M, Whittaker M T, Muenchausen R E and Sze R C 2000 *J. Appl. Phys.* **88** 7360
[91] Pauwels D, Le Masson N, Viana B, Kahn-Harari A, van Loef E V D, Dorenbos P and van Eijk C W E 2000 *IEEE Trans. Nucl. Sci.* **47** 1787
[92] Kawamura S, Higuchi M, Kaneko J H, Nishiyama S, Haruna J, Saeki S, Ueda S, Kurashige H, Ishibashi H and Furusaka M. 2009 *Cryst. Growth Des.* **9** 1470
[93] Kramer K W, Dorenbos P, Gudel H U and van Eijk C W E 2006 *J. Mater. Chem.* **6** 2773
[94] Chen H, Yang P, Zhou C, Jiang C and Pan J 2006 *Cryst. Growth Des.* **6** 809
[95] Kamada K, Yanagida T, Pejchal J, Nikl M, Endo T, Tsutumi K, Usuki Y, Fujimoto Y, Fukabori A and Yoshikawa A 2012 *J. Cryst. Growth* **352** 35
[96] Kamada K, Endo T and Tsutumi K 2011 *Cryst. Growth Des.* **11** 4484
[97] Kamada K, Yanagida T, Pejchal J, Nikl M, Endo T, Tsutumi K, Fujimoto Y, Fukabori A and Yoshikawa A 2011 *J. Phys. D: Appl. Phys.* **44** 505104
[98] Wu Y T and Ren G H 2013 *Opt. Mater.* **35** 2146
[99] Raukas M, Basun S A, van Schaik W, Yen W M and Happek U 1996 *Appl. Phys. Lett.* **69** 3300

[100] Nikl M, Vedda A, Fasoli M, Fontana I, Laguta V V, Mihokova E, Pejchal J, Rosa J and Nejezchleb K 2007 *Phys. Rev.* B **76** 195121

[101] Chewpraditkul W, Swiderski L, Moszynski M, Szczesniak T, Syntfeld-Kazuch A, Wanarak C and Limsuman P 2009 *IEEE Trans. Nucl. Sci.* **56** 3800

[102] Li X D, Sun X D, Li J-G, Xiu Z M, Gao T, Liu Y N and Hu X Z 2010 *Int. J. Appl. Ceram. Technol.* **7** E1

[103] Yanagida T *et al* 2007 *Nucl. Instrum. Methods Phys. Res.* A **579** 23

[104] Li X D, Li J-G, Xiu Z M, Huo D and Sun X D 2010 *J. Am. Ceram. Soc.* **93** 2229

[105] Cherepy N J *et al* 2009 *IEEE Trans. Nucl. Sci.* **56** 873

[106] Munoz-Garcia A B, Artacho E and Seijo L 2009 *Phys. Rev.* B **80** 014105

[107] Zorenko Y, Voznyak T, Gorbenko V V, Doroshenko A, Tolmachev A, Yavetskiy R, Petrusha I and Turkevich V 2013 *Opt. Mater.* **35** 2049

[108] Haven DT, Dickens P T, Weber M H and Lynn K G 2013 *J. Appl. Phys.* **114** 043102

10

Preferred growth orientation and microsegregation behaviors of eutectic in a nickel-based single-crystal superalloy

Fu Wang, Dexin Ma and Andreas Bührig-Polaczek

Foundry Institute, RWTH Aachen University, D-52072 Aachen, Germany

E-mail: d.ma@gi.rwth-aachen.de

Abstract

A nickel-based single-crystal superalloy was employed to investigate the preferred growth orientation behavior of the $(\gamma + \gamma')$ eutectic and the effect of these orientations on the segregation behavior. A novel solidification model for the eutectic island was proposed. At the beginning of the eutectic island's crystallization, the core directly formed from the liquid by the eutectic reaction, and then preferably grew along [100] direction. The crystallization of the eutectic along [110] always lagged behind that in [100] direction. The eutectic growth in [100] direction terminated on impinging the edge of the dendrites or another eutectic island. The end of the eutectic island's solidification terminates due to the encroachment of the eutectic liquid/solid interface at the dendrites or another eutectic island in [110] direction. The distribution of the alloying elements depended on the crystalline axis. The degree of the alloying elements' segregation was lower along [100] than [110] direction with increasing distance from the eutectic island's center.

Keywords: crystal growth, orientation, microstructure, segregation

1. Introduction

To meet the enhanced requirement of the operating temperature of turbines, advanced single-crystal Ni-based superalloys have elevated contents of refractory elements (Ta, W, and Re), which improves their high-temperature performance [1]. However, the refractory elements segregate during crystallization [2]. During solidification, the dendritic γ phase is first frozen and is mainly enriched in W and Re, whereas Al, Ta and Ti segregate into the remaining liquid which forms non-equilibrium eutectic islands in the interdendritic regions at the final stage of the solidification. The large $(\gamma + \gamma')$ eutectic island is undesirable for the as-cast microstructure in the single-crystal components because, to dissolve it, a longer solution treatment time is subsequently required.

Over the past few decades, a number of experimental and numerical modeling studies have been carried out on the solidification sequence of the $(\gamma + \gamma')$ eutectic island [2–8] as well as on the effect of processing parameters on the eutectic in Ni-based superalloys [9–13]. However, until now no work has been published on the preferential orientation growth behavior of the $(\gamma + \gamma')$ eutectic. In the present study, the preferred growth orientation behavior of the $(\gamma + \gamma')$ eutectic island is presented and analyzed in detail. In addition to this, the segregation profile is characterized as a function of its crystallographic orientation.

2. Experiments

The material used in this study was Mar-M 247 LC superalloy (Ni-8.21Cr-0.51Mo-9.57 W-9.26Co-0.7Ti-5.62Al-3.18Ta-1.48Hf-0.075C, wt%). Single-crystal cylindrical bars having a height and diameter of 150 and 20 mm, respectively, were solidified by using the selector method in an ALD Vacuum Technologies, Inc. Bridgman furnace. The shell mold was produced by a standard investment casting procedure. The

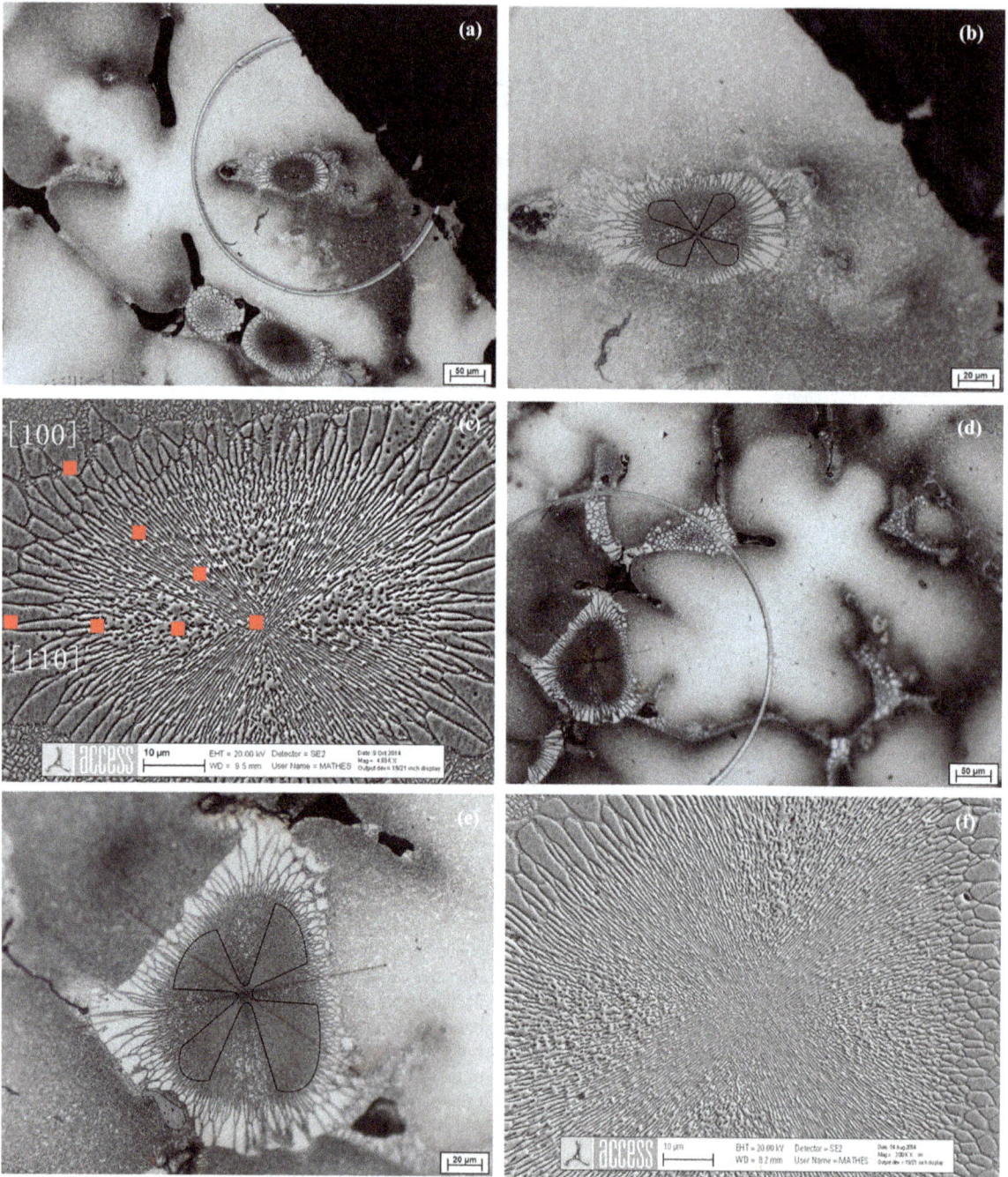

Figure 1. (a) and (d) The morphologies of cruciform dendrites and $(\gamma + \gamma')$ eutectic island; (b) and (e) magnified photos of the eutectic island; (c) and (f) SEM micrographs of the eutectic island and the measuring paths of the EDX scans along [100] and [110] directions.

wall thickness of the mold was 7–8 mm. In each shell mold cluster, the bars were assembled around a central rod. To measure the temperature development in the bars, thermocouples were placed at the top positions where the metallographic samples were taken. For the casting experiment, the shell mold cluster was placed on the copper chill plate in the Bridgman furnace and preheated, poured with superalloy melt, and then withdrawn out of the heating zone through the baffle into the cooling zone. In this experiment, a heater and pouring temperature was 1773 K (1500 °C) was used. A withdrawal rate of $V = 1 \text{ mm min}^{-1}$ was applied. The measured temperature gradient was 1.7 K mm^{-1}.

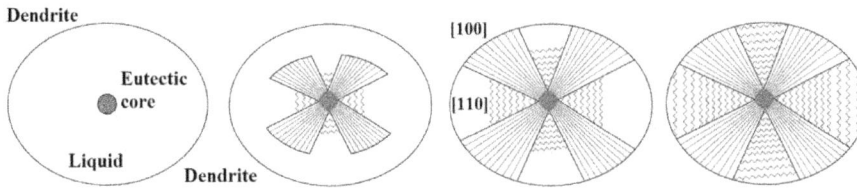

Figure 2. Solidification model of the $(\gamma+\gamma')$ eutectic island.

After solidification, the shell mold was cooled down in the furnace. When the temperature in the furnace reached under 473 K (200 °C), the vacuum was released and the casting mold was removed. The casting part was then knocked out of the mold and the bars were separated appropriately from the casting cluster. Finally, the bars were sectioned transversely at the top position (perpendicular to the [001] direction) and samples were mounted and polished for microstructural analysis. The microstructures were characterized using optical microscopy and scanning electron microscopy (SEM) after etching with 60 ml $C_2H_5OH + 40$ ml $HCl + 2$ g $(Cu_2Cl \cdot 2H_2O)$. The concentrations of the alloying elements of different microstructures were measured locally using energy dispersive x-ray spectrometer (EDX).

3. Results and discussion

Figures 1(a) and (d) show a typical dendritic microstructure and morphologies of the $(\gamma+\gamma')$ eutectic islands. Since [001] is the preferred direction for the crystal growth in cubic metals and the direction of the primary dendrite arm is [001], the transverse sections of the dendrite are cruciform with orthogonal secondary dendrite arms aligned in [100] and [010] directions. The details of the $(\gamma+\gamma')$ island shown in figures 1(b) (c) (e) and (f) demonstrate different microstructures depending on the crystallographic orientation.

A regularly arranged $(\gamma+\gamma')$ microstructure was observed in [100] direction, and γ and γ' phases were approximately parallel to [100] direction, whereas a disorderly structure was found in [110] direction. This suggests that anisotropy also exists in the growth of the eutectic. Both γ and γ' phases have a face-centered cubic structure and the preferred direction is also [100]. Therefore, an orthogonal cruciform of the transverse sections of the $(\gamma+\gamma')$ eutectic similar to the dendrite was exhibited in figures 1(b) and (e). The difference in the morphology of the eutectic between [100] and [110] directions depends on the sectioned crystallographic orientation and plane. In this analysis, the sample was sectioned and examined on the (001) plane in which the eutectic lamellar structure along the [100] direction was relatively orderly arranged. In addition to this, it can be seen that the $(\gamma+\gamma')$ eutectic's microstructure gradually diverges from the center to the periphery.

Based on the examination and the previous studies [6, 14], a novel formation model of the eutectic island can be depicted in figure 2. On commencing, the $(\gamma+\gamma')$ eutectic core

forms directly from the remaining liquid by eutectic reaction. At this stage, the anisotropy has no effect on the formation of the core. As this reaction proceeds, the eutectic prefers to grow in [100] direction, and the advancing velocity of the solidification front is the higher than [110] direction. When the liquid/solid interface of the eutectic in [100] direction impinges on the peripheral dendrites or another eutectic island, and there is no more free space for further growth, the eutectic reaction will cease in this direction, and the eutectic growth continues in [110] direction until the end of solidification. That is, the eutectic growth in [110] direction always lags behind that in [100] direction. Previous studies [15, 16] suggest that a long-range solute diffusion boundary layer established ahead of the eutectic solid/liquid interface may destabilize the morphology of the eutectic interface as a whole. During the solidification of the $(\gamma+\gamma')$ eutectic, owing to the crystallization of γ dendrite around the eutectic, the positive segregation elements such as Al, Ta, Ti and Hf are rejected in the interdendritic remaining liquid, and some of them can enrich ahead of the eutectic interface by a long-distance diffusion. Consequently, a compositional gradient of these elements perpendicular to the interface is built as shown in figure 3. In the present investigation, the longitudinal compositional gradient causes a concave eutectic liquid/solid interface (isotherm) to the remaining liquid. In addition to this, due to the formation of the $(\gamma+\gamma')$ core, the latent heat is gradually released, which causes a reduction in the nucleating undercooling and retards the eutectic reaction. The γ' and γ lamellas have sufficient time to thicken in the final solidification, which will gradually widen the eutectic liquid/solid interface. Since the growth directions of the eutectic lamellas are usually vertical to the isotherm, the eutectic lamella emanates with increasing distance from its center.

As indicted by the red squares in figure 1(c), the concentration of the alloying elements was measured along both [100] and [110] directions in the $(\gamma+\gamma')$ eutectic island. In figure 2 the measured solute profiles are plotted versus the distance from the eutectic center, along [100] and [110] directions, respectively. According to the sizes in both of the $(\gamma+\gamma')$ eutectic island's directions, two different examined lengths (shown in figure 1(c)) were chosen. These were 39.5 μm (L_1) and 40 μm (L_2) for [100] and [110] directions, respectively. To plot the results onto one graph for the same alloying element in both directions, and facilitate the comparison, the center of the eutectic island was set as zero, and the distance X to each examined point were then normalized

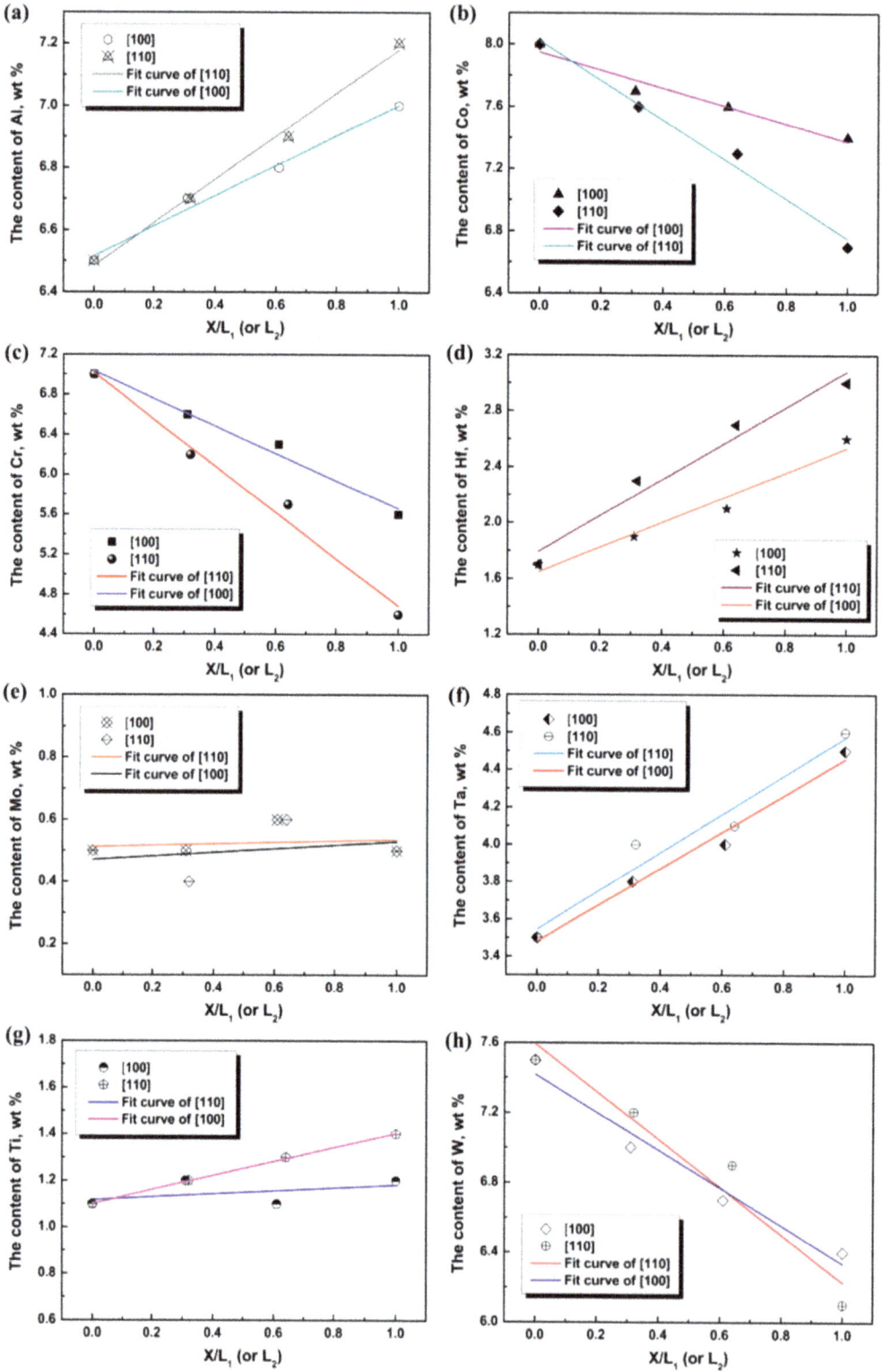

Figure 3. Measured solute distribution from the center of the $(\gamma + \gamma')$ eutectic island, along [100] and [110] directions.

with respect to L_1 or L_2. These dimensionless numbers were used as the abscissa.

Figure 3 shows that the concentrations of Al, Hf, Ta and Ti increase with increasing distance from the eutectic center along [100] and [110] directions, but reduction trends are observed for Co, Cr and W. The concentration of Mo is approximately constant. In addition to this, compared to [110] direction, the concentration gradient of Al, Co, Cr, Hf, W and Ti is smaller in [100] direction. Similar change is found for Ta along the both directions. It indicates that the solute inhomogeneity along [110] direction is more pronounced than that along [100] direction.

In the superalloy system, Al, Ta, Ti, and Hf segregate to the interdendritic regions during solidification, while Co, Cr and W segregate to the dendrite core. Mo does not consistently segregate to either side [17, 18]. At the beginning of solidification, Al, Ta, Ti, and Hf require long-distance diffusion to segregate to the center of the interdendritic region where the eutectic core forms. For this reason, the concentrations of these elements are small in the early crystallized eutectic core. As the solidification proceeds, the eutectic growth approaches the edge of dendrites. A short diffusion distance leads to increasing concentrations of these alloying elements. On the other hand, owing to the thickening of γ dendrite, Co, Cr and W are depleted. The concentrations of these elements decrease with increasing distance from the center of the eutectic. In addition to this, due to the lagging eutectic growth in [110] direction, the remaining liquid contains more segregated elements for forming the eutectic in [110] direction. Consequently, the degree of segregation along [100] direction is lower than that along [110] direction. These segregation profiles provide evidence for the preferred growth model of the $(\gamma + \gamma')$ eutectic islands, as stated above.

4. Conclusions

In this paper, a preferred growth orientation behavior of the $(\gamma + \gamma')$ eutectic and the effect of these orientations on the segregation were investigated. At the beginning of the eutectic's crystallization, the core directly formed from the liquid by the eutectic reaction, and then preferably grew along [100] direction. The crystallization of the eutectic along [110] always lagged behind that in [100] direction. The eutectic growth in [100] direction terminated on impinging the edge of the dendrites or another eutectic island. Eutectic solidification stopped due to the encroachment of the eutectic liquid/solid interface at the dendrites or another eutectic island in the [110] direction. The degree of the alloying elements'

segregation was lower along [100] direction than that along [110] direction.

Acknowledgments

This research was supported by German Research Foundation through Grant No. MA 2505/3-1. One of the authors (FW) would like to acknowledge the China Scholarship Council for supporting his stays in Germany. The authors would like to acknowledge the aid of Maria Schaarschmidt, Elke Schaberger-Zimmermann and Elke Breuer.

References

[1] Yeh A and Tin S 2005 *Scr. Mater.* **52** 519
[2] Long F, Yoo Y S, Seo S M, Jin T, Hu Z Q and Jo C Y 2011 *J. Mater. Sci. Technol.* **27** 101
[3] D'Souza N and Dong H B 2008 *Superalloys 2008* ed R C Reed, K A Green, P Caron, T P Gabb, M G Fahrmann, E S Huron and S A Woodard (TMS) p 261
[4] Zhu Y X, Zhang S N, Xu L Y, Bi J, Hu Z Q and Shi C X 1988 *Superalloy 1988* ed S Reichman, D N Duhl, G Maurer, S Antolovich and C Lund (TMS) p 703
[5] D'Souza N and Hong H B 2007 *Scr. Mater.* **56** 41
[6] Heckel A, Rettig R, Cenanovic S, Göken M and Singer R F 2010 *J. Cryst. Growth* **312** 2137
[7] Pang H T, Dong H B, Beanland R, Stone H J, Rae C M F, Midgley P A, Brewster G and D'Souza N 2009 *Metall. Mater. Trans.* A **40** 1660
[8] Seo S M, Lee J H, Yoo Y S, Jo C Y, Miyahara H and Ogi K 2008 *Superalloys 2008* ed R C Reed, K A Green, P Caron, T P Gabb, M G Fahrmann, E S Huron and S A Woodard (TMS) p 277
[9] Seo S M, Lee J H, Yoo Y S, Jo C Y, Miyahara H and Ogi K 2011 *Metall. Mater. Trans.* A **42** 3150
[10] Wang F, Ma D, Zhang J, Liu L, Bogner S and Bührig-Polaczek A 2014 *J. Alloy. Compd.* **616** 102
[11] Zhang J, Li J, Jin T, Sun X and Hu Z 2010 *Adv. Mater. Res.* **97-101** 1016
[12] Liu G, Liu L, Ai C, Ge B, Zhang J and Fu H 2011 *J. Alloy. Compd.* **509** 5866
[13] Wang F, Ma D, Zhang J, Liu L, Hong J, Bogner S and Bührig-Polaczek A 2014 *J. Cryst. Growth* **389** 47
[14] Wang F, Ma D, Zhang J, Bogner S and Bührig-Polaczek A 2015 *Mater. Charact.* **101** 20
[15] Zhao S, Li J F, Liu L and Hou Y H 2009 *Mater. Charact.* **60** 519
[16] Kurz W and Fisher D J 1998 *Fundamentals of Solidification* 4th edn (Aedermansdorf: Trans. Tech. Publications Ltd) p 108
[17] Zeisler-Mashl K L and Pletka B J 1992 *Superalloys 1992* ed S D Antolovich, R W Stusrud, R A MacKay, D L Anton, T Khan, R D Kissinger and D L Klarstrom (TMS) p 175
[18] Wang F, Ma D X, Zhang J, Bogner S and Bührig-Polaczek A 2014 *J. Mater. Process. Technol.* **214** 3112

11

Combination of supported bimetallic rhodium–molybdenum catalyst and cerium oxide for hydrogenation of amide

Yoshinao Nakagawa, Riku Tamura, Masazumi Tamura and
Keiichi Tomishige

Department of Applied Chemistry, School of Engineering, Tohoku University, 6-6-07, Aoba, Aramaki,
Aoba-ku, Sendai 980-8579, Japan

E-mail: yoshinao@erec.che.tohoku.ac.jp and tomi@erec.che.tohoku.ac.jp

Abstract

Hydrogenation of cyclohexanecarboxamide to aminomethylcyclohexane was conducted with silica-supported bimetallic catalysts composed of noble metal and group 6–7 elements. The combination of rhodium and molybdenum with molar ratio of 1:1 showed the highest activity. The effect of addition of various metal oxides was investigated on the catalysis of $Rh–MoO_x/SiO_2$, and the addition of CeO_2 much increased the activity and selectivity. Higher hydrogen pressure and higher reaction temperature in the tested range of 2–8 MPa and 393–433 K, respectively, were favorable in view of both activity and selectivity. The highest yield of aminomethylcyclohexane obtained over $Rh–MoO_x/SiO_2 + CeO_2$ was 63%. The effect of CeO_2 addition was highest when CeO_2 was not calcined, and CeO_2 calcined at >773 K showed a smaller effect. The use of CeO_2 as a support rather decreased the activity in comparison with $Rh–MoO_x/SiO_2$. The weakly-basic nature of CeO_2 additive can affect the surface structure of $Rh–MoO_x/SiO_2$, i.e. reducing the ratio of $Mo–OH/Mo–O^-$ sites.

Keywords: heterogeneous catalysis, rhodium, ceria, amide, amine

1. Introduction

Heterogeneous catalysis is one of the important applications of inorganic materials [1, 2]. Reduction reactions are one class of target reactions for heterogeneous catalysis and are widely used in both laboratory-scale organic synthesis and industrial processes [3–5]. Hydrogenation of amides produces amines that have been used in various fields such as pharmaceutical industries (equation (1)) [6, 7]. However, this reaction is rather difficult than other reduction reactions of carbonyl compounds, because the π-electrons in the carbonyl groups of amides are stabilized by the conjugation with nitrogen (equation (2)). Possible formation of various by-products such as alcohols, ammonia and secondary amines is another difficulty.

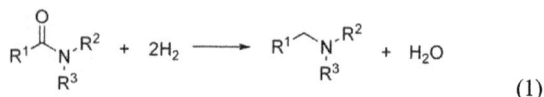

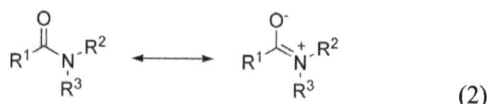

$$R^1\overset{\displaystyle O}{\underset{\underset{\displaystyle R^3}{|}}{\overset{||}{C}}-N-R^2} + 2H_2 \longrightarrow R^1\overset{R^2}{\underset{\underset{\displaystyle R^3}{|}}{N}} + H_2O \tag{1}$$

$$R^1\overset{\displaystyle O}{\underset{\underset{\displaystyle R^3}{|}}{\overset{||}{C}}-N-R^2} \longleftrightarrow R^1\overset{\displaystyle O^-}{\underset{\underset{\displaystyle R^3}{|}}{\overset{||}{C}=\overset{+}{N}-R^2}} \tag{2}$$

Conventionally, reduction of amides has been conducted non-catalytically with strongly reductive reagents such as $LiAlH_4$. Problems with this conventional method include the difficult handling and cost of the reactive reagents and complex workup. Therefore, development of catalytic systems for hydrogenation of amides with molecular hydrogen has been intensively carried out. Homogeneous systems using Ru complex catalysts have been reported to be effective [8–12]; however, homogeneous systems have difficulty in removal of catalyst from the reaction mixture after use. Heterogeneous catalysts are favorable in view of workup. Several research groups have reported that unsupported bimetallic catalysts which consist of noble metal and group 6 or 7 elements are effective such as Rh–Mo, Rh–Re, Ru–Re and Pt–Re [13–16]. However, reports of effective supported catalysts are limited [13, 17–19], while for other reduction reactions supported catalysts are more common than unsupported ones. Recently, we have showed that silica- or carbon-supported bimetallic catalysts composed of noble metal and group 6–7 elements are very effective for various reduction reactions such as hydrogenation of unsaturated aldehydes [20–22], hydrogenolysis of poly-alcohols [23–27] and hydrogenation of carboxylic acids [28–30]. We have also showed that the performance of some catalysts in this category is affected by addition of solid metal oxide in the reaction media [31–33]: for example, hydrogenolysis activity of Ir–ReO$_x$/SiO$_2$ catalyst is promoted by addition of solid acid such as H-ZSM-5 zeolite or silica-alumina and decreased by addition of basic oxides such as CeO$_2$ [31]. In this study, we applied various silica-supported bimetallic catalysts combined with metal oxides to hydrogenation of amide. We found that the addition of CeO$_2$ much increases the activity of Rh–MoO$_x$/SiO$_2$ catalyst.

2. Experimental

M^1–M^2O$_x$/SiO$_2$ catalysts (M^1 = noble metal; M^2 = Mo, W and Re) were prepared by sequential impregnation method as reported previously [23–26]. First, M^1/SiO$_2$ catalysts were prepared by impregnating SiO$_2$ (Fuji Silysia G-6; BET surface area 535 m^2 g^{-1}) with an aqueous solution of noble metal precursor (RhCl$_3$ · 3H$_2$O, H$_2$PtCl$_6$ · 6H$_2$O, RuCl$_3$ · nH$_2$O, PdCl$_2$ and H$_2$IrCl$_6$). The loading amount of M^1 was 4 wt%. After impregnation, they were dried at 383 K overnight. And then the second impregnation was conducted with an aqueous solution of M^2 precursor ((NH$_4$)$_6$Mo$_7$O$_{24}$ · 4H$_2$O, (NH$_4$)$_{10}$W$_{12}$O$_{41}$ · 5H$_2$O and NH$_4$ReO$_4$) to prepare M^1–M^2O$_x$/SiO$_2$. The loading amount of M^2 was set to M^2/M^1 = 1 in molar basis unless noted. After impregnation, the bimetallic catalysts were dried at 383 K overnight and calcined at 773 K for 3 h. Monometallic catalysts were also calcined at 773 K for 3 h when used for catalytic reaction.

Activity tests were performed in a 190 mL stainless steel autoclave with an inserted glass vessel. Typically, catalyst (100 mg), cyclohexanecarboxamide (0.25 g; 2 mmol), 1,2-dimethoxyethane (solvent, 20 g) and CeO$_2$ (Daiichi Kigenso HS, 120 m^2 g^{-1}; 100 mg) were put into an autoclave together with a spinner. After sealing the reactor, the air content was quickly purged by flushing three times with 1 MPa hydrogen. The autoclave was then heated to reaction temperature (typically 413 K), and the temperature was monitored using a

thermocouple inserted in the autoclave. After the temperature reached the desired one, the H$_2$ pressure was increased to set value (typically 8 MPa). During the experiment, the stirring rate was fixed at 500 rpm (magnetic stirring). After appropriate reaction time (typically 4 or 24 h), the autoclave was quickly cooled to room temperature, and the gases were collected in a gas bag. n-dodecane (0.1 mL) was added to the liquid content as an internal standard material, and the catalyst was separated by filtration. The products in gas and liquid phases were analyzed with GC and GC-MS. A CP-Sil-5 capillary column was used for separation. The formation of gaseous products was always negligible. Selectivities were calculated based on the number of carbon atoms. The reproducibility of carbon balance in different runs with the same conditions was ±3%. The loss of carbon balance was included to 'others'. Other metal oxides used instead of CeO$_2$ were ZrO$_2$ (Daiichi Kigenso; 88 m^2 g^{-1}), TiO$_2$ (AEROXIDE P25; 50 m^2 g^{-1}), γ-Al$_2$O$_3$ (Sumitomo KHO-24; 140 m^2 g^{-1}), MgO (Ube 500 A; 33 m^2 g^{-1}), SiO$_2$–Al$_2$O$_3$ (JGC C&C and Catalysis Society of Japan, JRC-SAL-3; 504 m^2 g^{-1}), and H-ZSM-5 (Süd Chemie and Catalysis Society of Japan, JRC-25–90 H; 325 m^2 g^{-1}).

3. Results and discussion

3.1. Hydrogenation of cyclohexanecarboxamide over various catalysts

First, we applied various silica-supported bimetallic catalysts to hydrogenation of cyclohexanecarboxamide (CyCONH$_2$) (table 1). We chose cyclohexanecarboxamide as a representative substrate of primary amide [14, 18], and the target product of this reaction is aminomethylcyclohexane (CyCH$_2$NH$_2$). By-products include cyclohexanemethanol (CyCH$_2$OH) which can be formed by C–N dissociation of amide, cyclohexanecarboxylic acid (CyCOOH) which is produced by hydrolysis of cyclohexanecarboxamide, and bis(cyclohexylmethyl)amine ((CyCH$_2$)$_2$NH; secondary amine). The formation mechanism of bis(cyclohexylmethyl)amine is discussed in section 3.5. Significant loss of carbon balance was observed in many cases. We included the loss to the selectivity to 'others' because TG analysis confirmed the deposition of organic material on the catalyst. Rh–MoO$_x$/SiO$_2$ showed the highest activity and selectivity to aminomethylcyclohexane in M^1–MoO$_x$/SiO$_2$ catalysts (M^1 = noble metal) and Rh–M^2O$_x$/SiO$_2$ catalysts (M^2 = Mo, W and Re). Monometallic Rh/SiO$_2$ and MoO$_x$/SiO$_2$ catalysts showed almost no activity in amine formation. The effect of Mo addition to Rh/SiO$_2$ catalyst is more evident than in the reported case of unsupported Rh–Mo catalysts where monometallic Rh catalyst shows some activity [13]. Among Rh–MoO$_x$/SiO$_2$ catalysts with different Mo/Rh ratios, the catalyst with Mo/Rh = 1 showed the highest activity. The catalysts with lower Mo amount showed higher selectivity to secondary amine in addition to lower activity. This activity trend is different from that of the same catalysts in C–O hydrogenolysis [24, 25, 34] and amino acid hydrogenation [29].

Table 1. Hydrogenation of cyclohexanecarboxamide over various catalysts[a].

Entry	Catalyst	Molar ratio of M^2/noble metal	Conv. (%)	Selectivity (%)				
				$CyCH_2NH_2$	$CyCH_2OH$	$(CyCH_2)_2NH$	$CyCOOH$	Others
1	Rh–MoO$_x$/SiO$_2$	1	74	43	6	30	<1	21[b]
2	Pt–MoO$_x$/SiO$_2$	1	<1	—	—	—	—	—
3	Ru–MoO$_x$/SiO$_2$	1	1[c]	~15	~30	~55	<5	<5
4	Pd–MoO$_x$/SiO$_2$	1	1[c]	~10	~30	~60	<5	<5
5	Ir–MoO$_x$/SiO$_2$	1	12	<1	<1	<1	<1	>99[b]
6	Rh–WO$_x$/SiO$_2$	1	3[c]	~20	~5	~70	<5	<5
7	Rh–ReO$_x$/SiO$_2$	1	20	47	4	39	<1	10[b]
8	Rh–MoO$_x$/SiO$_2$	0.25	29	23	5	50	<1	21[b]
9	Rh–MoO$_x$/SiO$_2$	0.5	58	41	5	36	<1	19[b]
10	Rh–MoO$_x$/SiO$_2$	2	67	44	8	28	<1	21[b]
11	Rh/SiO$_2$	0	2[c]	~15	~30	~55	<5	<5
12	MoO$_x$/SiO$_2$[d]	—	24	<1	<1	<1	<1	>99[b]

[a] Reaction conditions: catalyst 100 mg (noble metal 4 wt%), CyCONH$_2$ 0.25 g (2 mmol), 1,2-dimethoxyethane 20 g, H$_2$ 8 MPa, 413 K, 24 h. Cy = cyclohexyl.
[b] Representing loss of carbon balance predominantly via formation of solid products on the catalyst surface.
[c] Selectivities in these entries are nominal ones as the low conversions preclude obtained data from being comparable with other entries.
[d] Mo 3.7 wt%.

Table 2. Hydrogenation of cyclohexanecarboxamide over Rh–MoO$_x$/SiO$_2$ catalyst + various metal oxides[a].

Entry	Metal oxide	Conv. (%)	Selectivity (%)				
			$CyCH_2NH_2$	$CyCH_2OH$	$(CyCH_2)_2NH$	$CyCOOH$	Others[b]
1	CeO$_2$	89	62	6	5	<1	27
2	ZrO$_2$	26	21	5	11	26	37
3	TiO$_2$	24	21	5	15	20	40
4	γ-Al$_2$O$_3$	61	53	4	8	<1	35
5	MgO	20	39	4	15	<1	42
6	SiO$_2$–Al$_2$O$_3$	27	28	5	14	<1	53
7	H-ZSM-5	34	17	3	11	<1	69
8	None	18	49	7	26	<1	19

[a] Reaction conditions: Rh–MoO$_x$/SiO$_2$ 100 mg (Rh 4 wt%, Mo/Rh = 1), CyCONH$_2$ 0.25 g (2 mmol), metal oxide 100 mg, 1,2-dimethoxyethane 20 g, H$_2$ 8 MPa, 413 K, 4 h. Cy = cyclohexyl.
[b] Representing loss of carbon balance predominantly via formation of solid products on the catalyst surface.

The catalyst with Mo/Rh = 1/8 shows the highest activity in C–O hydrogenolysis and amino acid hydrogenation. There may be difference in the active sites between amide hydrogenation and these reactions. We selected Rh–MoO$_x$/SiO$_2$ (Mo/Rh = 1) in the following studies.

3.2. Addition of metal oxides to the catalytic system of Rh–MoO$_x$/SiO$_2$

We investigated the effects of addition of metal oxides on the catalysis of Rh–MoO$_x$/SiO$_2$ (Mo/Rh = 1). The results are shown in table 2. The reaction time was set to be shorter than table 1 to compare the activities. The selectivities of Rh–MoO$_x$/SiO$_2$ were almost the same at different reaction times (table 1, entry 1; table 2, entry 8). Addition of weakly basic CeO$_2$ and γ-Al$_2$O$_3$ increased the activity (conversion of

substrate) and selectivity to $CyCH_2NH_2$. The formation of secondary amine (($CyCH_2)_2NH$) was significantly suppressed by the addition. The addition of CeO$_2$ showed the best effect. On the other hand, acidic additives such as H-ZSM-5, silica-alumina, ZrO$_2$ and TiO$_2$ showed little effect on the conversion and much decreased the selectivity to $CyCH_2NH_2$. Strongly basic MgO also decreased the selectivity to $CyCH_2NH_2$ and had little effect on the conversion. The selectivity to 'others', which comprised solid polymerized products deposited on the catalyst, was increased by addition of acidic or strongly basic oxides.

Figure 1 shows the time course of hydrogenation of CyCONH$_2$ over Rh–MoO$_x$/SiO$_2$ in combination with CeO$_2$. The selectivities were hardly changed until the total conversion of CyCONH$_2$, and then the selectivity to $CyCH_2NH_2$ was gradually decreased and that to ($CyCH_2)_2NH$ was

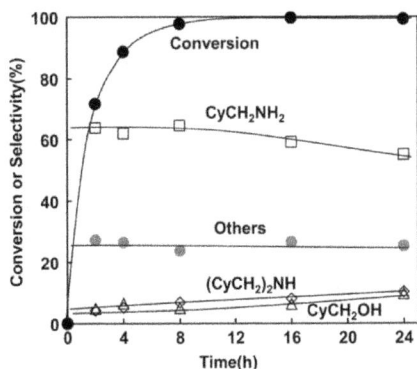

Figure 1. Time course of hydrogenation of cyclohexanecarboxamide (CyCONH$_2$) over Rh–MoO$_x$/SiO$_2$ + CeO$_2$. Reaction conditions: Rh–MoO$_x$/SiO$_2$ (Rh 4 wt%, Mo/Rh = 1) 100 mg, CeO$_2$ (uncalcined) 100 mg, 1,2-dimethoxyethane 20 g, H$_2$ 8 MPa, 413 K. Cy = cyclohexyl. 'Others' comprise unknown solid products leading to loss of carbon balance during catalysis.

gradually increased. The highest yield of CyCH$_2$NH$_2$ was 63% obtained at 8 h (equation (3)). Although the yield value was lower than that over unsupported Rh/Mo catalyst in the literature (87%), the activity was significantly higher (CyCONH$_2$/Rh$_{total}$ = 50 at 413 K, 8 h in this study; CyCONH$_2$/Rh$_{total}$ = 20 at 433 K, 16 h in the literature [13]).

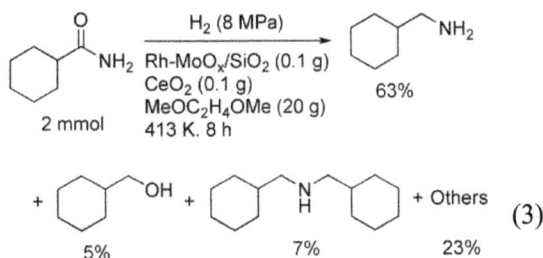

$$(3)$$

The life of the catalyst is also an issue. The reusability of Rh–Mo catalysts has been reported for reduction reactions [13, 24, 30], and good stability in the structure has been observed by XRD and EXAFS analyses [24, 30]. However, the deposition of organic material on the catalyst in this system clearly limits the long-term use. The development of effective regeneration method without aggregation of active metal particles is a target of further study.

3.3. Effect of amount and type of CeO$_2$ additive

The effect of amount of CeO$_2$ additive on the catalysis is shown in figure 2. The activity was increased with increasing CeO$_2$ amount; however, too much amount of CeO$_2$ decreased the selectivity to CyCH$_2$NH$_2$ and increased the selectivity to unknown by-products. 50–100 mg of CeO$_2$ was the best amount as additive to Rh–MoO$_x$/SiO$_2$, and we used 100 mg of CeO$_2$ in the other parts of this study.

While we used commercial CeO$_2$ without calcination pretreatment, it is well known that the crystallinity and the surface area of CeO$_2$ can be changed by calcination

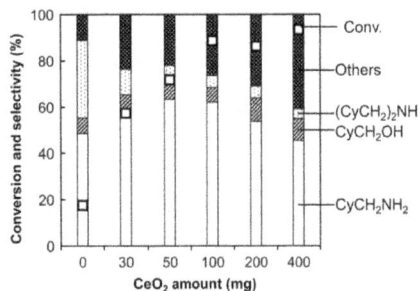

Figure 2. Hydrogenation of cyclohexanecarboxamide (CyCONH$_2$) over Rh–MoO$_x$/SiO$_2$ + various amounts of CeO$_2$. Reaction conditions: Rh–MoO$_x$/SiO$_2$ (Rh 4 wt%, Mo/Rh = 1) 100 mg, CeO$_2$ (uncalcined) 0–400 mg, 1,2-dimethoxyethane 20 g, H$_2$ 8 MPa, 413 K, 4 h. Cy = cyclohexyl. 'Others' comprise unknown solid products leading to loss of carbon balance during catalysis.

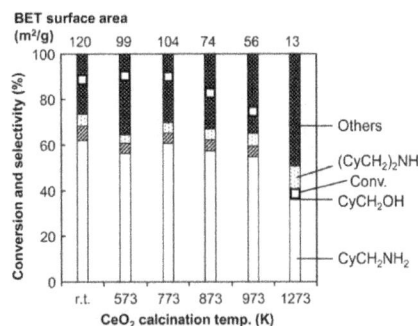

Figure 3. Hydrogenation of cyclohexanecarboxamide (CyCONH$_2$) over Rh–MoO$_x$/SiO$_2$ + calcined CeO$_2$. Reaction conditions: Rh–MoO$_x$/SiO$_2$ (Rh 4 wt%, Mo/Rh = 1) 100 mg, CeO$_2$ 100 mg, 1,2-dimethoxyethane 20 g, H$_2$ 8 MPa, 413 K, 4 h. Cy = cyclohexyl. 'Others' comprise unknown solid products leading to loss of carbon balance during catalysis. 'r.t.' stands for room temperature.

pretreatment [35, 36]. The surface area is reduced by calcination at higher temperature, and the surface of CeO$_2$ samples without calcination or calcined at lower temperature (<873 K) is partly amorphous [37]. Indeed, we have used CeO$_2$ catalysts after calcination at different temperatures for various CO$_2$ utilization reactions such as carbonate synthesis, and we have found that CeO$_2$ after 873 K calcination shows the highest activity probably because crystalline CeO$_2$ surface is the active site [37–40]. Figure 3 shows the results of hydrogenation of CyCONH$_2$ over Rh–MoO$_x$/SiO$_2$ and CeO$_2$ calcined at various temperatures. The addition effect of CeO$_2$ was highest when CeO$_2$ was not calcined or calcined at <773 K, and the effect became smaller when CeO$_2$ was calcined at higher temperature. This behavior shows that the addition effect was mostly determined by the surface area.

We also prepared CeO$_2$-supported Rh–MoO$_x$ catalyst. However, the catalytic activity was even lower than Rh–MoO$_x$/SiO$_2$ without CeO$_2$ addition, although the dimerization side-reaction was surely suppressed similarly to external addition of CeO$_2$ (figure 4). These data suggest that the direct interaction between CeO$_2$ and Rh (or Mo) is not important in the catalysis.

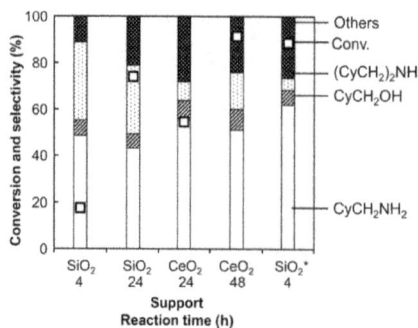

Figure 4. Hydrogenation of cyclohexanecarboxamide (CyCONH$_2$) over Rh–MoO$_x$ catalyst with different supports. Reaction conditions: Rh–MoO$_x$/support (Rh 4 wt%, Mo/Rh = 1) 100 mg, 1,2-dimethoxyethane 20 g, H$_2$ 8 MPa, 413 K, 4–48 h. Cy = cyclohexyl. 'Others' comprise unknown solid products leading to loss of carbon balance during catalysis. *: CeO$_2$ (100 mg).

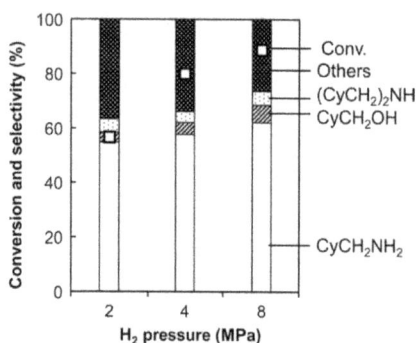

Figure 6. Effect of reaction temperature on hydrogenation of cyclohexanecarboxamide (CyCONH$_2$) over Rh–MoO$_x$ catalyst + CeO$_2$. Reaction conditions: Rh–MoO$_x$/SiO$_2$ (Rh 4 wt%, Mo/Rh = 1) 100 mg, CeO$_2$ (uncalcined) 100 mg, 1,2-dimethoxyethane 20 g, H$_2$ 8 MPa, 393–433 K, 4 h. Cy = cyclohexyl. 'Others' comprise unknown solid products leading to loss of carbon balance during catalysis. *: half amount of catalyst and CeO$_2$ was used (50 mg each).

After all, higher hydrogen pressure and higher reaction temperature are favorable in this reaction.

3.5. Reaction mechanism

We have characterized Rh–MoO$_x$/SiO$_2$ catalysts with various Mo/Rh ratios (0.13–0.5) in the previous papers [24, 34], where the catalysts were used for C–O hydrogenolysis reactions. According to the data of temperature-programmed reduction, CO adsorption, XRD and XAFS, the catalyst with larger Mo amount contains Rh metal particles with size of ~3 nm, and MoO$_x$ species with average valence of around 4 are present on the surface of Rh metal particles under reductive conditions. It should be noted that we obtained essentially the same characterization results for different lots of Rh–MoO$_x$/SiO$_2$ (Mo/Rh = 1/8) catalysts [24, 30, 34], suggesting the good reproducibility in preparation of Rh–MoO$_x$/SiO$_2$ catalysts. Similar structures of unsupported Rh/Mo catalysts have been reported in the literature [13]: aggregates of Rh metal particles with the size of 2–4 nm and molybdenum oxide species whose valence is predominantly 4. The reaction mechanism over Rh–MoO$_x$/SiO$_2$ catalyst can be the same as that over unsupported Rh/Mo catalysts.

Several literature studies [6] proposed the reaction mechanism of hydrogenation of amides over bimetallic catalysts as follows: first, the carbonyl group of the amide is hydrogenated (equation (4)), and then dehydration occurs to form imine intermediate (equation (5)). Hydrogenation of imine gives amine product (equation (6)). The hydrogenation of deactivated carbonyl group (equation (4)) is the rate-determining step.

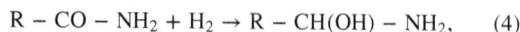

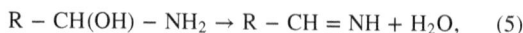

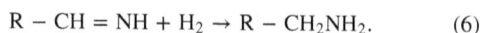

Figure 5. Effect of H$_2$ pressure on hydrogenation of cyclohexanecarboxamide (CyCONH$_2$) over Rh–MoO$_x$ catalyst + CeO$_2$. Reaction conditions: Rh–MoO$_x$/SiO$_2$ (Rh 4 wt%, Mo/Rh = 1) 100 mg, CeO$_2$ (uncalcined) 100 mg, 1,2-dimethoxyethane 20 g, H$_2$ 2–8 MPa, 413 K, 4 h. Cy = cyclohexyl. 'Others' comprise unknown solid products leading to loss of carbon balance during catalysis.

3.4. Effect of reaction conditions

The effect of hydrogen pressure on the catalysis of Rh–MoO$_x$/SiO$_2$ + CeO$_2$ was examined (figure 5). It should be noted that comparison of selectivities at different conversion level is possible because selectivities are hardly changed on reaction time until complete conversion (figure 1). Higher activity was observed under higher hydrogen pressure. The selectivity to CyCH$_2$NH$_2$ was also slightly increased with increasing hydrogen pressure, and instead the formation of unknown by-products was suppressed.

Figure 6 shows the effect of reaction temperature. Higher temperature increased the activity. The selectivity to CyCH$_2$NH$_2$ became slightly higher with increasing the temperature, as clearly seen up to 423 K. At 433 K, the conversion level was too high to compare selectivity in the standard reaction conditions. Therefore we further conducted reaction tests at 433 K (and 413 K for comparison) with smaller amount of Rh–MoO$_x$/SiO$_2$ catalyst and CeO$_2$. The selectivity to aminomethylcyclohexane was higher at 433 K than 413 K.

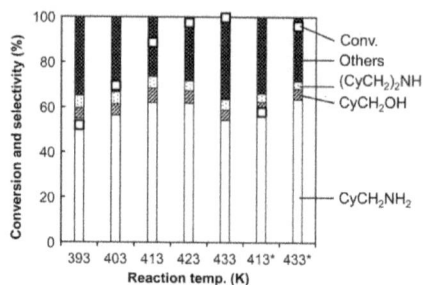

$$R - CO - NH_2 + H_2 \rightarrow R - CH(OH) - NH_2, \quad (4)$$

$$R - CH(OH) - NH_2 \rightarrow R - CH = NH + H_2O, \quad (5)$$

$$R - CH = NH + H_2 \rightarrow R - CH_2NH_2. \quad (6)$$

There is another reaction mechanism proposed in the literature: first amide is dehydrated to form nitrile (equation (7)), and then nitrile is hydrogenated to amine (equation (8)) [6, 14].

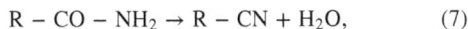

$$R - CO - NH_2 \rightarrow R - CN + H_2O, \quad (7)$$

$$R - CN + 2H_2 \rightarrow R - CH_2NH_2. \quad (8)$$

The present data agree with the former mechanism with imine intermediate (equations (4)–(6)). The positive reaction order with respect to hydrogen pressure (figure 5) indicates that the rate-determining step involves hydrogen species. For the mechanism with imine intermediate, the reaction order corresponded with that the step of equation (4) is rate determining. On the other hand, for the mechanism with nitrile intermediate (equations (7) and (8)), the reaction order means that the dehydration step (equation (7)) is fast. However, cyclohexaneacetonitrile, which is the dehydration product of cyclohexanecarboxamide, was not detected in the hydrogenation of cyclohexanecarboxamide. Although the concurrent participation of both mechanisms is not ruled out, the main reaction route should be the former mechanism.

Now we discuss the mechanism of addition effects of CeO_2. As shown in section 3.2, two promoting effects were present: increase in the catalytic activity (substrate conversion) and increase in the selectivity to primary amine (target product). The former effect can be explained by the increase of the number of active site. According to the reported density functional calculation for $Pt–ReO_x/TiO_2$-catalyzed hydrogenation of amide [17], the amide substrate is first adsorbed on the Re^{n+} center with the carbonyl group, and then the carbonyl group is reduced. In contrast, as shown in our previous papers, the active sites of $M^1–M^2O_x/SiO_2$ catalysts ($M^1 = $ Rh, Ir; $M^2 = $ Mo, Re) for activation of alcohols in C–O hydrogenolysis are M^2–OH sites [25, 26, 41–44], and the addition of solid acid to $Ir–ReO_x/SiO_2$ increases the number of Re–OH sites by protonation of Re–O^- [31]. The addition of solid base (CeO_2) to $Rh–MoO_x/SiO_2$ may well decrease the number of acidic Mo–OH sites (equation (9)).

$$Mo - OH + Ce - O \rightleftarrows Mo - O^- + Ce - OH^+. \quad (9)$$

The Mo–OH site has Brønsted acidity and thus the amide substrate can be adsorbed on the proton rather than the Mo^{4+} center which activates carbonyl group. Therefore, the addition of CeO_2 to $Rh–MoO_x/SiO_2$ can increase the number of site for activation of carbonyl group of amide to increase the activity.

The latter effect (increase in the selectivity) is accompanied by the suppression of secondary amine formation. According to the literature, secondary amine is mainly produced by addition reaction of imine intermediate with primary amine (equation (10)) and addition of ammonia to the reaction media is effective to suppress secondary amine formation [13]. The reaction of equation (10) competes with the hydrogenation of imine (equation (6)). Suppression of the reaction of equation (10) and/or promotion of the reaction of equation (6) increases the selectivity to primary amine.

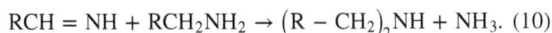

$$RCH = NH + RCH_2NH_2 \rightarrow (R - CH_2)_2NH + NH_3. \quad (10)$$

In the present system, ammonia was generated by alcohol ($CyCH_2OH$) formation. One explanation of addition effect of CeO_2 to increase selectivity is that basic CeO_2 reduces the acidity of $Rh–MoO_x/SiO_2$ surface to increase the concentration of free ammonia in the reaction media (equation (11)).

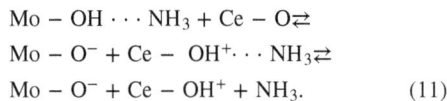

$$Mo - OH \cdots NH_3 + Ce - O \rightleftarrows$$
$$Mo - O^- + Ce - OH^+ \cdots NH_3 \rightleftarrows$$
$$Mo - O^- + Ce - OH^+ + NH_3. \quad (11)$$

Another explanation for the increase in selectivity is that the step of imine hydrogenation (equation (6)) is accelerated by the CeO_2 addition. From table 2, the systems that showed higher yield of reduction products ($CyCH_2NH_2 + CyCH_2OH + (CyCH_2)_2NH$) tend to show higher selectivity ratio of $CyCH_2NH_2/(CyCH_2)_2NH$. Further investigation is necessary to clarify the mechanism of increasing selectivity to amine.

4. Conclusions

The addition of CeO_2 to $Rh–MoO_x/SiO_2$ increases the catalytic activity in hydrogenation of cyclohexanecarboxamide to aminomethylcyclohexane. The selectivity to aminomethylcyclohexane is also increased by the addition of CeO_2. The activity of this combined catalyst system is higher than that of unsupported Rh/Mo catalyst system, which has been reported in the literature, although the aminomethylcyclohexane yield is still lower. The crystallinity of CeO_2 does not affect the addition effect, suggesting that only the weakly-basic nature of CeO_2 surface induces the addition effect. The addition effect of CeO_2 can be related to the ratio of Mo–O^- to Mo–OH sites on the surface of $Rh–MoO_x/SiO_2$.

Acknowledgment

This work was supported by JSPS KAKENHI grant number 26249121.

References

[1] Shi J 2013 *Chem. Rev.* **113** 2139–81
[2] Bell A T 2003 *Science* **299** 1688–91
[3] Nakagawa Y, Tamura M and Tomishige K 2013 *ACS Catal.* **3** 2655–68
[4] Nakagawa Y, Tamura M and Tomishige K 2014 *J. Mater. Chem. A* **2** 6688–702
[5] Tomishige K, Tamura M and Nakagawa Y 2014 *Chem. Rec.* **14** 1041–54
[6] Smith A M and Whyman R 2014 *Chem. Rev.* **114** 5477–510
[7] Dub P A and Ikariya T 2012 *ACS Catal.* **2** 1718–41
[8] Werkmeister S, Junge K and Beller M 2014 *Org. Process Res. Dev.* **14** 289–302
[9] John J M and Bergens S H 2011 *Angew. Chem., Int. Edn Engl.* **50** 10377–80
[10] Balaraman E, Gnanaprakasam B, Shimon L J W and Milstein D 2010 *J. Am. Chem. Soc.* **132** 16756–8
[11] Ikariya T and Kayaki Y 2014 *Pure Appl. Chem.* **86** 933–43

[12] Coetzee J, Dodds D L, Klankermayer J, Brosinski S, Leitner W, Slawin A M Z and Cole-Hamilton D J 2013 *Chem. Eur. J.* **19** 11039–50

[13] Beamson G, Papworth A J, Philipps C, Smith A M and Whyman R 2010 *J. Catal.* **269** 93–102

[14] Beamson G, Papworth A J, Philipps C, Smith A M and Whyman R 2011 *J. Catal.* **278** 228–38

[15] Hirosawa C, Wakasa N and Fuchikami T 1996 *Tetrahedron Lett.* **37** 6749–52

[16] Maj A M, Suisse I, Pinault N, Robert N and Agbossou-Niedercorn F 2014 *ChemCatChem* **6** 2621–5

[17] Burch R, Paun C, Cao X, Crawford P, Goodrich P, Hardacre C, Hu P, McLaughlin L, Sá J and Thompson J M 2011 *J. Catal.* **283** 89–97

[18] Stein M and Breit B 2013 *Angew. Chem., Int. Edn Engl.* **52** 2231–4

[19] Coetzee J, Manyar H G, Hardacre C and Cole-Hamilton D J 2013 *ChemCatChem* **5** 2843–7

[20] Tamura M, Tokonami K, Nakagawa Y and Tomishige K 2013 *Chem. Commun.* **49** 7034–6

[21] Liu S, Amada Y, Tamura M, Nakagawa Y and Tomishige K 2014 *Green Chem.* **16** 617–26

[22] Liu S, Amada Y, Tamura M, Nakagawa Y and Tomishige K 2014 *Catal. Sci. Technol.* **4** 2535–49

[23] Koso S, Furikado I, Shimao A, Miyazawa T, Kunimori K and Tomishige K 2009 *Chem. Commun.* 2035–7

[24] Koso S, Ueda N, Shinmi Y, Okumura K, Kizuka T and Tomishige K 2009 *J. Catal.* **267** 89–92

[25] Shinmi Y, Koso S, Kubota T, Nakagawa Y and Tomishige K 2010 *Appl. Catal.* B **94** 318–26

[26] Nakagawa Y, Shinmi Y, Koso S and Tomishige K 2010 *J. Catal.* **272** 191–4

[27] Chen K, Koso S, Kubota T, Nakagawa Y and Tomishige K 2010 *ChemCatChem* **2** 547–55

[28] Takeda Y, Nakagawa Y and Tomishige K 2012 *Catal. Sci. Technol.* **2** 2221–3

[29] Tamura M, Tamura R, Takeda Y, Nakagawa Y and Tomishige K 2014 *Chem. Commun.* **50** 6656–9

[30] Tamura M, Tamura R, Takeda Y, Nakagawa Y and Tomishige K 2014 *Chem. Eur. J.* at press (doi: 10.1002/chem.201405769)

[31] Nakagawa Y, Ning X, Amada Y and Tomishige K 2012 *Appl. Catal.* A **433–434** 128–34

[32] Chen K, Tamura M, Yuan Z, Nakagawa Y and Tomishige K 2013 *ChemSusChem* **6** 613–21

[33] Liu S, Tamura M, Nakagawa Y and Tomishige K 2014 *ACS Sustainable Chem. Eng.* **2** 1819–27

[34] Koso S, Watanabe H, Okumura K, Nakagawa Y and Tomishige K 2012 *Appl. Catal.* B **111–112** 27–37

[35] Tomishige K, Yasuda H, Yoshida Y, Nurunnabi M, Li B and Kunimori K 2004 *Green Chem.* **6** 206–14

[36] Tamura M, Honda M, Nakagawa Y and Tomishige K 2014 *J. Chem. Technol. Biotechnol.* **89** 19–33

[37] Yoshida Y, Arai Y, Kado S, Kunimori K and Tomishige K 2006 *Catal. Today* **115** 95–101

[38] Honda M, Kuno S, Sonehara S, Fujimoto K, Suzuki K, Nakagawa Y and Tomishige K 2011 *ChemCatChem* **3** 365–70

[39] Honda M, Sonehara S, Yasuda H, Nakagawa Y and Tomishige K 2011 *Green Chem.* **13** 3406–13

[40] Honda M, Tamura M, Nakagawa Y and Tomishige K 2014 *Catal. Sci. Technol.* **4** 2830–45

[41] Koso S, Nakagawa Y and Tomishige K 2011 *J. Catal.* **280** 221–9

[42] Amada Y, Shinmi Y, Koso S, Kubota T, Nakagawa Y and Tomishige K 2011 *Appl. Catal.* B **105** 117–27

[43] Nakagawa Y, Mori K, Chen K, Amada Y, Tamura M and Tomishige K 2013 *Appl. Catal.* A **468** 418–25

[44] Chen K, Mori K, Watanabe H, Nakagawa Y and Tomishige K 2012 *J. Catal.* **294** 171–83

12

Synthesis and characterization of 3D topological insulators: a case $TlBi(S_{1-x}Se_x)_2$

Kouji Segawa

Institute of Scientific and Industrial Research, Osaka University 8-1 Mihogaoka, Ibaraki, Osaka 567-0047, Japan

E-mail: segawa@sanken.osaka-u.ac.jp

Abstract

In this article, practical methods for synthesizing Tl-based ternary III-V-VI$_2$ chalcogenide $TlBi(S_{1-x}Se_x)_2$ are described in detail, along with characterization by x-ray diffraction and charge transport properties. The $TlBi(S_{1-x}Se_x)_2$ system is interesting because it shows a topological phase transition, where a topologically nontrivial phase changes to a trivial phase without changing the crystal structure qualitatively. In addition, Dirac semimetals whose bulk band structure shows a Dirac-like dispersion are considered to exist near the topological phase transition. The technique shown here is also generally applicable for other chalcogenide topological insulators, and will be useful for studying topological insulators and related materials.

Keywords: topological insulator, Dirac semimetal, topological phase transition

1. Introduction: Tl-based topological insulators

Topological insulators have attracted great interest because of their peculiar surface state, which hosts spin-polarized Dirac electrons [1–3]. Possible motivations for studying topological insulators are utilization of the surface state, such as transparent electrodes [4], and/or exploring exotic phenomena. To date, one of the necessary conditions for a material to be a topological insulator is occurrence of band inversion due to strong spin-orbit interaction, and therefore topological insulators must contain heavy elements like Bi. The prototypical topological insulator is tetradymite Bi_2Se_3 [5, 6], which many three-dimensional (3D) topological insulators are rooted on. Bi_2Se_3 has a so-called quintuple-layer structure, which consists of Se–Bi–Se–Bi–Se layers coupled to one another with a van der Waals gap. In many 3D topological insulators like $PbBi_2Te_4$, $GeBi_2Te_4$, $GeBi_4Te_7$, etc [7–12], the unit structure is a similar kind of multiple layer; for example, septuple

layers of Te–Bi–Te–Ge–Te–Bi–Te are formed in $GeBi_2Te_4$. The symmetry of the crystal structure is rhombohedral, and it is essentially unchanged from Bi_2Se_3. The unit structure of Bi_2Se_3 tetradymite is understood as a distorted rock-salt-like structure that is elongated along the (111) direction, and many brother systems take similar structures. Reduction of symmetry from cubic to rhombohedral is likely to be necessary for band inversion at odd numbers of time-reversal symmetry momenta, which causes nontriviality [13].

To date, there are not many variations of the crystallographic systems of 3D topological insulators characterized by a Z_2 invariant (Z_2 topological insulators), although more than 20 systems have been discovered as Z_2 topological insulators [3]. However, there is a very different and intriguing system among them: a series of Tl-based ternary topological insulators such as $TlBiSe_2$. There are many interesting features to these materials. (1) They have a topologically *trivial* compound whose crystallographic structure is identical and continuously connected to nontrivial phase. In other words, the topological phase transition is realized in these systems, which is rarely seen in Z_2-topological-insulator systems. (2) A 3D Dirac semimetal is expected to be realized, in relation with the topological phase transition [14, 15]. (3)

There is a system in which the surface Dirac cone has a gap, although there is no factor which breaks the time-reversal symmetry [16]. (4) In particular for $TlBiSe_2$, the band structure of the surface state is quite simple and the Dirac cone is placed at the $\bar{\Gamma}$ point. (5) The bulk band gap is ~ 0.35 eV, which is the largest among known topological insulators and important for application at high temperatures. (6) No van der Waals gap takes place in their structure. This can be a demerit since scanning tunneling microscope/spectroscopy cannot be applied to these systems.

The first time that the systems attracted interest as topological insulators was in February 2010. Two reports of the theoretical prediction of the realization of the topologically nontrivial phase in Tl-based ternary III–V–VI_2 compounds were posted independently [17, 18]. At first, $TlBiTe_2$ had been considered to show superconductivity at low temperatures [19], but now people realize that the superconductivity is due to an impurity in TlTe [20]. Although superconductivity has nothing to do with this, interest in a possible new topological insulator system remained. The first report of an observation of a Dirac cone in $TlBiSe_2$ was posted in June 2010 by Sato et al [21], and it was followed by two independent groups [22, 23].

The nontriviality of $TlBiSe_2$ and $TlBiTe_2$ was certainly predicted [17, 18]; however, reality turned out to be different from the prediction on $TlBiS_2$, as the real structure is completely different from the basis of the prediction in $TlSbS_2$ and $TlSbSe_2$. Experimentally $TlBiS_2$ is revealed to be topologically *trivial* by angle-resolved photoemission spectroscopy (ARPES), and this fact has been reported by two groups independently [16, 24]. The structure of the trivial $TlBiS_2$ is qualitatively identical to that of the nontrivial $TlBiSe_2$, and thus it is obvious that there is some point at which a phase transition from trivial to nontrivial takes place. Indeed, the topological phase transition is observed [16, 24], and the transition point turns out to be close to $x = 0.5$ of $TlBi(S_{1-x}Se_x)_2$ [24]. This feature gives us significant help in determining whether some phenomenon originates from the topological surface state. If such a phenomenon is not observed in $TlBiS_2$ but observed in $TlBiSe_2$, we can easily say that that has a topological origin. Therefore, the system $TlBi(S_{1-x}Se_x)_2$ provides a useful platform to study the topological surface state experimentally. Furthermore, between the trivial and nontrivial phases, a Dirac semimetal whose 3D band structure has a Dirac-cone dispersion is considered to realize [25, 26].

Another striking feature is the Dirac gap in $TlBi(S_{1-x}Se_x)_2$ for $0.5 < x < 1.0$. Sato et al reported that a gap opens in the surface Dirac cone if the composition is changed from $TlBiSe_2$ toward $TlBiS_2$ while keeping the composition within that of the nontrivial phases. This feature is quite unusual because breaking the time reversal symmetry is necessary to open the gap in the Dirac cone. One may suspect that the gap is due to hybridization of some imperfect cleavage of the sample, but this is not likely because the gap opening is very reproducible in experiments by Sato et al [16]. The origin of the gap has been proposed by theorists [27], but it is still puzzling.

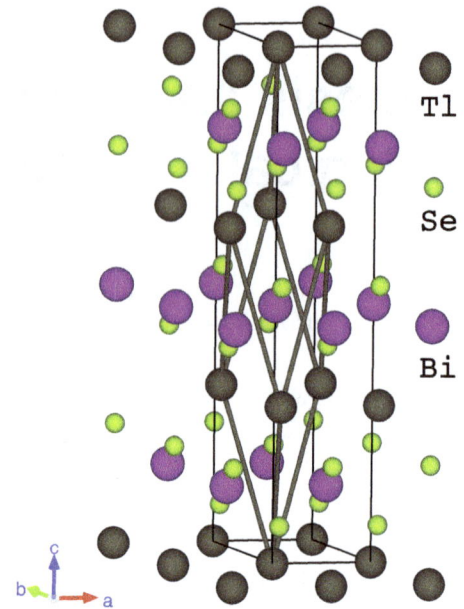

Figure 1. Crystal structure of $TlBiSe_2$. Hexagonal unit cell is shown by black thin lines, and the rhombohedral unit cell is shown by gray thick lines [34].

In this paper, I describe the synthesis and characterization of the intriguing system, $TlBi(S_{1-x}Se_x)_2$, which is useful for studying the topological phase transition, Dirac semimetals, and the mechanism of the unexpected opening of the gap in the topological surface-state dispersion.

2. Single crystals of $TlBi(S_{1-x}Se_x)_2$

2.1. Structure

Figure 1 shows the structure of $TlBiSe_2$, which is called an $NaFeO_2$-type. The crystallographic structure was first reported for $TlBiTe_2$ in 1961 [28], and the structures of $TlBiSe_2$ and $TlBiS_2$ are essentially identical. The space group is R $\bar{3}$ m (No. 166), which is the same as for other Z_2 topological insulators such as $Bi_{1-x}Sb_x$, Bi_2Se_3, and tetradymite families such as $GeBi_2Te_4$, etc. $Bi_{1-x}Sb_x$ takes a face-centered cubic (fcc)-like lattice structure distorted along the (111) direction. The unit structure of tetradymite is similar to that of $TlBiSe_2$ and $TlBiS_2$, but it continues for only five layers like Se-Bi-Se-Bi-Se because of the charge neutrality. In the $TlBiSe_2$ and $TlBiS_2$, Tl^+ and Bi^{3+} align alternately, the averaged valence of the cations is +2, and charge neutrality is naturally conserved. Therefore, the structure of $TlBiSe_2$ and $TlBiS_2$ can be understood as a distorted NaCl-type, in which Tl^+ and Bi^{3+} are alternately placed at the cation sites. The lattice constants were given in multiple papers [30–33], and the results are essentially consistent. Please note that the number of independent parameters is two, and thus the parameters of the hexagonal unit cell can be converted into a rhombohedral

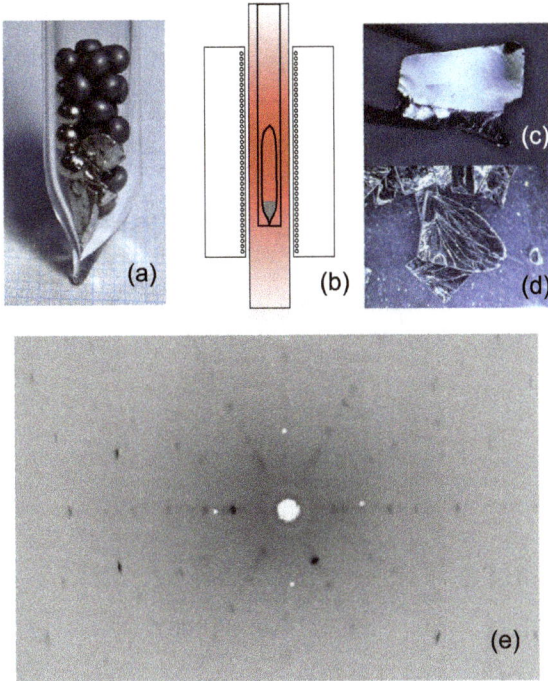

Figure 2. (a) Raw materials put in a sealed quartz tube. Round black shots are selenium, round silvery shots are bismuth, and irregularly shaped pieces are thallium. The outer and inner diameters of the quartz tube are 9 mm and 7 mm, respectively. (b) Schematic of the growth, where the sealed quartz tube is put at a slightly lower position in the furnace. (c), (d) Pictures of cleaved crystals of TlBiSe$_2$. (e) Laue diffraction pattern of TlBiSe$_2$ on the cleaved surface.

unit-cell expression, and vice versa. The conversion formulae are shown in the appendix.

2.2. Synthesis

Single crystals of most chalcogenide topological insulators are made by a melt-grown method. The Bridgman method is one kind of melt-grown method. Since the compound melts congruently, the growth process consists of slowly cooling down a melt under a certain temperature gradient. The samples must be sealed in a quartz tube, because the vapor pressure of chalcogenides is usually high at high temperatures. For TlBiSe$_2$ and TlBiS$_2$ the melting temperature is low, and the reaction between the melt and a quartz tube is negligible. Synthesis of those systems has already been reported [30–33]. A ternary phase diagram was also investigated for both TlBiSe$_2$ and TlBiS$_2$ [35–37], and it is known that these compounds melt congruently. The growth method is essentially the same in all reports, and thus I introduce our method for synthesizing crystals as an example.

A picture of the starting materials sealed in a quartz tube is shown in figure 2(a). The tip of the quartz tube is narrowed to restrict nucleation of the growth at the beginning. This is useful for obtaining crystals of larger domains. The purity of starting materials of Tl, Bi, Se, and S is 99.999%(5N), 99.9999%(6N), 5N, and 5N, respectively. To remove the oxidized surface of the raw materials, Tl shots are annealed in

a hydrogen atmosphere at 270–300 °C and Bi shots are washed with a diluted HNO$_3$ solution. The amount of the raw materials is carefully controlled in the glove box, and the materials are mixed into the quartz-glass tube, which is sealed after evacuating with a diffusion pump and then filled with a small amount of pure argon. The ratio of the materials is stoichiometric in the present experiment, but recently, the shifted composition of starting materials turned out to be useful for synthesizing bulk-insulating samples [38, 39]. Before the main growth procedure, the raw materials are slowly warmed up at the rate of 100 °C/h and kept at 900 °C to complete the reaction. At 900 °C, the quartz tube is shaken to homogenize the melt and eliminate bubbles.

A schematic picture of the growth setup is shown in figure 2(b). In the present experiment, I used a vertical tube furnace with a single temperature control. A natural temperature gradient is used for growth, and it is measured with thermocouples to be ∼10 °C/cm. The vacant part of the quartz tube is placed at a higher temperature position in the furnace to prevent condensation of vapor into any other unexpected phase. For growing TlBiSe$_2$, the temperature-sweep rates of 0.5, 1, and 2 °C/h are tried, and the best value among them is determined to be 1 °C/h by evaluating the transport properties (details are shown below). As for the solid-solution compounds, the temperature sweep rate of 2 °C/h is applied to grow TlBi(S$_{1-x}$Se$_x$)$_2$ crystals, including TlBiS$_2$, because comparable quality is achieved in the growth of TlBiSe$_2$ at the rate of 2 °C/h. The melting points of TlBiSe$_2$ and TlBiS$_2$ are slightly different: 720 °C [40] and 740 °C [32], respectively. Therefore, the growth conditions of TlBi(S$_{1-x}$Se$_x$)$_2$ are changed for x: The temperature sweep range for solid-solution compositions is from 860 °C to 660 °C for $x \leqslant 0.3$, and from 840 °C to 640 °C for $x \geqslant 0.4$.

3. Characterization: x-ray diffraction

Figures 2(c) and (d) show photographs of TlBiSe$_2$ crystals. The obtained crystals are easily cleavable, and the grain size is as long as ∼5 mm along the cleavage plane. Figure 2(e) shows the Laue picture taken with respect to the cleavage plane. This clearly shows the three-fold symmetry of the system, and the result is consistent with the structure of TlBiSe$_2$. The lattice constants of TlBi(S$_{1-x}$Se$_x$)$_2$ are obtained by both single-crystal and powder x-ray diffraction (XRD), as shown in figure 3. If I take a hexagonal lattice, $a_{\text{hex}} = 0.425$–0.410 nm and $c_{\text{hex}} = 2.23 - 2.19$ nm for $x = 1.0$ to 0.0. It is important to note that the lattice parameters change *continuously* with changing x, and the crystallographic structure does not change.

Figures 4(a) and (b) show the single-crystal x-ray profile of TlBiSe$_2$ and TlBiS$_2$. Apparently the intensity of $(0\ 0\ 6n+3)_{\text{hex}}$ peaks with integer n is much weaker in TlBiS$_2$. Figure 4(c) shows $(0\ 0\ 9)_{\text{hex}}$ peaks normalized by those of $(0\ 0\ 6)_{\text{hex}}$ for various compositions of TlBi(S$_{1-x}$Se$_x$)$_2$. The intensity decreases with a change in the composition from TlBiSe$_2$ to TlBiS$_2$. As described above, the crystal structure of a series of the Tl-based ternary compound can be understood as an fcc structure distorted along the (111) direction [17, 29]. However, the displacement of chalcogen sites from

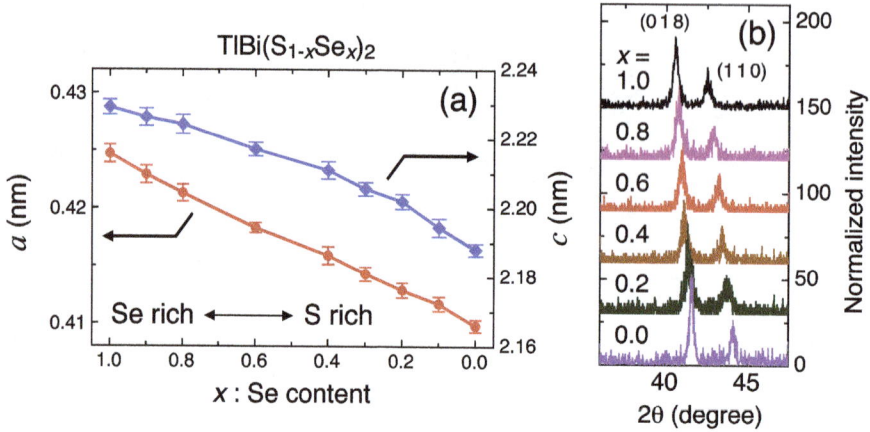

Figure 3. (a) Lattice constants determined from power XRD are plotted as functions of the Se content. (b) Powder XRD patterns. Normalized intensity profiles of the (018) and (110) peaks in the powdered $TlBi(S_{1-x}Se_x)_2$ single crystals for a series of Se concentrations; the (018) peak reflects both the in-plane and out-of-plane periodicities, whereas the (110) peak reflects solely the in-plane periodicity. Taken from [16].

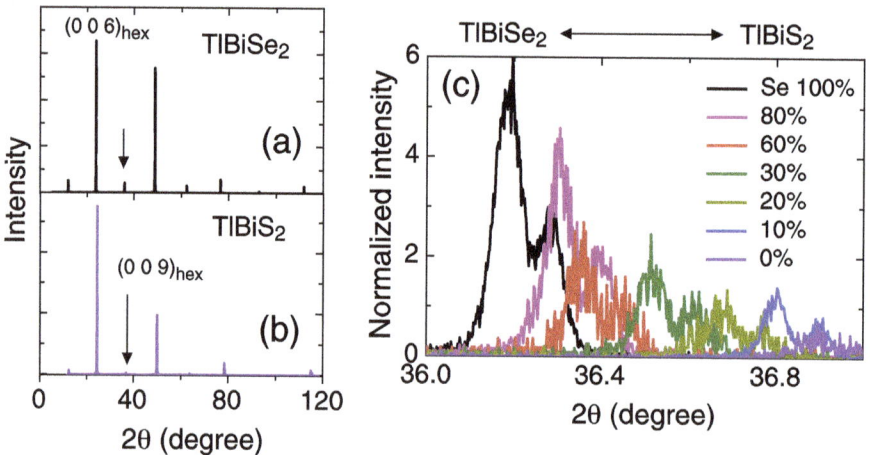

Figure 4. (a, b) Single-crystal XRD patterns of $TlBiSe_2$ and $TlBiS_2$ taken on the c-plane. Arrows show peaks at $(0\ 0\ 9)_{hex}$. (c) Profiles of $TlBi(S_{1-x}Se_x)_2$ around the $(0\ 0\ 9)_{hex}$ peak normalized by the intensity of $(0\ 0\ 6)_{hex}$ peaks. The normalized intensity shows a clear decrease with decreasing x.

the fcc position and the alternate order of Tl and Bi causes longer periodicity and produces $(0\ 0\ 6n+3)_{hex}$ peaks in the XRD profile. Therefore, the smaller intensity of $(0\ 0\ 6n+3)_{hex}$ peaks suggests that the displacement of chalcogens in $TlBiS_2$ is smaller than that in $TlBiSe_2$. In the calculation by Lin *et al*, the displacement of chalcogens is a key for producing band inversion [17]; this may be an origin of the absence of band inversion in $TlBiS_2$, along with weakness of the spin-orbit interaction due to the lighter element, S.

4. Characterization: transport properties

4.1. Electrical contacts

Measurements of transport properties are performed by the conventional six-probe method. Usually, for many chalcogenide topological insulators, electrical contacts are attached with silver paste such as 4922N by Dupont, which can be

cured at room temperature. However, if Tl is included as a constituent element, then the contact resistance becomes higher than $\sim 15\,\Omega$ and it quickly degrades $\sim 1\ k\Omega$ after several hours. Indium solder can be used to make contacts with the contact resistance as low as $\sim 1\,\Omega$, but the superconductivity of indium solder sometimes produces significant signals at low temperatures. To date, the spot-welding of gold wires is the best method for making electrical contacts, which become as low as $\sim 1\,\Omega$ and remain low for at least a few weeks. In the present experiment, 30-μm-diameter gold wires were used as lead wires (figure 5), and the typical voltage for welding the wire was 5–9 V with a capacitor of 100 μF.

4.2. Transport properties of TlBiSe₂: optimization of synthesis conditions

Figures 6(a)–(c) show the temperature dependences of (a) resistivity, (b) the Hall coefficient, and (c) Hall mobility for

Figure 5. A sample with six welded contacts, which is mounted on a copper plate with insulation.

three samples of TlBiSe$_2$ single crystals grown in different conditions. Samples A, B, and C are grown at a temperature sweep rate of 0.5, 1, and 2 °C/h, respectively. The temperature dependences of the resistivity of all the three samples show a metallic behavior (figure 6(a)), and this indicates that the samples are degenerate semiconductors. Figure 6(b) shows that the temperature dependences of the Hall coefficient are not strong, and 4.2–6.2·10^{19} cm^{-3} $\left[= \frac{1}{eR_\mathrm{H}} \right]$ of n-type carriers are doped in this system (the Hall resistivity, ρ_{yx}, shows a linear B-dependence up to ±7 T). The growth-rate dependences of the transport properties are not simple; figure 6(a) shows that the resistivity of sample A grown by the slowest rate shows the highest resistivity of the three, but the fast-grown sample C does not show the lowest resistivity. The Hall coefficient also depends on the growth conditions, and there is a tendency for the absolute value of the Hall coefficient to decrease with increasing resistivity. This suggests that higher resistivity is not caused by a reduction in the charge carriers, but by an increase in disorder. Indeed, the sample A (B) of highest (lowest) resistivity shows the lowest (highest) Hall mobility (figure 6(c)). Apparently sample B is the best among the three samples, because it has lowest carrier concentration and disorder. On the other hand, I applied the growth conditions of sample C (the temperature sweep rate of 2 °C/h) to other TlBi(S$_{1−x}$Se$_x$)$_2$ systems. The quality of sample C is certainly not the best among the three, but it is reasonably good because the mobility is more than three times of that of sample A. The data for lowest-mobility sample A are consistent with that by Mitsas *et al* [31], and thus in the present experiment the quality of samples is apparently improved. This is probably because of both the higher purity of the raw materials and the optimization of the growth conditions.

Now let us consider the origin of the growth-condition dependence of transport properties. Inductively-coupled plasma atomic-emission spectroscopy (ICP-AES) analysis is performed on samples A and B grown by the slow- and the middle-temperature sweep rates, respectively, to obtain the resultant chemical composition of crystals (table 1). The sum of the Tl and Bi contents is fixed to be 2. In both samples, bismuth composition is slightly larger than thallium, and this shows the substitution of Bi^{3+} for Tl$^+$. The fraction of the substitution is higher in sample A than in sample B, and thus it is naturally expected that this difference causes an increase in n-type carriers and imperfections by Bi^{3+} partially substituting Tl$^+$ in sample A. In other words, it is important to suppress substitution of Bi for Tl to obtain better samples with less disorder.

4.3. Transport properties of TlBiSe$_2$

In this section, let me discuss the magneto transport of the best sample of TlBiSe$_2$ (sample B in figure 6). Figure 7(a) shows the magnetic-field dependence of the Hall resistivity, ρ_{yx}, up to ±14 T. The solid line shows the linear fitting, and the lack of non-linearity (deviation from the linear fit is less than 1%) is indicative of the single-channel electronic transport. The carrier concentration of this sample is $n = 4.31 \times 10^{19}$ cm^{-3} [$R_\mathrm{H} = -145$ (10^{-3} cm^3/C)], and this result is in agreement with low-field data in figure 6(b). Please note that the bulk conduction channel must govern the transport properties of the present sample, and thus the topological surface channel does not play an important role, unfortunately. Figure 7(c) shows the field dependence of the magnetoresistance up to 14 T. The data are symmetrized with respect to the magnetic field. At very low fields, no anomaly is observed, and thus weak antilocalization behavior is missing in this system. This result is indicative of the dominance of the bulk-channel transport. At high fields, magnetoresistance shows a clear oscillation. Since the oscillatory component of the magnetoresistance shows a periodicity of $1/B$, this behavior is Shubnikov-de-Haas (SdH) oscillation. The frequency of the SdH oscillation is $f = 296$ T, and this gives $k_\mathrm{F} = 9.48 \times 10^6$ cm^{-1} according to the Onsager relation, $f = (\hbar c/2\pi e)\pi k_\mathrm{F}^2$. The k_F value is in good agreement with the ARPES result [21]. If a spherical Fermi surface (FS) is assumed, $k_\mathrm{F}^3/3\pi^2$ gives $n = 2.88 \times 10^{19}$ cm^{-3}. This value is smaller than n obtained from the Hall coefficient, but it can be understood if the FS is elongated along the c_hex^*-axis. Figure 3(b) shows the temperature dependence of the Seebeck coefficient, S. The negative sign of S is consistent with n-type doping in the present system.

4.4. Transport properties of TlBi(S$_{1−x}$Se$_x$)$_2$

From the transport properties at high temperatures, one may extract the energy gap if an activation behavior is observed [3]. In the present experiment I could not observe any activation behavior either in the resistivity or the Hall coefficient. However, the temperature dependence of resistivity shows some difference between Se-rich and Se-poor samples (figure 8). The resistivity above 300 K and the temperature dependences in $x = 0.1$ and 0.4 samples show a significant increase up to 500 K, whereas the increase is modest in $x = 0.6$ and 1.0 samples. In all these samples, the charge

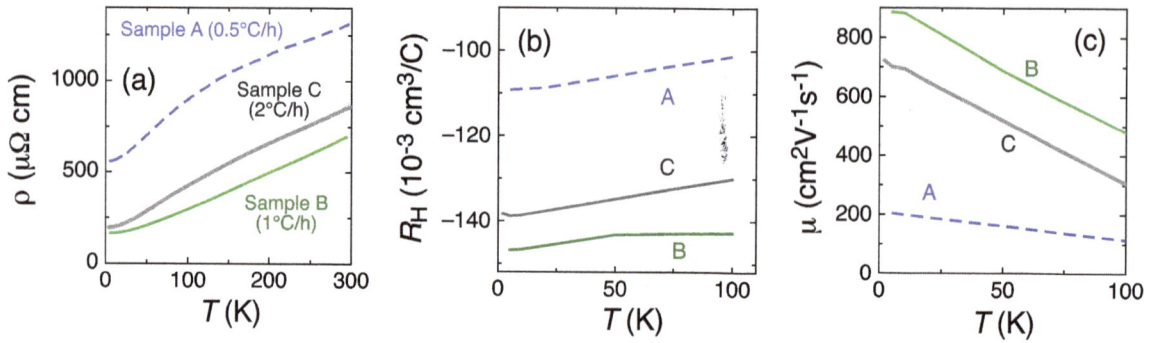

Figure 6. Transport properties of TlBiSe$_2$ samples. Samples A, B, and C are grown at a temperature sweep rate of 0.5, 1, and 2 °C/h, respectively. Panels (a)–(c) present the temperature dependences of (a) ρ_{ab}, (b) R_H, and (c) μ_H.

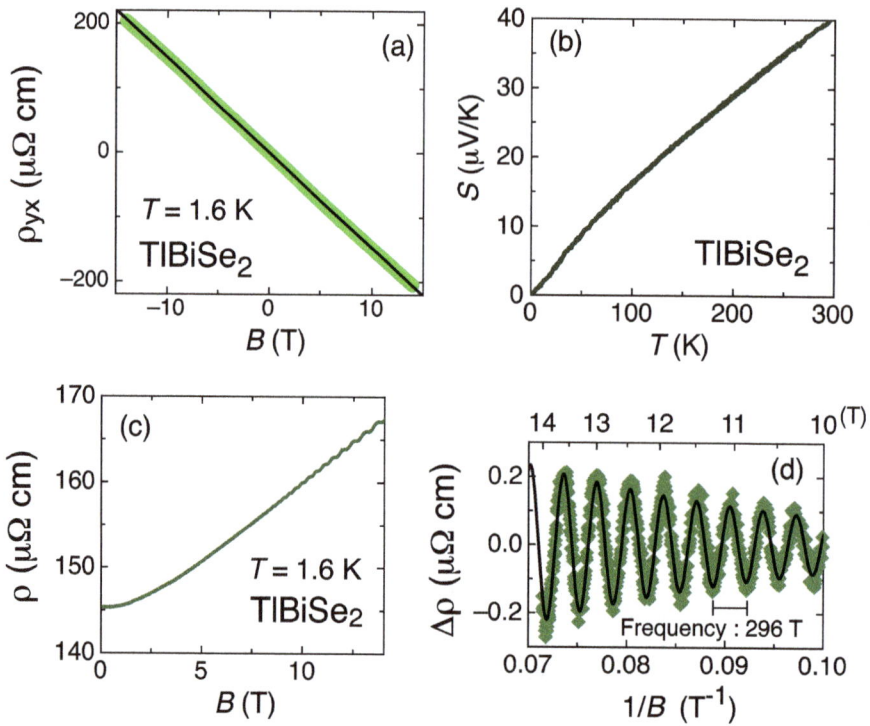

Figure 7. Transport properties of the best TlBiSe$_2$ sample. (a) Field dependence of the Hall resistivity, ρ_{yx}. (b) Temperature dependence of the Seebeck coefficient. (c) Field dependence of the resistivity at 1.6 K. (d) Oscillatory component of magnetoresistance versus inverse of the magnetic fields. The solid line is a fit to data.

Table 1. The data are obtained from ICP-AES analyses. The sum of thallium and bismuth contents is fixed to be 2.

TlBiSe$_2$ Sample	Growth rate (°C/h)	Tl	Bi	Se
A	0.5	0.976 ± 0.008	1.024 ± 0.008	2.010 ± 0.003
B	1	0.985 ± 0.008	1.015 ± 0.008	2.002 ± 0.006

carriers are electrons, and thus the chemical potential is located in the conduction band. The observed behavior suggests that the responsible bands for the Se-rich composition and the other are different from each other, and thus it is suggested that band inversion occurs when x changes from 0.4 to 0.6 in the present system.

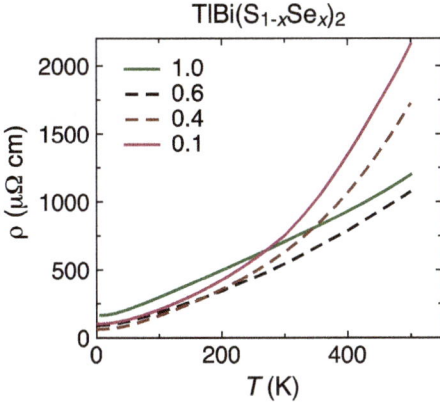

Figure 8. Temperature dependences of the resistivity for TlBi(S$_{1-x}$Se$_x$)$_2$, x = 1.0, 0.6, 0.4, and 0.1, up to 500 K.

5. Summary

This article provides detailed information on the synthesis and characterization of TlBi(S$_{1-x}$Se$_x$)$_2$. The growth condition of TlBiSe$_2$ is optimized to obtain higher mobility using transport properties. XRD analysis confirmed that the change in the crystallographic structure is gradual in TlBi(S$_{1-x}$Se$_x$)$_2$, and this system is useful for studying the topological phase transition and related Dirac semimetals.

Acknowledgments

The present work was performed in collaboration with Yoichi Ando, T Minami, K Eto, and S Sasaki. I acknowledge the Comprehensive Analysis Center of Institute of Scientific and Industrial Research at Osaka University for assistance in the quantitative analysis with ICP-AES. This work was supported by JSPS (KAKENHI 24540320) and MEXT (Innovative Area 'Topological Quantum Phenomena' KAKENHI).

Appendix A. Parameter and vector conversion of Tl-based ternary system and tetradymite topological insulators between hexagonal and rhombohedral unit-cell systems

The crystallographic structure of the Tl-based ternary system and tetradymite topological insulators can be expressed by either hexagonal or rhombohedral unit cells, which specify the structure by two independent lattice parameters. In the following, I show the conversion formulae from hexagonal to rhombohedral, and vice versa:

$$a_{\text{rhm}} = \frac{1}{3}\sqrt{a_{\text{hex}}^2 + \frac{c_{\text{hex}}^2}{3}},$$

$$\alpha_{\text{rhm}} = \arccos\left(1 - \frac{a_{\text{hex}}^2}{2c_{\text{hex}}^2}\right),$$

$$a_{\text{hex}} = \sqrt{2a_{\text{rhm}}^2(1 - \cos\alpha_{\text{rhm}})},$$

$$c_{\text{hex}} = \sqrt{3a_{\text{rhm}}^2(1 + 2\cos\alpha_{\text{rhm}})},$$

where a_{hex} and c_{hex} (a_{rhm} and α_{rhm}) are the lattice constants for hexagonal (rhombohedral) unit cells. In the hexagonal system, the c-axis is perpendicular to the layer-structure planes.

In the real space, the formulae for converting unit vectors between hexagonal and rhombohedral unit-cell systems are given as follows: Here, the unit vectors of a rhombohedral (hexagonal) system are expressed as $\vec{a}_1$, $\vec{a}_2$, and $\vec{a}_3$ ($\vec{a}_{\text{hex}}$, $\vec{b}_{\text{hex}}$, and $\vec{c}_{\text{hex}}$), and

$$\vec{c}_{\text{hex}} = \vec{a}_1 + \vec{a}_2 + \vec{a}_3$$

is always valid. But, $\vec{a}_{\text{hex}}$ and $\vec{b}_{\text{hex}}$ depend on how to choose the rhombohedral unit-cell vectors. If the unit vectors of a rhombohedral system are taken as

$$\vec{a}_{\text{hex}} = \vec{a}_1 - \vec{a}_2,$$

then $\vec{b}_{\text{hex}}$ is given by

$$\vec{b}_{\text{hex}} = \vec{a}_2 - \vec{a}_3.$$

The reversal conversion formulae are given by

$$\vec{a}_1 = \frac{2}{3}\vec{a}_{\text{hex}} + \frac{1}{3}\vec{b}_{\text{hex}} + \frac{1}{3}\vec{c}_{\text{hex}},$$

$$\vec{a}_2 = -\frac{1}{3}\vec{a}_{\text{hex}} + \frac{1}{3}\vec{b}_{\text{hex}} + \frac{1}{3}\vec{c}_{\text{hex}},$$

and

$$\vec{a}_3 = -\frac{1}{3}\vec{a}_{\text{hex}} - \frac{2}{3}\vec{b}_{\text{hex}} + \frac{1}{3}\vec{c}_{\text{hex}}.$$

By using these formulae, conversion of an axis is easily performed as follows.

$$(1, 0, 0)_{\text{hex}} = (1, -1, 0)_{\text{rhm}}$$

$$(0, 0, 1)_{\text{hex}} = (1, 1, 1)_{\text{rhm}}$$

$$(1, 0, 0)_{\text{rhm}} = \left(\frac{2}{3}, \frac{-1}{3}, \frac{-1}{3}\right)_{\text{hex}}$$

The above conversion is in the real space, and here let us move to the reciprocal space. Here I show the conversion formulae for converting Miller indices as follows.

$$h_{\text{rhm}} = \frac{1}{3}(-h_{\text{hex}} + 2k_{\text{hex}} + l_{\text{hex}})$$

$$k_{\text{rhm}} = \frac{1}{3}(2h_{\text{hex}} - k_{\text{hex}} + l_{\text{hex}})$$

$$l_{\text{rhm}} = \frac{1}{3}(-h_{\text{hex}} - k_{\text{hex}} + l_{\text{hex}})$$

$$h_{\text{hex}} = k_{\text{rhm}} - l_{\text{rhm}}$$

$$k_{\text{hex}} = h_{\text{rhm}} - l_{\text{rhm}}$$

$$l_{\text{hex}} = h_{\text{rhm}} + k_{\text{rhm}} + l_{\text{rhm}}.$$

Note that the conversion formulae are quite different from those of the real space, because the direction of the reciprocal vector, $\vec{a}^*$ and $\vec{b}^*$, changes from that of the real space, unlike rectangular structures, such as cubic or orthorhombic. Those formulae naturally tell us the rule of absent reflection: if $-h_{hex} + 2k_{hex} + l_{hex} \neq 3n$ (n is an integer), no reflection is observed at the position of the Miller index, because h_{rhm} does not become an integer.

References

[1] Hasan M Z and Kane C L 2010 *Rev. Mod. Phys.* **82** 3045
[2] Qi X L and Zhang S C 2011 *Rev. Mod. Phys.* **83** 1057
[3] Ando Y 2013 *J. Phys. Soc. Jpn.* **82** 102001
[4] Peng H, Dang W, Cao J, Chen Y, Wu D, Zheng W, Li H, Shen Z X and Liu Z 2012 *Nat. Chem.* **4** 281
[5] Zhang H, Liu C X, Qi X L, Dai X, Fang Z and Zhang S C 2009 *Nat. Phys.* **5** 438
[6] Xia Y *et al* 2009 *Nat. Phys.* **5** 398
[7] Souma S, Eto K, Nomura M, Nakayama K, Sato T, Takahashi T, Segawa K and Ando Y 2012 *Phys. Rev. Lett.* **108** 116801
[8] Kuroda K *et al* 2012 *Phys. Rev. Lett.* **108** 206803
[9] Xu S Y *et al* 2010 arXiv:1007.5111
[10] Neupane M *et al* 2012 *Phys. Rev. B* **85** 235406
[11] Okamoto K *et al* 2012 *Phys. Rev. B* **86** 195304
[12] Eremeev S V *et al* 2012 *Nat. Commun.* **3** 635
[13] Fu L and Kane C L 2007 *Phys. Rev. B* **76** 045302
[14] Burkov A A and Balents L 2011 *Phys. Rev. Lett.* **107** 127205
[15] Singh B, Sharma A, Lin H, Hasan M Z, Prasad R and Bansil A 2012 *Phys. Rev. B* **86** 115208
[16] Sato T, Segawa K, Kosaka K, Souma S, Nakayama K, Eto K, Minami T, Ando Y and Takahashi T 2011 *Nat. Phys.* **7** 840
[17] Lin H, Markiewicz R S, Wray L A, Fu L, Hasan M Z and Bansil A 2010 *Phys. Rev. Lett.* **105** 036404
[18] Yan B, Liu C X, Zhang H J, Yam C Y, Qi X L, Frauenheim T and Zhang S C 2010 *Eur. Phys. Lett.* **90** 37002
[19] Hein R A and Swiggard E M 1970 *Phys. Rev. Lett.* **24** 53
[20] Popovich N S, Diakonov V K S V P, Fita I M and Levchenko G G 1984 *Solid State Commun.* **50** 979
[21] Sato T, Segawa K, Guo H, Sugawara K, Souma S, Takahashi T and Ando Y 2010 *Phys. Rev. Lett.* **105** 136802
[22] Kuroda K *et al* 2010 *Phys. Rev. Lett.* **105** 146801
[23] Chen Y L *et al* 2010 *Phys. Rev. Lett.* **105** 266401
[24] Xu S Y *et al* 2011 *Science* **332** 560
[25] Neupane M *et al* 2014 *Nat. Commun.* **5** 3786
[26] Novak M, Sasaki S, Segawa K and Ando Y 2015 *Phys. Rev. B* **91** 041203
[27] Isobe H and Nagaosa N 2012 *Phys. Rev. B* **86** 165127
[28] Hockings E F and White J G 1961 *Acta Crystallogr.* **14** 328
[29] Gitsu D V, Kantser V G, Malkova N M and Tofan V A 1990 *J. Phys.: Condens. Matter* **2** 1129
[30] Toubektsis S N and Polychroniadis E K 1987 *J. Cryst. Growth* **84** 316
[31] Mitsas C L, Siapkas D I, Polychroniadis E K and Paraskevopoulos O V K M 1993 *Phys. Stat. Sol.* (a) **136** 483
[32] Ozer M, Paraskevopoulos K M, Anagnostopoulos A N, Kokkou S and Polychroniadis E K 1996 *Semicond. Sci. Technol.* **11** 1405
[33] Ozer M, Paraskevopoulos K M, Anagnostopoulos A N, Kokkou S and Polychroniadis E K 1998 *Semicond. Sci. Technol.* **13** 86
[34] Momma K and Izumi F 2011 *J. Appl. Crystallogr.* **44** 1272
[35] Babanly M B, Kesamanly M F and Kuliev A A 1988 *Rus. J. Inorg. Chem.* **33** 1546
[36] Babanly M B, Kesamanly M F and Kuliev A A 1988 *Rus. J. Inorg. Chem.* **33** 2371
[37] Babanly M B, Popovkin B A, Zamani I S and Guseinova R R 2003 *Rus. J. Inorg. Chem.* **48** 1932
[38] Kuroda K *et al* 2013 arXiv:1308.5521v2
[39] Eguchi G, Kuroda K, Shirai K, Kimura A and Shiraishi M 2014 *Phys. Rev. B* **90** 201307
[40] Mitsas C L, Polychroniadis E K and Siapkas D I 1999 *Thin Solid Films* **353** 85

13

Preparation of macroporous zirconia monoliths from ionic precursors via an epoxide-mediated sol-gel process accompanied by phase separation

Xingzhong Guo[1]**, Jie Song**[1]**, Yixiu Lvlin**[1]**, Kazuki Nakanishi**[2]**, Kazuyoshi Kanamori**[2] **and Hui Yang**[1]

[1] School of Materials Science and Engineering, Zhejiang University, Hangzhou, 310027, People's Republic of China
[2] Department of Chemistry, Graduate School of Science, Kyoto University, Kitashirakawa, Sakyo-ku, Kyoto 606-8502, Japan

E-mail: msewj01@zju.edu.cn (X Guo) and yanghui@zju.edu.cn

Abstract
Monolithic macroporous zirconia (ZrO_2) derived from ionic precursors has been successfully fabricated via the epoxide-mediated sol-gel route accompanied by phase separation in the presence of propylene oxide (PO) and poly(ethylene oxide) (PEO). The addition of PO used as an acid scavenger mediates the gelation, whereas PEO enhances the polymerization-induced phase separation. The appropriate choice of the starting compositions allows the production of a macroporous zirconia monolith with a porosity of 52.9% and a Brunauer–Emmett–Teller (BET) surface area of $171.9\,m^2 \cdot g^{-1}$. The resultant dried gel is amorphous, whereas tetragonal ZrO_2 and monoclinic ZrO_2 are precipitated at 400 and 600 °C, respectively, without spoiling the macroporous morphology. After solvothermal treatment with an ethanol solution of ammonia, tetragonal ZrO_2 monoliths with smooth skeletons and well-defined mesopores can be obtained, and the BET surface area is enhanced to $583.8\,m^2 \cdot g^{-1}$.

Keywords: porous materials, zirconia monoliths, sol-gel, phase separation, heat-treatment, solvothermal treatment

1. Introduction

As an important structural and functional material, zirconia (ZrO_2) has attracted considerable attention because of superior thermal, mechanical, chemical stability and electrical properties [1–3]. Zirconia has been extensively applied as an advanced material in various areas such as electronics, optics, catalysis and high-temperature structural engineering [4–10]. The tetragonal phase of ZrO_2 has both acidic and basic properties and gives the most active catalyst for several

catalytic reactions [11, 12]. Porous zirconia with precisely designed pore structures has become the new development of zirconia materials for catalyst supports, heat insulation, particle filters and gas membranes under severe conditions such as high-temperature and corrosive environments [13–18].

A lot of effort has been made to obtain porous zirconia such as pore-forming agents, tape-casting, gel casting, templating, impregnation and injection molding [19–23]. Sakka *et al* [24] reported a template-assisted method to fabricate macroporous ceramics consisting of TiO_2 and ZrO_2. The sol-gel method accompanied by phase separation is known as a promising technique for fabricating monolithic materials with a hierarchical porous structure [25, 26]. This method has been used to fabricate porous SiO_2 [26], TiO_2 [27–30] and Al_2O_3 [31]. In our previous works, we have demonstrated the

Table 1. Starting compositions of the samples.

Sample No.	$ZrOCl_2 \cdot 8H_2O$ g^{-1}	H_2O (V_{H2O}) mL^{-1}	EtOH (V_{EtOH}) mL^{-1}	PO (V_{PO}) mL^{-1}	PEO (W_{PEO}) g^{-1}
P1	1.610	2.0	2.4	0.50	0.100
P2	1.610	2.0	2.4	0.52	0.100
P3	1.610	2.0	2.4	0.54	0.100
P4	1.610	2.0	2.4	0.56	0.100
P5	1.610	2.4	2.4	0.52	0.060
P6	1.610	2.4	2.4	0.52	0.075
P7	1.610	2.4	2.4	0.52	0.090
P8	1.610	2.4	2.4	0.52	0.105
P9	1.610	2.4	2.4	0.52	0.120
P10	1.610	2.4	2.4	0.52	0.135
P11	1.610	1.8	3.0	0.52	0.115
P12	1.610	2.0	2.8	0.52	0.115
P13	1.610	2.2	2.6	0.52	0.115
P14	1.610	2.4	2.4	0.52	0.115
P15	1.610	2.6	2.2	0.52	0.115

preparation of monolithic macroporous mayenite [32], mullite [33], cordierite [34] and AlPO$_4$ [35] by this approach. Konishi *et al* [36] reported the preparation of monolithic zirconia gels by this route. However, they used highly reactive and expensive zirconium propoxide (Zr(O^{n}Pr)$_4$) as the starting materials under a strongly acidic condition, and the process and the pore structure are too hard to control on account of the rapid reaction rate.

In this work, we demonstrate the novel and facile preparation of a porous zirconia monolith via the sol-gel process, accompanied by phase separation. The ionic precursors (zirconium oxychloride, ZrOCl$_2 \cdot$ 8H$_2$O) are utilized to synthesize a zirconia monolith in the presence of propylene oxide (PO) and poly(ethylene oxide) (PEO). The gelation of the system is mediated by PO, whereas PEO is added as a phase separation inducer. The resulting zirconia monoliths possess precisely controllable macropores at suitable starting compositions, and the solvothermal treatment of the resultant monoliths can further yield a high density of mesopores in crystallized skeletons, leading to a remarkable increase of surface area. The synthesis process has several advantages compared to the traditional alkoxide-derived sol-gel route, which allows the technique to involve various transition metals.

2. Experimental details

2.1. Materials

Zirconium oxychloride (ZrOCl$_2 \cdot$ 8H$_2$O) was used as a zirconium source. Mixtures of distilled water (H$_2$O) and ethanol (EtOH, Sinopharm Chemical Reagent Co., Ltd, China, AR) were used as the solvents. PO (Sigma-Aldrich Co., USA, 99.5%) was used to initiate gelation, and PEO (Aladdin Co., China, AR) having an average molecular weight (M_w) of 1×10^6 was used as the polymer to induce phase separation. All chemicals were used as received.

2.2. Preparation of monoliths

The starting compositions were listed in table 1. In a typical synthesis, 1.610 g of ZrOCl$_2 \cdot$ 8H$_2$O and W_{PEO} g of PEO were dissolved in a mixture of V_{H2O} mL of distilled water and V_{EtOH} mL of ethanol. After continuously stirring for 120 min, V_{PO} mL of PO was then added to the transparent solution under ambient conditions (25 °C). After stirring for 1 min, the resultant homogeneous solution was transferred into a container. The container was sealed and kept at 60 or 80 °C for gelation. After gelation, the obtained wet gels were aged at 40 °C for 72 h and then evaporation-dried at 60 °C for 72 h. Some of the resultant xerogels were subsequently heat-treated at various temperatures up to 1000 °C for 2 h with a heating rate of 2–3 °C · min^{-1}. Some of the wet gels were immersed in 0.5, 1.0 and 2.0 mol L^{-1} NH$_4$OH solution at 40 °C three times, respectively. Then, the wet gels with solvent were together transferred into a stainless-steel autoclave vessel with a Teflon inner lining and additionally aged at 180 °C for 12 h under an autogeneous pressure in order to tailor the micro- and mesoporous structures. The solvothermally aged gels thus obtained were solvent-exchanged with distilled water and 2-propanol and were evaporation-dried at 40 °C for 5 d.

2.3. Characterization of monoliths

Morphologies of the monolithic gels after drying, heat treatment and solvothermal treatment were observed by a field-emission scanning electron microscope (FESEM: SIRION-100, FEI Co., the Netherlands, with Au coating). Differential thermal analysis/thermogravimetry (DTA/TG; CRY-2P and WRT-3P, INESA, China) was performed at temperatures up to 1100 °C at a heating rate of 10 °C · min^{-1} while continuously supplying air at a rate of 100 mL · min^{-1}. Chemical bonding in the dried gels was investigated with Fourier transform infrared spectroscopy (FTIR: Nicolet 5700, ThermoFisher Co., USA) using the potassium bromide (KBr) pellet technique. The crystal structure was confirmed by powder x-ray diffraction (XRD: XRD-6000, Shimadzu

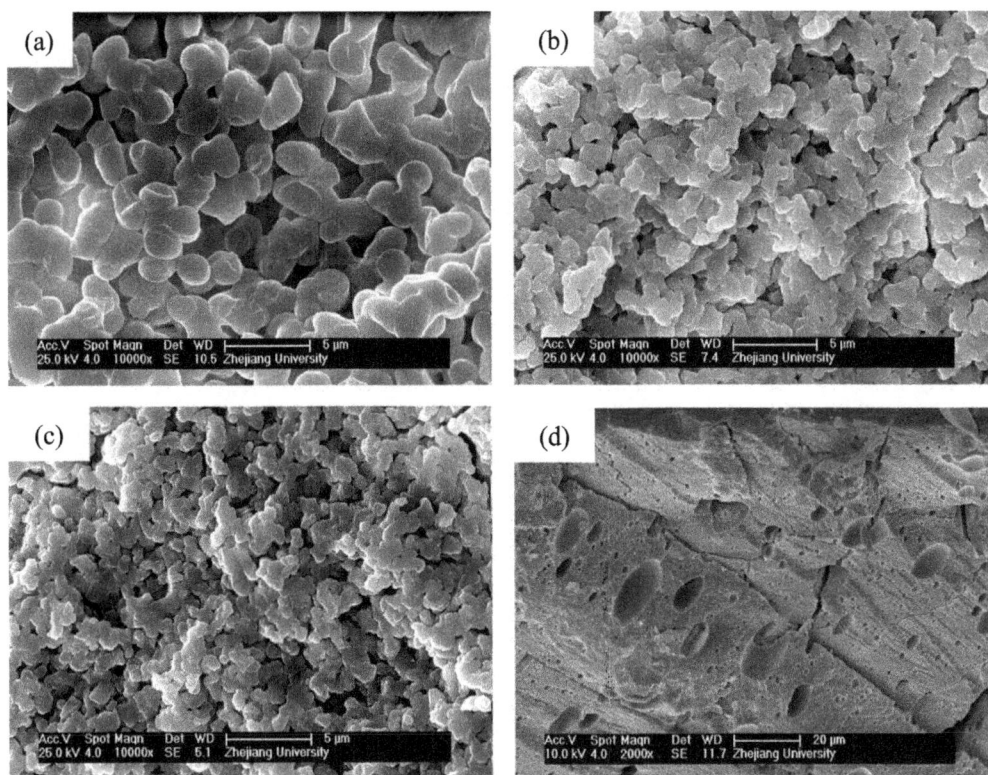

Figure 1. SEM images of dried ZrO_2 gels prepared with various V_{PO}: (a) 0.50 mL (P1), (b) 0.52 mL (P2), (c) 0.54 mL (P3) and (d) 0.56 mL (P4).

Corporation, Japan) using Cu $K\alpha$ ($\lambda = 0.154$ nm) as an incident beam. Macropore size distributions over the diameter from 20 nm to 100 μm were evaluated by mercury porosimetry (AutoPore IV 9510, Micromeritics Instruments, USA). Meso- and micropores were characterized by N_2 adsorption–desorption isotherms (Autosorb-1-C, Quantachrome Instruments, USA). Before the N_2 adsorption–desorption measurement, the samples were degassed at 100 °C under vacuum. The pore size distribution was calculated from them by the Barrett–Joyner–Halenda (BJH) method, and the surface area was obtained by the Brunauer–Emmett–Teller (BET) method. The bulk density of each sample was measured by mercury porosimetry. The porosity (%) of each sample was calculated as $[(1 - \rho_b)/\rho_s] \times 100$, where ρ_b and ρ_s refer to the bulk and skeletal densities, respectively.

3. Results and discussion

3.1. Formation of macroporous monoliths

In this system, zirconium oxychloride is used as the ionic precursor, PEO is used as the phase separation inducer and PO is used to mediate the sol-gel process. The epoxide-mediated sol-gel reaction was firstly reported by Gash *et al* [37–39]. The mechanism can be illustrated as follows. (1)

Epoxide added in the homogeneous solution is protonated by H^+. (2) An irreversible ring-opening reaction is brought about by nucleophilic groups such as H_2O, Cl^- and NO_3^-. The processes of this system are shown as the following reaction equations. The overall reaction can raise the pH of the solution quickly and uniformly, which promotes the hydrolysis and condensation.

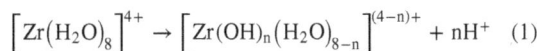

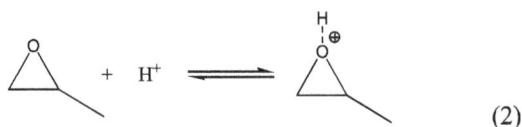

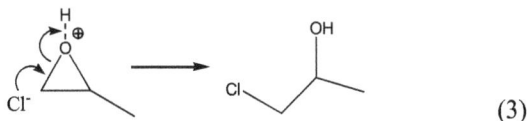

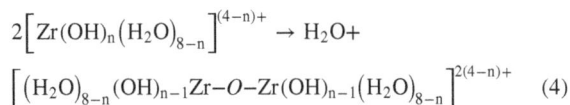

$$\left[Zr(H_2O)_8\right]^{4+} \rightarrow \left[Zr(OH)_n(H_2O)_{8-n}\right]^{(4-n)+} + nH^+ \quad (1)$$

$$(2)$$

$$(3)$$

$$2\left[Zr(OH)_n(H_2O)_{8-n}\right]^{(4-n)+} \rightarrow H_2O +$$
$$\left[(H_2O)_{8-n}(OH)_{n-1}Zr-O-Zr(OH)_{n-1}(H_2O)_{8-n}\right]^{2(4-n)+} \quad (4)$$

The phase separation tendency of the polymeric system can be evaluated by the Flory–Huggins theory [40–42]. The

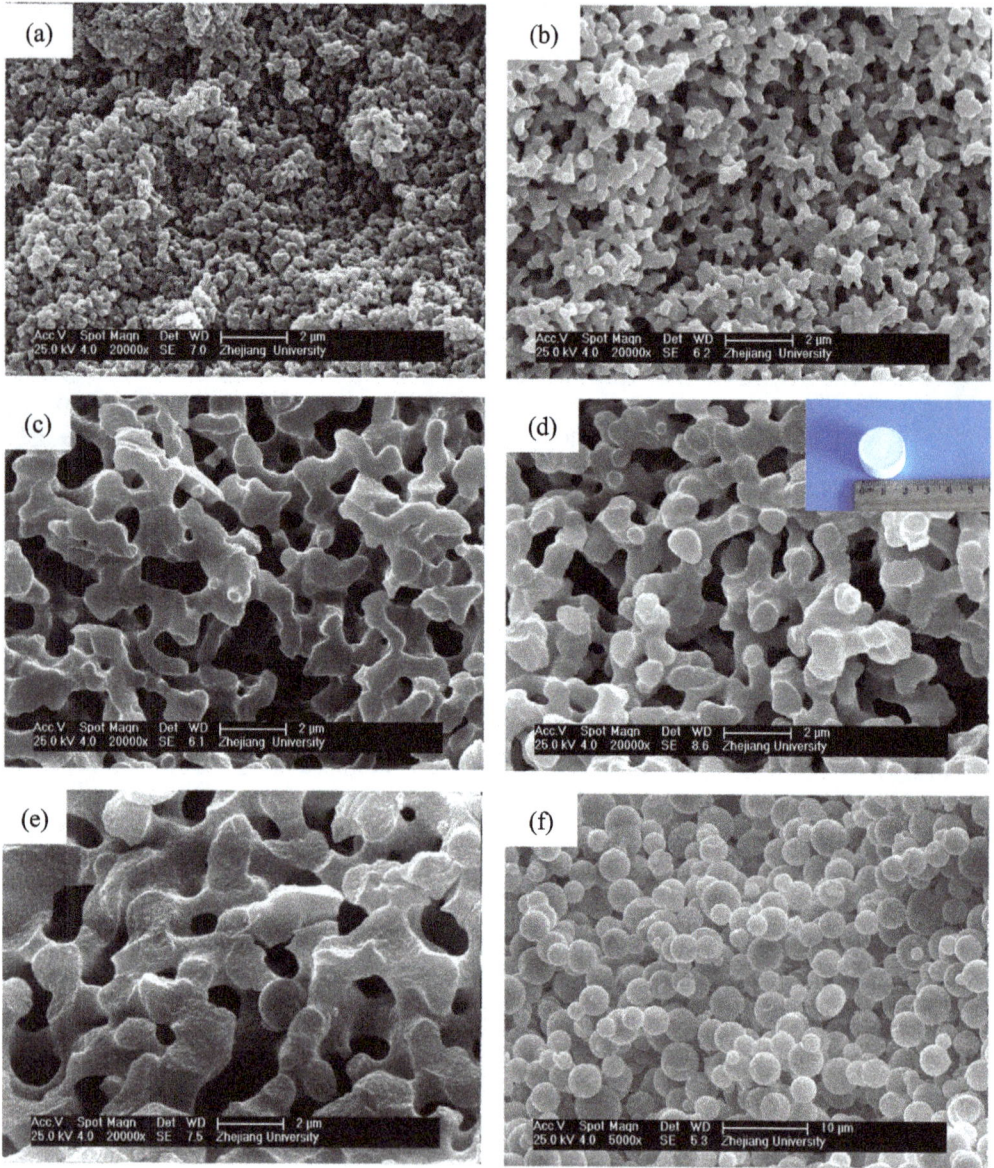

Figure 2. SEM images of dried ZrO_2 gels prepared with various W_{PEO}: (a) 0.060 g (P5), (b) 0.075 g (P6), (c) 0.090 g (P7), (d) 0.105 g (P8), (e) 0.120 g (P9) and (f) 0.135 g (P10); the insect picture in (d) is the appearance of the dried monolith of the P8 sample.

Gibbs free energy change of mixing, ΔG, can be described as follows:

$$\Delta G \propto RT \left(\frac{\varphi_1}{p_1} \ln \varphi_1 + \frac{\varphi_2}{p_2} \ln \varphi_2 + \chi \varphi_1 \varphi_2 \right) \qquad (5)$$

$$\chi \propto \frac{(\delta_1 - \delta_2)^2}{k_b T}, \qquad (6)$$

where φ_i, P_i and δ_i are the volume fraction, degree of polymerization and solubility parameter of component i ($i = 1$ or 2), respectively. k_b is the Kauri–Butanol value, R is the gas constant and T is the absolute temperature. The former two terms in parentheses represent the entropic contribution, and the last term is the enthalpic contribution. The effects of the main starting compositions (PO and PEO contents) on the phase separation and macrostructure formation of ZrO_2 dried gels will be investigated as follows:

The SEM microphotographs of the dried gels with different PO contents are shown in figure 1. It can be seen that the V_{PO} has a great impact on the macroporous structure of ZrO_2 monoliths. The morphology of gels changes from fine aggregated particles to isolated macropores. When V_{PO} is small, the sol-gel process freezes the late stage morphology of phase

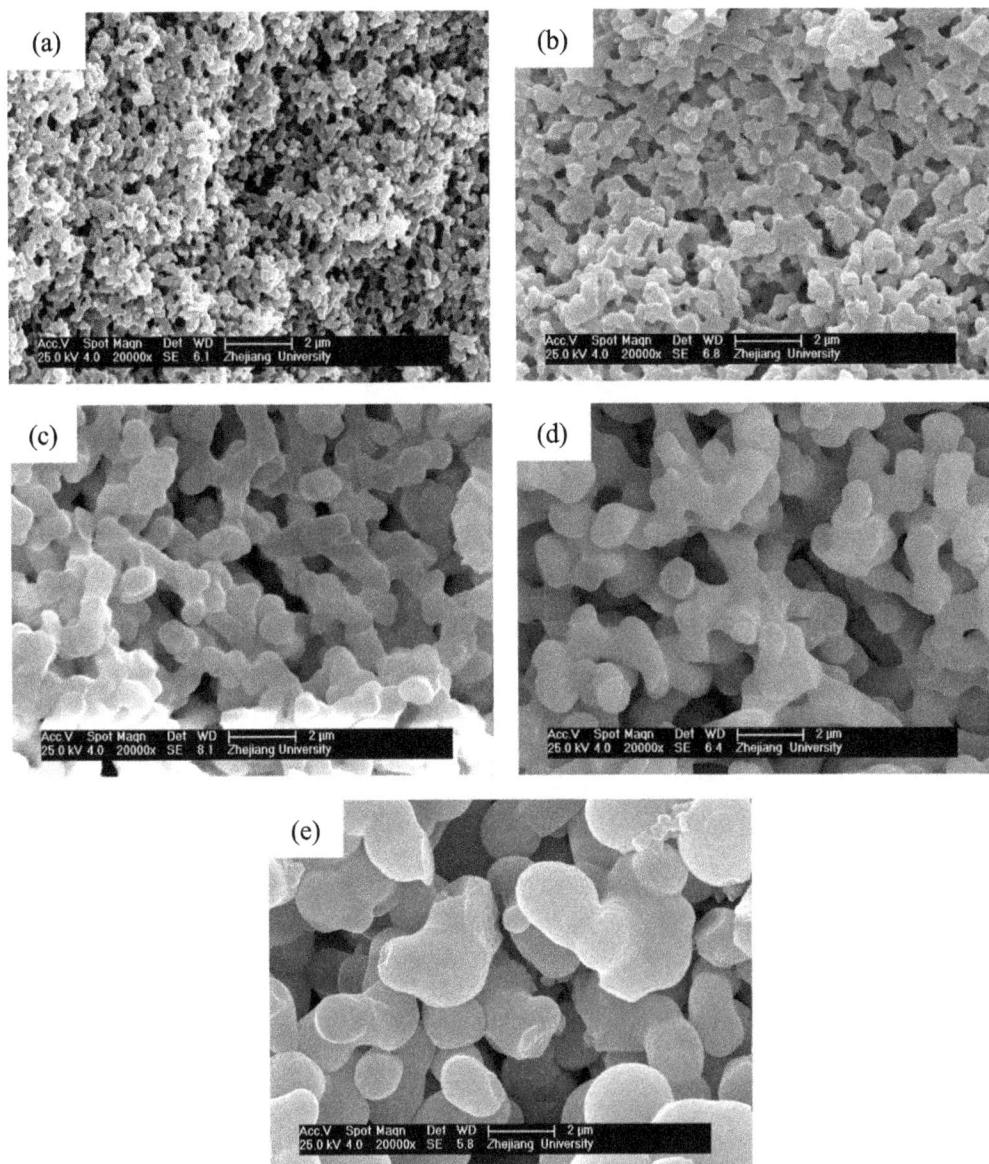

Figure 3. SEM images of dried ZrO_2 gels prepared with various solvent ratios (V_{H2O}/V_{EtOH}): (a) 1.8/3.0 (P11), (b) 2.0/2.8 (P12), (c) 2.2/2.6 (P13), (d) 2.4/2.4 (P14) and (e) 2.6/2.2 (P15).

separation; only aggregated particles are obtained (figure 1(a)). The increasing V_{PO} can dramatically shorten the gelation time of the system. Therefore, the early stage morphology of phase separation is acquired when too much PO is added (figure 1(d)).

Figure 2 presents the morphologies of dried gels prepared with different W_{PEO}. With the increasing W_{PEO}, the morphology of gels changes from nanopores (figure 2(a)), through co-continuous skeletons and pores (figures 2(b)–(e)), to particles (figure 2(f)), indicating the rising phase separation tendency. An appropriate proportion of PO and PEO can lead to a concurrent sol-gel transition and phase separation, producing co-continuous porous structures. The system is

divided into the co-continuous gels phase and the fluid phase. The fluid phase is transformed into macropores after evaporation drying. As a result, monolithic dried gels with interconnected macropores and co-continuous skeletons are obtained.

From the above equations (5) and (6), it can be seen that the increasing χ, which can be caused by the enlarging polarity difference of the gelation phase and mixed solvents, can also alter ΔG from negative to positive. As we have known, the polarity of the gel phase gradually decreases during the poly-condensation because of the consumption of high polarity hydroxy groups, and the polarity of the mixed

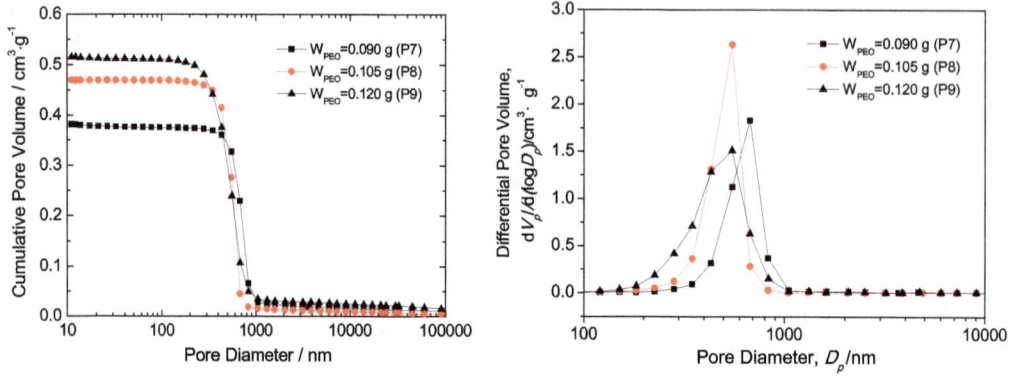

Figure 4. Macropore size distribution of the dried ZrO_2 monolith with various PEO contents evaluated by mercury porosimetry.

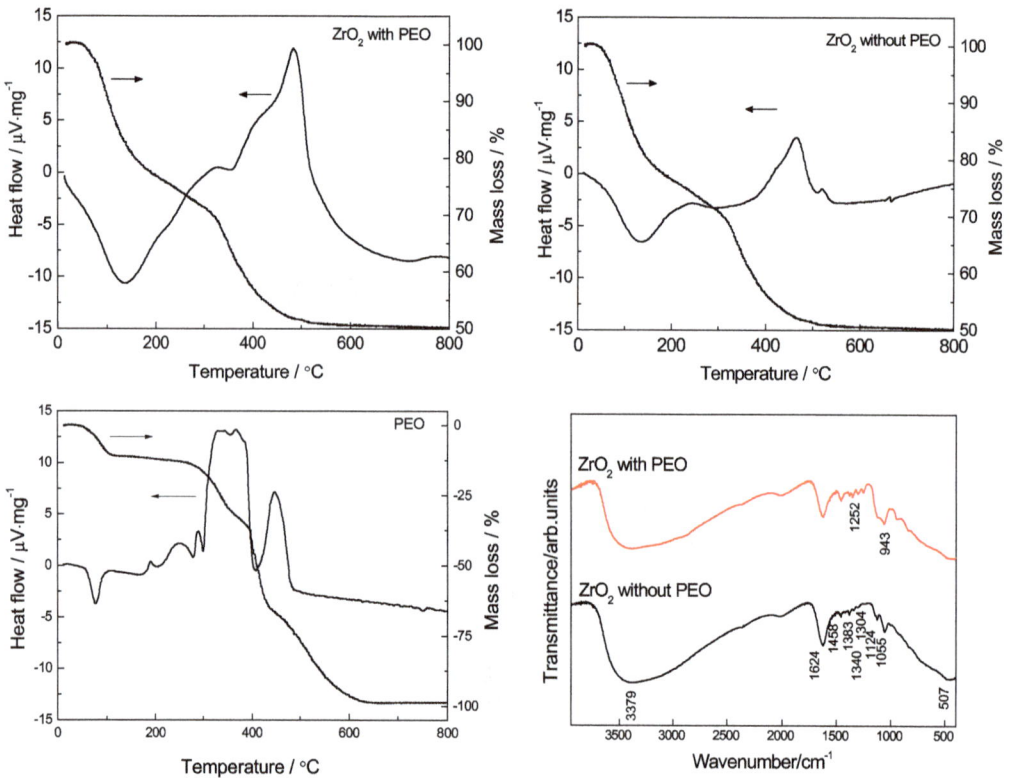

Figure 5. DTA/TG curves and FTIR spectra of dried ZrO_2 gels with and without PEO.

solvents rises due to the enlarging proportion of water. We also investigate the influence of different solvent proportions on the morphology of dried gels, as presented in figure 3. The morphologies of the dried gels change from nanopores, through co-continuous skeletons, to broken bulky skeletons with the increase of water proportion. When the proportion of water in the mixed solvents is low, the polarity difference between the gel phase and mixed solvents becomes small and leads to a weak phase separation tendency (figure 3(a)). In contrast, the phase separation tendency in the system becomes

stronger with a higher proportion of water, and the broken bulky skeletons are obtained (figure 3(e)).

The macropore size distributions of P7, P8 and P9 dried gels with interconnected macropores and co-continuous skeletons are shown in figure 4. Mercury porosimetry analysis indicates that the gels possess a narrow pore size distribution, reflecting the macrostructure formed via the spinodal decomposition [25]. The macropore size of the dried gels is distributed roughly between 0.3 and 1 μm, and the median macropore sizes of P7, P8 and P9 dried gels are 0.68, 0.55

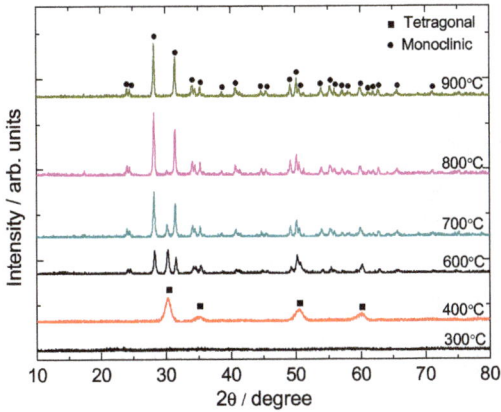

Figure 6. XRD patterns of a ZrO_2 monolith heat-treated at various temperatures.

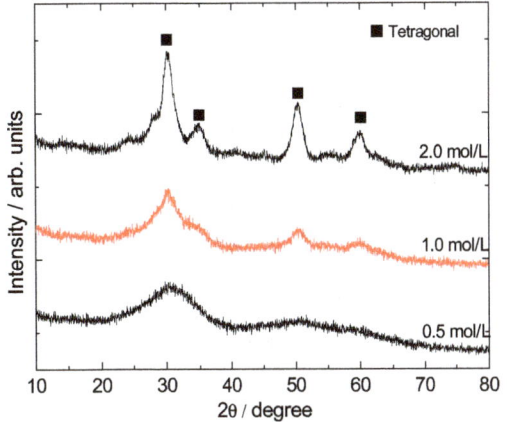

Figure 7. SEM image of ZrO_2 monoliths after heat treatment at 700 °C

and 0.55 μm, respectively. The macropore size distribution of the sample with 0.105 g PEO (P8) is narrower than the two others, indicating ideal macroporous structure. The bulk densities of the three dried gels are 1.21, 1.08 and 1.01 $g \cdot cm^{-3}$, corresponding to 20.5%, 18.3% and 17.1% of the theoretical density (5.89 $g \cdot cm^{-3}$); the total pore volumes are 0.399, 0.473 and 0.524 $cm^3 \cdot g^{-1}$; and the total porosities are 48.4%, 51.2% and 52.9%, respectively. It indicates that

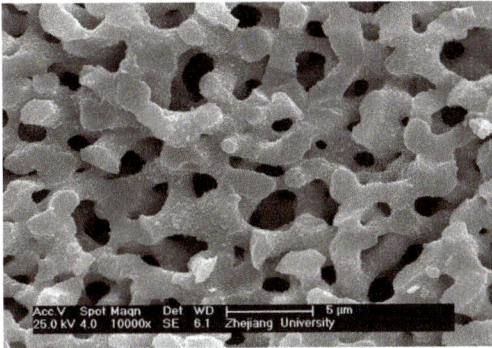

Figure 9. XRD patterns of ZrO_2 monoliths after solvothermal treatment with various ammonia concentrations.

the pore size, pore volume and porosity increase, and the bulk density of dried gels decreases with the increase of PEO content. The results agree well with the SEM morphology.

To clarify the distribution of PEO between the gel phase and liquid phase, thermal and infrared analyses were carried out, as shown in figure 5. There is an exothermal peak that resulted from the decomposition of PEO between 300 and 400 °C in the gels prepared with PEO; this peak does not appear in the DTA curve of the gels prepared without PEO. In the FTIR spectra two new peaks appear around 1252 and 943 cm^{-1} in the gel prepared with PEO. They correspond to the asymmetric torsional vibration and to the rocking vibration of the CH_2 group [43, 44], respectively. The peak around 1124 cm^{-1}, which originates from stretching of the C–O–C bond [45], is stronger in the gel prepared with PEO than in the PEO-free gel. These results confirm the presence of PEO in the dried gel fabricated with PEO. They suggest that in this ZrO_2 system, similar to the PEO-incorporated alkoxy-derived SiO_2 sol-gel [26], PEO is absorbed on the surface of ZrO_2 oligomers through hydrogen bonds [46–48], which can increase the hydrophobic–hydrophilic repulsive interaction with solvent mixtures and finally cause phase separation.

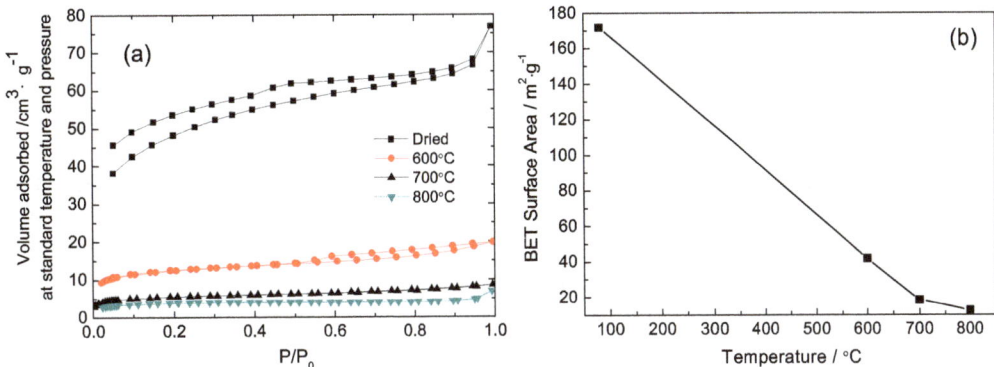

Figure 8. N_2 adsorption–desorption isotherms (a), BET surface area (b) of ZrO_2 monoliths heat-treated at various temperatures.

Figure 10. SEM images of ZrO_2 monoliths before (a) and after solvothermal treatment with various ammonia concentrations: (b) 0.5 mol L^{-1}, (c) 1.0 mol L^{-1} and (d) 2.0 mol L^{-1}.

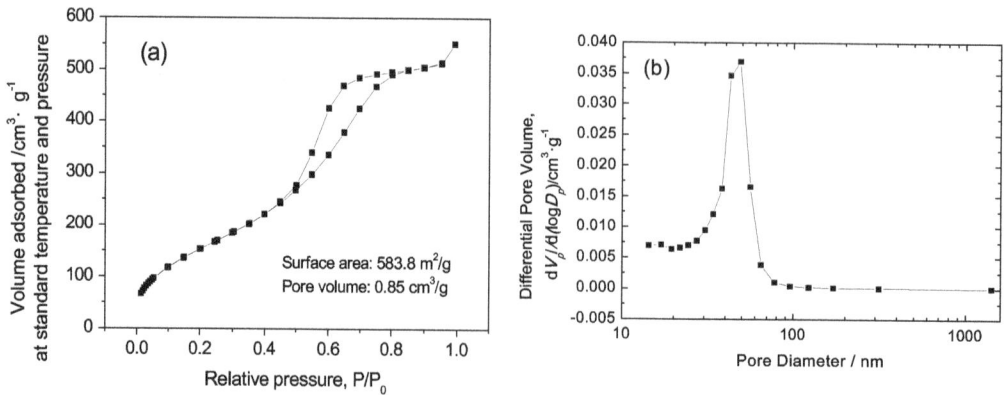

Figure 11. N_2 adsorption–desorption isotherm (a) and BJH pore size distribution (b) of a ZrO_2 monolith after solvothermal treatment with an ammonia concentration of 2.0 mol L^{-1}.

3.2. Heat treatment of porous monoliths

According to the DTA curves of the dried gels, the heat treatment of the dried gel (P8 sample) was carried out between 800 and 1100 °C for 8 h with a heating rate of 2–3 °C · min^{-1}. The impact of heat treatment on the crystallization and porous structure is examined below.

Figure 6 displays the XRD patterns of the gels heat-treated at various temperatures. No peaks are observed for the gels heat-treated at 300 °C, which manifests an amorphous state. After calcination at 400 °C, broad diffraction peaks appear due to the precipitation of tetragonal ZrO_2, and all tetragonal ZrO_2 turn to monoclinic ZrO_2 after calcination at 900 °C. From the Scherrer's equation the crystallite size of tetragonal ZrO_2 obtained at 400 °C is 7.6 nm, and the monoclinic crystallite size acquired at 900 °C is 35.4 nm. This means that increasing the heat-treated temperature can enlarge the tetragonal ZrO_2 crystallite and finally transform it into monoclinic ZrO_2 crystallite, which is more stable at a low temperature. The result is consistent with Garvie's theory [49], which believes that tetragonal ZrO_2 can exist in low temperatures when the nanoparticle size is less than 30 nm. Figure 7 presents the SEM image of gels heat-treated at 700 °C. It can be seen that the well-defined macroporous morphology is basically retained, signifying that the heat treatment does not destroy the macropore structure of ZrO_2 monoliths.

Figure 8 shows the N_2 adsorption–desorption isotherms and BET surface area of gels heat-treated at various temperatures. The as-dried and 600 °C heat-treated gels exhibit isotherms of type-IV, while the gels heat-treated at 700 and 800 °C show isotherms of type-I (figure 8(a)). The results indicate that the elevating temperature of heat treatment can exterminate the mesopores of monoliths. It is also confirmed by the corresponding BET surface area (figure 8(b)). The BET surface area decreases from 172 to 13 $m^2 \cdot g^{-1}$ after being heat-treated at 800 °C due to the disappearance of mesopores, caused by phase transformation, and the aggregation of nanoparticles due to sintering.

3.3. Solvothermal treatment of porous monoliths

The solvothermal treatment of gels is introduced to study the effects on the crystallization and modification of ZrO_2 skeletons. The choice of appropriate organic solvents plays a key role in the solvothermal synthesis, such as redox, polarity, complexation, viscosity, and so forth, and strongly influences the heterogeneous liquid–solid reactions [36]. In this study, an ethanol solution of ammonia was chosen as the solvent, and ZrO_2 monoliths were solvothermally treated with various ammonia concentrations at 180 °C for 12 h. The XRD patterns shown in figure 9 demonstrates that the peaks of tetragonal ZrO_2 crystallite become increasingly sharp when the ammonia concentration increases from 0.5 to 2.0 mol L^{-1}. This is on account of the higher solubility of ZrO_2 gel particles in the solvent with a higher concentration of ammonia and with the reaction of dissolution-precipitation, which will lead to the formation of the products with higher crystallinity at a rapid rate [26]. Figure 10 shows the macroporous morphology of ZrO_2 monoliths before and after being solvothermally treated with different ammonia concentrations. It is observed that ammonia concentration does not much affect the tailoring on the skeletons of gels. The scale of the co-continuous skeleton slightly increases, and the surface of the skeletons just become smooth.

During the drying stage of wet gels without solvothermal treatment, many micropores or/and mesopores in gel skeletons gradually decrease due to the capillary force, causing a large decrease in surface area. During solvothermal treatment of wet gels, the mesopores in the gel skeletons are regenerated via a process of dissolution/reprecipitation (Ostwald ripening). On the other hand, ammonia used as the solvent in the solvothermal treatment can improve the nucleation and growth of crystalline nanoparticles by providing more OH^- and can produce more mesopores constructed by nanoparticles. Moreover, the mesopores are not spoiled in the

subsequent drying stage. The N_2 adsorption–desorption isotherm of the ZrO_2 monolith after solvothermal treatment exhibits isotherms of type-IV with an H1 hysteresis loop, indicating the existence of mesopores (figure 11(a)). Also, the BJH pore size distribution (figure 11(b)) also verifies that the pore size concentrates in the mesopore range. According to the calculation, the BET surface area of the monolith is as high as 583.8 $m^2 \cdot g^{-1}$, the average pore diameter is 58.3 nm and the average pore volume is 0.8508 $cm^3 \cdot g^{-1}$. Compared to the as-dried gels, solvothermal treatment could generate a large amount of mesopores and significantly increase the average surface area and pore volume.

4. Conclusions

ZrO_2 gels derived from a low-cost metal salt precursor have been synthesized by an epoxide-mediated sol-gel route, accompanied by phase separation. The appropriate choice of the amounts of epoxides, solvents and polymers allowed the formation of a gel with controlled macroporous morphology. Heat treatment of the dried gels at 400 and 600 °C results in the formation of tetragonal ZrO_2 and monoclinic ZrO_2, respectively, without spoiling the macrostructure. Solvothermal treatment with an ethanol solution of ammonia increases the number and size of the micropores and mesopores, thereby increasing the BET surface area from 171.9 to 583.8 $m^2 \cdot g^{-1}$. The presented synthesis process thus enables us to produce ZrO_2 monoliths having bimodal meso-macroporous structures with adjustable pore sizes.

Acknowledgments

This work is supported by the National Natural Science Foundation of China (51372225) and Zhejiang Provincial Natural Science Foundation of China (LY13B010001).

References

[1] Tanabe K 1985 *J. Mater. Chem. Phys.* **13** 347
[2] Mercera P, Vanommen J, Doesburg E, Burggaaf A and Ross J 1991 *J. Appl. Catal.* **71** 363
[3] Wirth H and Hearn M A 1995 *J. Chromatogr.* **711** 223
[4] Ortiz-Landeros J, Contreras-García M E and Pfeiffer H 2009 *J. Porous Mater.* **16** 473
[5] Wu C H, Chen S Y and Shen P 2013 *J. Solid State Chem.* **200** 170
[6] Jung W, Hertz J L and Tuller H L 2009 *Acta Mater.* **57** 1399
[7] Gionco C, Paganini M C, Giamello E, Burgess R, DiValentin C and Pacchioni G 2014 *J. Phys. Chem. Lett.* **5** 447
[8] Larsen G, Lotero E, Petkovic L M and Shobe D S 1997 *J. Catal.* **169** 67
[9] Diaz-Torres L A, dela Rosa E, Salas P, Romero V H and Angeles-Chavez C 2008 *J. Solid State Chem.* **181** 75
[10] Fonseca F C, deFlorio D Z and Muccillo R 2009 *Solid State Ion.* **180** 822
[11] Yamaguchi T 1994 *J. Catal. Today* **20** 199

[12] Centi G, Cerrato G, Dangelo S, Finardi U, Giamello E, Morterra C and Perathoner S 1996 *J. Catal. Today* **27** 265

[13] Chuach G K 1999 *J. Catal. Today* **49** 131

[14] Chuach G K and Jaenicke S 1996 *J. Appl. Catal.* A **145** 267

[15] Minh N Q 1993 *J. Am. Ceram. Soc.* **76** 563

[16] Kim J D, Hana S, Kawagoe S, Sasaki K and Hata T 2001 *J. Thin Solid Films* **385** 293

[17] Laurent M, Schreiner U and Langjahr P A 2001 *J. Eur. Ceram. Soc.* **21** 1495

[18] Lee H L, Kim J T and Hong G G 1988 *J. Korean Ceram. Soc.* **25** 117

[19] Ortiz-Landeros J, Contreras-García M E, Gómez-Yáñez C and Pfeiffer H 2011 *J. Solid State Chem.* **184** 1304

[20] Byrappa K and Adschiri T 2007 *Prog. Cryst. Growth Char. Mater.* **53** 117

[21] Liang X and Patel R L 2014 *Ceram. Int.* **40** 3097

[22] Bluthardt C, Fink C, Flick K, Hagemeyer A, Schichter M and Volpe A Jr 2008 *Catal. Today* **137** 132

[23] Prete F, Ruzzuti A, Esposito L, Tucci A and Leonelli C 2011 *J. Am. Ceram. Soc.* **94** 3587

[24] Sakka Y, Tang F Q, Fudouzi H and Uchikoshi T 2005 *Sci. Technol. Adv. Mater.* **6** 915

[25] Nakanishi K 1997 *J. Porous Mater.* **4** 67

[26] Nakanishi K and Soga N 1991 *J. Am. Ceram. Soc.* **74** 2518

[27] Hasegawa G, Kanamori K, Nakanishi K and Hanada T 2010 *J. Am. Ceram. Soc.* **93** 3110

[28] Konishi J, Fujita K, Nakanishi K and Hirao K 2006 *Chem. Mater.* **18** 6069

[29] Konishi J, Fujita K, Nakanishi K and Hirao K 2004 *Mater. Res. Soc. Symp. Proc.* **788** 391

[30] Fujita K, Konishi J, Nakanishi K and Hirao K 2006 *Sci. Technol. Adv. Mater.* **7** 511

[31] Tokudome Y, Fujita K, Nakanishi K, Miura K and Hirao K 2007 *Chem. Mater.* **19** 3393

[32] Guo X Z, Cai X B, Song J, Zhu Y, Nakanishi K, Kanamori K and Yang H 2014 *New J. Chem.* **38** 5832

[33] Guo X Z, Li W Y, Nakanishi K, Kanamori K, Zhu Y and Yang H 2013 *J. Eur. Ceram. Soc.* **33** 1967

[34] Guo X Z, Nakanishi K, Kanamori K, Zhu Y and Yang H 2014 *J. Eur. Ceram. Soc.* **34** 817

[35] Li W Y, Zhu Y, Guo X Z, Nakanishi K, Kanamori K and Yang H 2013 *Sci. Technol. Adv. Mater.* **14** 045007

[36] Konishi J, Fujita K, Oiwa S, Nakanishi K and Hirao K 2008 *J. Chem. Mater.* **20** 2165

[37] Gash A E, Tillotson T M, Satcher J H Jr, Poco J F, Hmbesh L W and Simpson R L 2001 *J. Chem. Mater.* **13** 999

[38] Gash A E, Tillotson T M, Satchel J H Jr, Hrubesh L W and Simpson R L 2001 *J. Non-Cryst. Solids* **285** 22

[39] Gash A E, Satchel J H and Simpson R L 2004 *J. Non-Cryst. Solids* **350** 145

[40] Huggins M L 1942 *J. Am. Chem. Soc.* **64** 2716

[41] Huggins M L 1942 *J. Phys. Chem.* **46** 151

[42] Flory P J 1942 *J. Chem. Phys.* **10** 51

[43] McLachlan R D and Nyquist R A 1968 *Spectrochim. Acta A: Molec. Spectrosc.* **24** 103

[44] Durig J R, Zhen M, Heusel H L, Joseph P J, Groner P and Little T S 1985 *J. Phys. Chem.* **89** 2877

[45] Liang C Y and Marchessault R H 1959 *J. Polym. Sci.* **37** 385

[46] Tokudome Y, Fujita K, Nakanishi K, Miura K and Hirao K 2007 *Chem. Mater.* **19** 3393

[47] Saravanan L and Subramanian S 2005 *J. Colloid Interface Sci.* **284** 363

[48] Siffert B and Li J F 1989 *J. Colloids Surf.* **40** 207

[49] Garvie R 1978 *J. Phys. Chem.* **82** 218

14

Dynamic probe of ZnTe(110) surface by scanning tunneling microscopy

Ken Kanazawa, Shoji Yoshida, Hidemi Shigekawa and Shinji Kuroda

Faculty of Pure and Applied Sciences, University of Tsukuba, Tsukuba 305-8573, Japan

E-mail: kanazawa@ims.tsukuba.ac.jp

Abstract

The reconstructed surface structure of the II–VI semiconductor ZnTe (110), which is a promising material in the research field of semiconductor spintronics, was studied by scanning tunneling microscopy/spectroscopy (STM/STS). First, the surface states formed by reconstruction by the charge transfer of dangling bond electrons from cationic Zn to anionic Te atoms, which are similar to those of IV and III–V semiconductors, were confirmed in real space. Secondly, oscillation in tunneling current between binary states, which is considered to reflect a conformational change in the topmost Zn–Te structure between the reconstructed and bulk-like ideal structures, was directly observed by STM. Third, using the technique of charge injection, a surface atomic structure was successfully fabricated, suggesting the possibility of atomic-scale manipulation of this widely applicable surface of ZnTe.

Keywords: scanning tunneling microscopy, atom manipulation, ZnTe, semiconductor surface

1. Introduction

The characterization of semiconductor surfaces is important not only for fundamental scientific research but also for the application of nanotechnology. Group IV (Si and Ge) and III–V (GaAs, GaP and InSb) semiconductor surfaces have been intensively studied, particularly from the viewpoint of developing novel electronic devices. ZnTe, which we focus on in this study, is a II–VI semiconductor with a relatively wide direct band gap and is expected to be applied for optical devices such as green light-emitting diodes [1] and THz light generators and detectors [2, 3]. In addition, ZnTe is also commonly used as a substrate for CdTe self-assembled quantum dots [4]. Furthermore, it is used as a host material of diluted magnetic semiconductors (DMS), which are promising materials in the research field of semiconductor spintronics [5, 6]. For example, (Zn,Cr)Te is well known for its intrinsic room-temperature ferromagnetism, which is realized when the Cr composition is higher than 20% ($Zn_{0.8}Cr_{0.2}Te$).

Comparing to group IV or III–V semiconductors, however, the ZnTe surface has not been well comprehended. Moreover, reported studies are mainly concerned with static surface properties such as atomic structure and electronic ground state, but dynamic phenomena on the surface have been largely uncertain. ZnTe(110) surface states, which are the topic of this paper, have been studied by low-energy electron diffraction [7], scanning tunneling microscopy (STM) [8, 9], angle-resolved photoelectron spectroscopy (ARPES) [10], electron energy loss spectroscopy (EELS) [11] and density functional theory (DFT) calculations [12]. These results have revealed that the topmost cationic and anionic atoms are slightly displaced toward the inside of the crystal and vacuum from the ideal (110) structures of truncated bulk crystal, respectively, and the atomic arrangement of ZnTe (110) is fundamentally similar to that of III–V semiconductor surfaces (hereafter referred to as the reconstructed surface). The STM observations, however, showed that this surface is unstable and friable, i.e., line defects tend to be generated along the ⟨110⟩ direction (figure 1(a)) when a clean surface is obtained by cleavage [8, 9]. In addition, when the surface is scanned by STM with a positive sample bias voltage, surface atoms are often desorbed from the surface [8]. These characteristics are unique and markedly different from those of IV and III–V semiconductor surfaces.

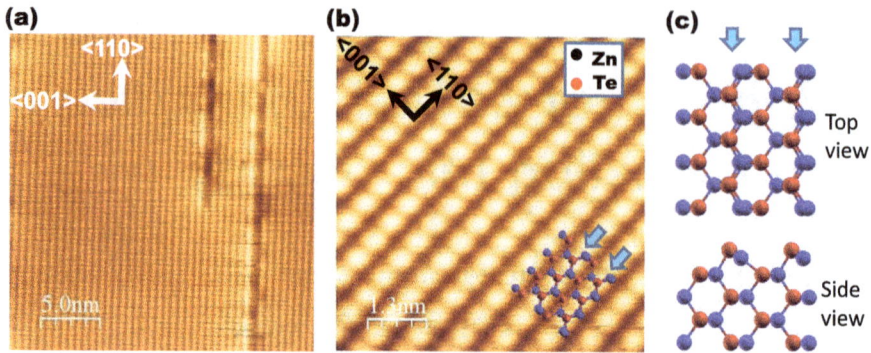

Figure 1. (a) STM image of ZnTe(110) surface ($V_s = -1.5$ V, $I_t = 60$ pA). (b) High-resolution image of ZnTe(110) surface ($V_s = -2.0$ V, $I_t = 20$ pA). (c) Schematic structures of ZnTe(110) surface.

With the reduction of device size, the development of composite materials with multiple functions is required. Control of the surface atomic and electronic structures, particularly that of atomic-scale defects, has become a key technology to meet the requirements. Therefore, to exploit the characteristics of ZnTe, it is important to clarify the structures and the associated phenomena of its surface. For this purpose, we have carried out scanning tunneling microscopy and spectroscopy (STM/STS) analysis of this surface. In addition to confirming the charge transfer on this surface, oscillation in tunneling current, which is considered to reflect the positional fluctuation of the topmost Zn atom induced by charge injection was observed. Furthermore, by applying this mechanism, we have succeeded in fabricating atomic structures on the surface. Such fabrication is expected to play a role, for example, in modulating the local potential of a surface to design functional devices with various electronic properties.

2. Experimental methods

A p-type ZnTe single crystal doped with phosphorus of a concentration of 10^{18} cm^{-3} was used as a sample in this study. All STM/STS measurements were performed with an Omicron low-temperature scanning tunneling microscope under ultrahigh-vacuum conditions ($<10^{-8}$ Pa) at 8 K and 77 K using an electrochemically sharpened W tip ($\phi = 0.3$ mm). A ZnTe(110) surface was obtained by cleaving the sample in a high vacuum ($\sim 1 \times 10^{-5}$ Pa).

3. Results and discussion

Figure 1(a) shows a typical STM image obtained at a sample bias voltage of $V_s = -1.5$ V. Since occupied lone pair states are formed on the topmost Te atoms due to charge transfer from topmost Zn atoms, the Te atoms can be imaged as bright points at a negative sample bias voltage. In addition to the bright lines of Te atoms along the $\langle 110 \rangle$ direction, several dark lines along the same direction were observed. According to the previous study [8], these dark lines are line defects

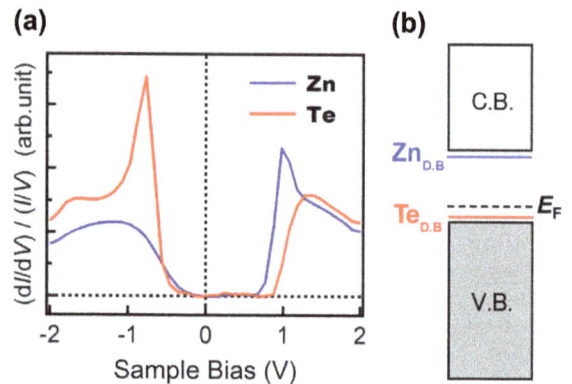

Figure 2. (a) Normalized differential conductance (NDC)—sample bias voltage spectra obtained above a Zn atom (blue) and a Te atom (red). The setpoint voltage and current were $V_s = -2.0$ V and $I_t = 20$ pA, respectively. The peaks shown in the blue (+1.0 eV) and red (−0.8 eV) lines indicate the energy levels of the dangling bond states of the surface Zn and Te atoms, respectively. (b) Schematic of the ZnTe band structure derived from the NDC-V_s spectra.

caused by the absence of topmost atoms and are intrinsic defects of the ZnTe(110) surface, as mentioned above. From the different brightness on the both sides of the dark-line defects, we can determine the indexes of the in-plane crystallographic axes on the surface (the distinction between [110] and [−110] axes), as shown in figure 1(a) [9]. And we could determine the correspondence of the STM image to the atomic arrangement on the surface (figure 1(c)) as shown in figure 1(b).

To investigate the surface electronic structure in more detail, we carried out STS measurement. Figure 2 shows the results of normalized differential conductance (NDC) analysis, which reflects the local density of states (LDOS). NDC-V_s curves were measured on topmost Zn (blue) and Te (red) atoms, respectively. There are differences between the two curves, particularly at the rising edges from the semiconductive gap region near $V_s = 0$. At the rising edge at a positive (negative) sample bias voltage, the NDC-V_s spectrum obtained at the Zn (Te) position exhibits a higher peak than that obtained at the Te (Zn) position. These results can be comprehensively explained by the mechanism that the empty and occupied dangling bond states are formed through the charge transfer from cationic Zn to anionic Te associated with the surface reconstruction. Similar charge transfer is well known to occur on several semiconductor surfaces such as Si (100) dimer and GaAs(110) reconstructed surfaces, which has been confirmed by STM observation with atomic resolution [13, 14]. The STS analysis has confirmed the surface states of ZnTe(110) in the real space. The observed energy levels correspond to the surface states studied by ARPES [10], EELS [11], and DFT [12], which have suggested that the anionic Te (cationic Zn) dangling bond state is located near the valence band maximum (conduction band minimum), as shown in figure 2(b). Here, the energy gap of the host ZnTe was determined from the distance between the Zn (−1.0 V) and Te (+1.3 V) peaks in the spectra, because we have to remove the influence of their surface states. The gap energy obtained is about 2.3 eV, which is close to the previously reported energy gap of ZnTe (2.39 eV at 4.2 K).

In this STS measurement, a small current set point (−2 V, 20 pA), which was used to observe the STM image in figure 1(b), was chosen to reduce the effect of the surface instability reported in an STM measurement for the positive sample bias voltage region. In fact, no remarkable phenomenon related to the instability was observed during the STS measurements. However, with a larger tunneling current setpoint, the surface instability appeared. Therefore, next we investigated the instability observed during the STM measurement at a positive sample bias voltage in more detail. To observe the expected dynamics on the surface, we studied the effect of the electron-injection from the STM tip to the sample at 8 K with the tip position fixed at meshed points during the two-dimensional scan over the surface.

Figure 3(a) shows an STM image of the surface with an 8×8 mesh in each of which the STM electron-injection sequence shown in figure 3(b) was carried out as follows. (1) during an STM scan with the set-point conditions (sample bias and current) of $V_s = -2.0$ V and $I_t = -1.0$ nA, the tip was held at a certain line crossing in the mesh ($t = 0$). (2) After opening the STM feedback loop, the sample bias voltage was changed to a positive value of $V_s = +1.5$ V (red arrow 1), at which the tunneling current usually became very small reflecting the I–V character of this surface. (3) Then, the feedback loop was closed at the set point with a tunneling current of +0.03 nA (green arrow). In this step, the distance between the STM tip and the sample became small and electronic injection from the tip to the sample started to occur (injection). After a very short time, we open the feedback loop again to fix the distance between the tip and the surface and started measurement of tunneling current at the same time. (4) After an electron injection time of $t_i = 0.3$ s, the bias voltage was returned to $V_s = -2.0$ V (red arrow 2).

Figure 3(c) shows the tunneling current as a function of time measured in each 8×8 grid in figure 3(a). The origin, $t_{mes} = 0$, is the time when the tunneling current was set to +0.03 nA. Variations in the tunneling current can be seen. Figure 3(d) shows three typical examples of data selected from figure 3(c). The current appears to oscillate between the

original 'low-current' state and another 'high-current' state. No damage was observed in the STM image obtained after sequential measurements, as shown in figure 3(e), which indicates that, during the sequence, the charge injection of this condition induced only current oscillation without any destructive contact or scratching of the surface with the STM tip.

As a possible reason of the current oscillation, we may consider first the simple reflection of a temporal occupying event of electrons at the surface such as the Coulomb-blockade, which, however, is not consistent with our experimental results. If Coulomb-blockade was formed, we should observe the effect of the 'staying electron' as the drop of the tunneling current. However, we observed the jump-up of tunneling current from the original 'low-current' state at the start of the injection of the tunneling electron.

Comparing the current oscillation with the unstable STM observation with a positive sample bias voltage as reported by the earlier study [8], a possible phenomenon to be able to explain the current oscillation is a structural variation between the original surface-reconstructed structure and another structure, which causes the 'high-current' state, i.e., the STM sequence induces a structural change of the topmost Zn or Te atom toward the vacuum side from the reconstructed structure.

What causes the structural change? A possible mechanism to excite the structural change is the effect of inelastic tunneling which is widely observed. However, any current oscillation was not observed in a similar sequence with a negative sample bias voltage, which should appear in the case of inelastic tunneling: inelastic tunneling occurs at the same positive and negative bias voltages.

In order to establish a possible interpretation for the surface dynamics induced by the electron injection of the STM sequence, we carried out a DFT calculation using the Perdew–Burke–Ernzerhof generalized gradient approximation with the ABINIT code and plane-wave-based norm-conserving pseudopotentials. A supercell including a slab model with five ZnTe(110) layers was used for the calculation. The dangling bonds of atoms in the bottom layer were passivated by hydrogen atoms, as shown in figure 4(a). The energy cutoff of the calculation was 60 Ry. As the sampling k-points, we used $8 \times 8 \times 2$ grids within the first Brillouin zone. Here, we focused on a possibility of the atomic displacement of the topmost Zn atom, because the Zn atom is considered to be most easily desorbed from the surface, referring to the previous study [9], which reports that the defects on the cleaved ZnTe(110) surface mainly consist of the absences of Zn only or Zn–Te pairs and not Te only. As a possible computational model to describe the atomic displacement of the surface Zn atom toward the vacuum side from the reconstructed structure, we consider the ideal (110) surface structure consisting of a truncated bulk structure. The stability of the surface was analyzed by comparing the total energy difference between the ideal and reconstructed surface structures in the neutral state and in charged states with up to six additional electrons per supercell.

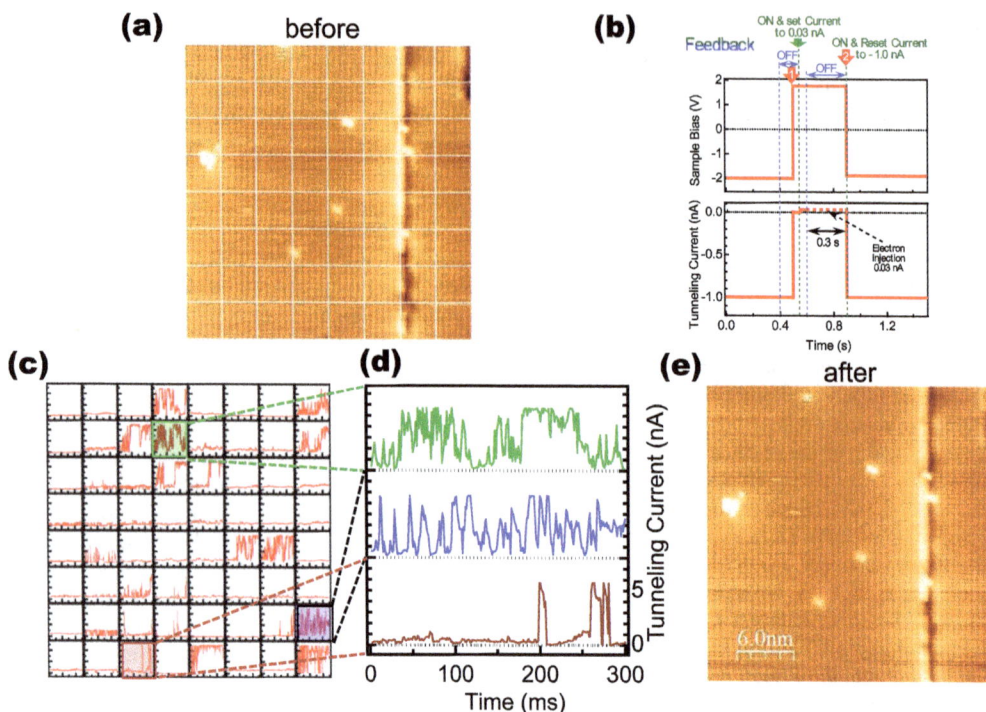

Figure 3. (a) STM image obtained at 8 K ($V_s = -3.0$ V, $I_t = 1.0$ nA) with 8×8 mesh in which time-dependent STM measurements were carried out. (b) Sequence used in measurement of temporal change of the tunneling current (t_i: electron injection time). At first we opened the STM feedback loop and the sample bias voltage was changed to a positive value of $V_s = +1.5$ V (red arrow 1). Then, we closed the feedback loop at the set point with a tunneling current of $+0.03$ nA (green arrow) to start injection of electrons from the tip to the sample (injection). Almost at the same time, we opened the feedback loop again to fix the tip and started measurement of tunneling current. After an electron injection, the bias voltage was returned to $V_s = -2.0$ V (red arrow 2). (c) Measurement of temporal change of the tunneling current obtained for the 8×8 mesh shown in (a). (d) Three typical plots in a magnified scale selected from those in (c). The spectra shown in green, blue and red respectively correspond to those in the squares in (c) indicated by the same colors. (e) STM image obtained after time-resolved STM measurements shown in (c) ($V_s = -3.0$ V, $I_t = 1.0$ nA).

Figure 4(b) shows the summary of the results of our DFT calculation, showing the total energy difference between the ideal and reconstructed surface structures ($E_{ide} - E_{rec}$) as a function of the number of additional electrons per supercell. As shown in the figure, the reconstructed surface is more stable than the ideal surface in the neutral state (without additional electrons). This result validates the surface reconstruction of this surface with the buckling of the Zn–Te structure, similar to the cases of the Si dimer and Ga–As structure appearing upon their surface reconstruction [13, 14]. On the other hand, the surface stability changes with the charge injection; the ideal configuration becomes more stable when the number of additional electrons is larger than 1.5. Considering the result of our theoretical study with the result of our experiment, in which the variations in the tunneling current was observed, these results may indicate that the electrons injected from an STM tip induce modulation of the surface structure; the charge injection into the surface region may induce an atomic oscillation of the surface Zn atoms between the reconstructed structure and a structure with displacement of the Zn atom toward the vacuum side as schematically shown in figure 4(c).

Furthermore, in order to discuss the effect of the electrons injected from the STM tip in more detail theoretically, we calculated partial density of states (PDOS) for the topmost Zn and Te atoms and the Zn atom located in the third (110) layer, which should correspond to the state far from the topmost surface, of the reconstructed surface with no and one additional electron. The results of the PDOS calculations show that the PDOS of the topmost Zn atom is the largest near the conduction band minimum in the neutral condition (figure 4(d)) and the state is occupied by one additional electron with the shift of the Fermi energy (E_F) (dotted lines), as indicated by a black arrow (figure 4(e)). In addition, the theoretical results of the DOS (figure 4(f)) and spatial distribution of the LDOS integrated over $E_F - 0.2$ eV $< E < E_F$ (figure 4(g)) calculated for the ideal surface structure with two additional electrons, which may cause the inversion of the surface stability, clearly indicate that the additional electrons are preferentially located at the topmost Zn atom of the ideal structure. These results are consistent with those of our STS measurement, in which we observed the Zn_{DB} state near the conduction band minimum, suggesting that the injected electrons tend to be located at the dangling bond states of the topmost Zn. Upon the occupation of the empty Zn_{DB} states by

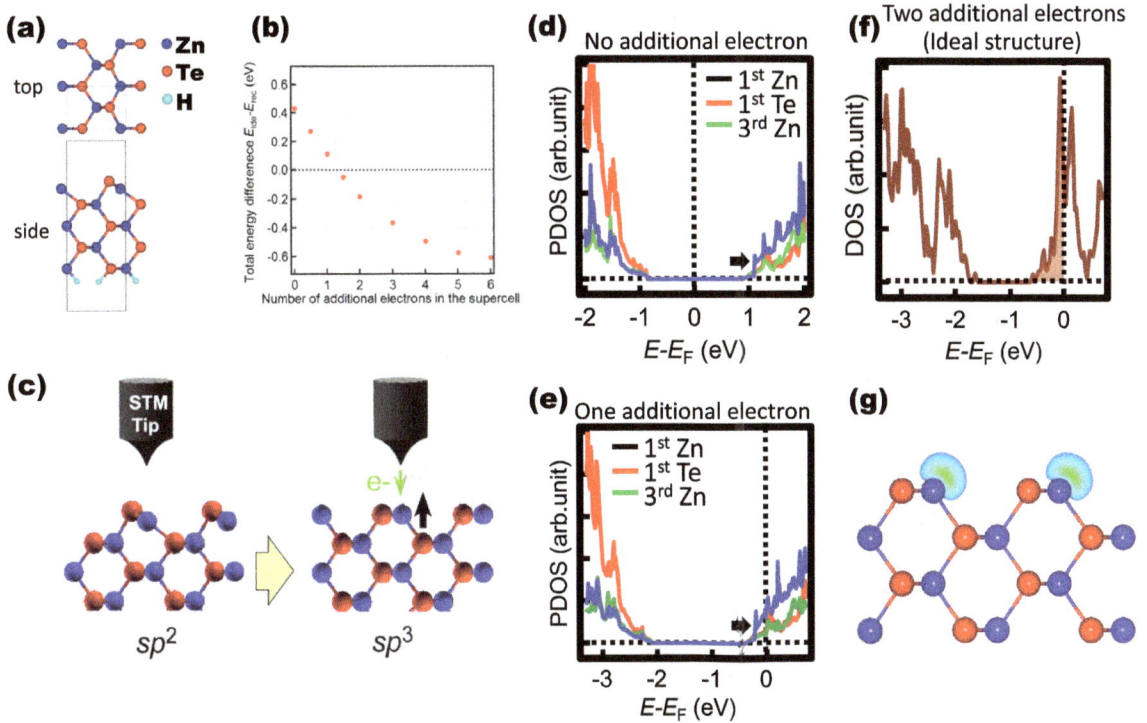

Figure 4. (a) Schematic illustration of the ZnTe structure used for the DFT calculation. (b) Total energy difference between the ideal and reconstructed surface structures (E_{ide}−E_{rec}) as a function of the number of injected electrons. (c) Schematic diagrams showing atom manipulation by charge injection. Electrons injected from the STM tip induce the displacement of the topmost Zn atoms toward the vacuum side. (d) and (e) Calculated partial density of states (PDOS) for the topmost Zn and Te atoms and the Zn atom located in the third (110) layer of the reconstructed surface with no (d) and one (d) additional electron. (f) DOS calculated for the ideal surface structure with two additional electrons. (g) Calculated spatial distributions of the LDOS integrated over E_F−0.2 eV < E < E_F (E_F: Fermi energy) for the ideal surface structure with two additional electrons.

the additional electrons, the bonding state of the Zn atom with neighboring Te atoms may change from the flat sp^2-like conformation to the sp^3-like tetrahedral coordination. The distance between the STM tip and the Zn atom decreases upon the conformational change from the reconstructed structure to the displaced structure with Zn atom moved toward the vacuum side, producing the high state observed in tunneling current. According to our theoretical study, the conformational change is considered to be a possible mechanism for producing the current oscillation observed during the STM electron-injection sequence.

If the STM tip and the surface Zn atom are closely located, the electrons are injected to the dangling bond with a high possibility, and, the displacement should occur frequently. In this case, the 'high-current' state is supposed to be observed continually. Actually, comparing the details of the current oscillations shown in figure 3(c), the tunneling current in the top spectrum of figure 3(d) tends to be high, for example, and in the bottom spectrum the oscillation occurs less often. This difference is considered to be due to the relative lateral positions of the STM tip to that of a Zn atom below the tip. Since the measurement in each mesh was started without checking the relative positions, the STM tip might be located only slightly above the Zn atom in some

cases, but at a much greater distance from the Zn atom in other cases.

Next, we consider the relationship between the structural oscillations and the atomic desorption from the surface induced by STM observation at a positive sample bias voltage in the previous study. No damage reflecting the atomic desorption was observed with the measurement condition of the STS measurement (figure 1) and during the above STM sequence (figure 3) in this study. However, the atomic desorption occurred when we used the setpoint tunneling current of ten times larger, I = 0.3 nA. Figures 5(a) and (c) show the STM images obtained before and after the electron injection sequence shown in figure 3, which was carried out at the grid points of the mesh drawn in figure 5(a). Formation of defects is clearly shown in figure 5(c). When the surface Zn atom is displaced toward to the vacuum by the electron injection, tunneling current increases. In fact, much larger tunneling current, which exceeded the detection limits of 5 nA was observed during the electron injection as shown in figure 5(b), which might induce the atomic desorption as a result of an enhancement of the atomic oscillation.

Here, the appearances of the atomic deficiencies generated by the desorption process are quite similar to the intrinsic line defects on the cleaved ZnTe(110) surface recognized as a

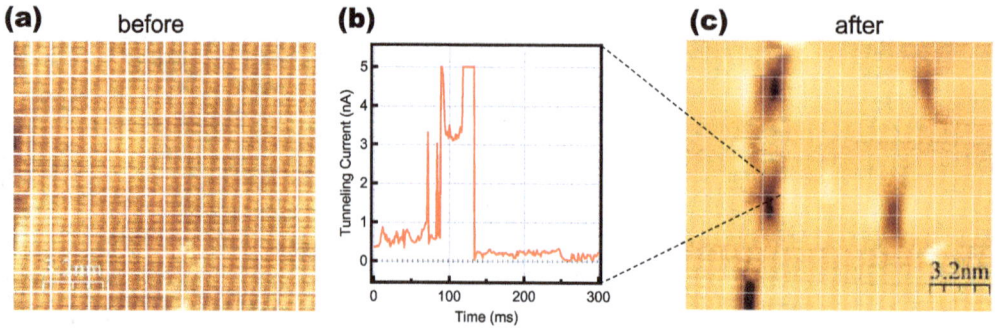

Figure 5. (a) STM image before the electron-injection sequence with $V_s = +1.5$ V and $I_t = 0.3$ nA ($V_s = -3.0$ V, $I_t = 1.0$ nA). (b) Temporal change of the current obtained during the electron-injection sequence carried out on the area indicated by dotted lines. (c) STM image of the surface defects generated by the electron-injection sequence ($V_s = -3.0$ V, $I_t = 1.0$ nA).

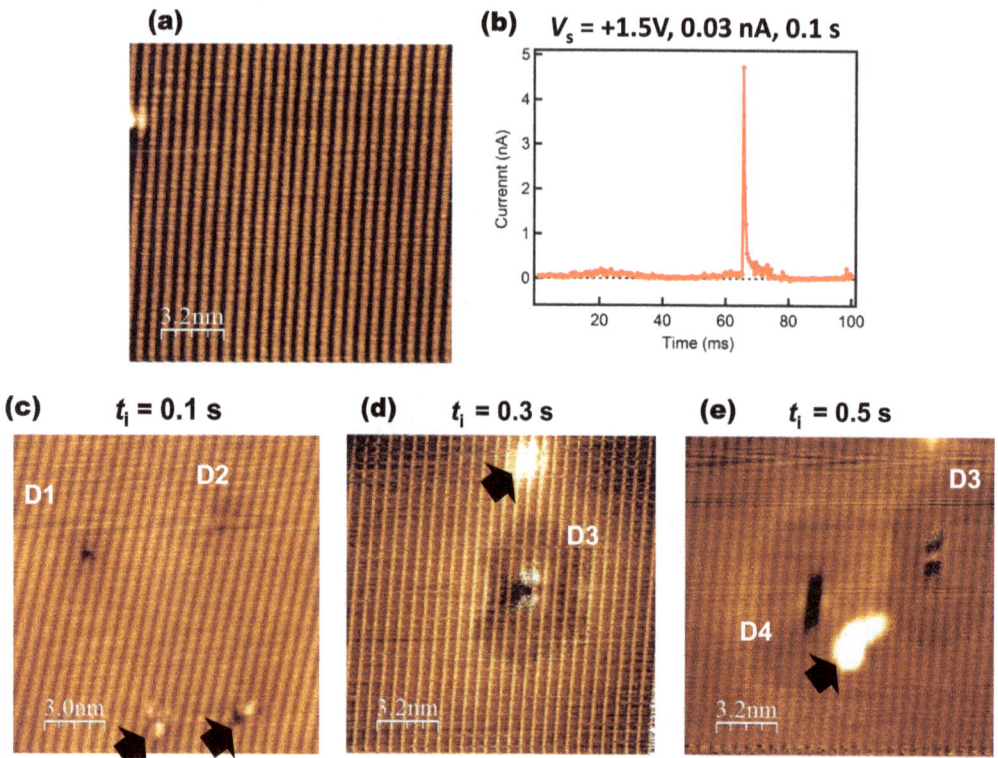

Figure 6. (a) STM image of a ZnTe surface obtained before the electron injection treatment ($V_s = -3.0$ V, $I_t = 1.0$ nA). (b) Typical temporal change of the current obtained during the treatment. (c), (d), (e) STM images after the treatment with three different electron injection times of t_i ((c) 0.1 s, (d) 0.3 s, (e) 0.5 s, $V_s = +1.5$ V, $I_t = 30$ pA). D1 to D4 are surface defects induced by each treatment. D3 in (d) and (e) indicate exactly the same defect. Black arrows indicate clusters of atoms desorbed by the manipulation.

missing row of Zn atoms or Zn–Te pair, though the length of the generated defects are quite short. Therefore, it is reasonable to consider that the short defects arise from the position of the surface Zn atom, corresponding to the theoretical calculation, which supports the model that the charge injection from the STM tip induces the displacement of the surface Zn atom.

This result suggests that we should consider the experimental parameters to enhance the interaction between the

STM tip and the surface atom during the electron injection, in order to induce the desorption of surface atoms. Specifically, possible reasons, which may influence the probability of the desorption, include the tunneling current, the distance, the electric field between the STM tip and the sample surface, and the local temperature.

To verify the mechanism, we carried out current injection under different conditions. Namely, we observed the surface and placed the STM tip immediately above a Zn atom, and then the sequence shown in figure 3(b) was carried out. The sample temperature was set at 77 K.

Figure 6(a) shows the surface observed before the treatment. Figure 6(b) shows a typical time-resolved spectrum, i.e., the tunneling current as a function of time, in which $t_{mes} = 0$ corresponds to the moment when the tunneling current was set to +0.03 nA, similarly to the case shown in figure 3(d). The tunneling current jumped from 0.03 to ~5 nA and then disappeared. Figure 6(c) shows the STM image after the treatment with the injection time of $t_i = 0.1$ s, in which an atomic-scale defect was generated. Namely, the jump in tunneling current shown in figure 6(b) is considered to correspond to the desorption of the surface Zn atom. To confirm this result, we carried out the same process with increased electron injection times t_i to 0.3 s and 0.5 s, the results of which are shown in figures 6(d) and (e), respectively. Larger defects were generated with increasing t_i, and clusters considered to be formed by the removal of atoms were observed, as indicated by arrows. After the desorption of the Zn atom immediately below the STM tip, the adjacent Zn atoms may have been desorbed depending on the injection time. Further analysis is necessary to determine in detail the conditions required for desorption. Although the lengths of the defects were short, their appearances were very similar to the intrinsic line defect on the ZnTe(110) surface shown in figure 1(a) in that they run along the ⟨110⟩ direction with single-atomic-row-width on the surface.

Here, the atomic manipulations were demonstrated at 77 K with the same parameter as that used in the previous experiment (figure 3) at 8 K. The higher environmental temperature of 77 K is only different experimental parameter from the former experiment. Considering the two results that the ten times larger tunneling current at 8 K and the higher environmental temperature of 77 K could induce the atomic desorption, environmental temperature may work as an additional driving force to overcome the potential hill of the change in state of the surface Zn atom located just below the STM tip [15], and the atomic desorption might have been induced easier at 77 K than at 8 K. Similar temperature dependence is known for the case of Si(001) surface dimer, where dimer flip-flop motion has temperature dependence [16]. The relationship between atomic desorption and STM parameters is yet unclear. It involves several factors, such as local heating by the tunneling current and the relative position between the tip and surface atoms, which affects the electric field distribution. Further study is necessary to understand the mechanism of the atomic oscillation and desorption.

Finally, we focus on the generated defects. An elliptical disk-shaped dark area was observed around the defects, as shown in figure 6, which is similar to that around the atomic-scale defects and surface steps observed on a GaAs surface [17]. Namely, the results suggest the modulation of the electronic structure in the region, such as charge doping to the host semiconductor [18]. Since a dark area is observed at a negative sample bias voltage, a positively charged dopant center is supposed to be formed here. Although the band structure must be examined in detail by taking into account a tip-induced band bending [18, 19], this atomic-scale

fabrication may give a possibility for controlling local electronic structures of this surface.

In fact, in (Ga,Mn)As, for example, activation energy of Mn acceptor was controlled by changing the condition of As vacancies around the Mn atom [20]. On the other hand, ZnTe is known to be a host material of (Zn,Cr)Te, one of the most promising DMSs with intrinsic room temperature ferromagnetism, as mentioned earlier. According to the results of previous studies [21, 22], the magnetic behavior of (Zn,Cr)Te and the electronic states of the host material ZnTe have a close relationship; therefore, the manipulation of the surface electronic states is a key technology for realizing future advances in semiconductor spintronics associated with, for example, nanomagnetism. Although there are many issues to overcome, such as removal of clusters of the desorbed atoms as indicated by black arrows in figures 6(c)–(e), for future application with further optimization and improvement of the manipulation techniques, we may be able to fabricate a desired atomic-scale electronic structure, for example, a local potential modulation, on a surface.

4. Conclusions

The charge transfer associated with the surface reconstruction of ZnTe(110) was confirmed in real space at the atomic scale. The tunneling current oscillation, which should be concerned with the positional change of a surface Zn atom between the reconstructed and bulk-like displaced structures as a result of charge injection, was directly observed using STM and was found to be in good agreement with the theoretical calculation of the structural stability. Analysis of the conformational change may provide information on the environmental conditions surrounding a Zn atom. In addition, although the conditions required to induce a conformational change must be clarified, atomic-scale defects were successfully generated below the STM tip using the atom manipulation process based on charge injection. Since the generated defects have similar characteristics to such as the charge dopant in a host semiconductor, the technique of precise atom manipulation may be a key factor in the future development of spintronics or nanomagnetics based on DMS materials.

Acknowledgment

Support from Japan Society for the Promotion of Science in the form of Grants-in-Aid for Scientific Research is acknowledged.

References

[1] Tanaka T, Saito K, Nishio M, Guo Q and Ogawa H 2009 *Appl. Phys. Express* **2** 122101
[2] Reimann K 2007 *Rep. Prog. Phys.* **70** 1597
[3] Wu Q, Litz M and Chang X-C 1996 *Appl. Phys. Lett.* **68** 2924

[4] Karczewski G, Maćkowski S, Kutrowski M, Wojtowicz T and Kossut J 1999 *Appl. Phys. Lett.* **74** 3011

[5] Ferrand D *et al* 2000 *J. Cryst. Growth* **214/215** 387

[6] Kuroda S, Nishizawa N, Takita K, Mitome M, Bando Y, Osuch K and Dietl T 2007 *Nat. Mater.* **6** 440

[7] Meyer R J, Duke C B, Paton A, So E, Yeh J L, Kahn A and Mark P 1980 *Phys. Rev.* B **22** 2875

[8] Wierts A, Ulloa J M, Celebi C, Koenraad P M, Boukari H, Maingault L, Andre R and Mariette H 2007 *Appl. Phys. Lett.* **91** 161907

[9] Çelebi C, Arı O and Senger R T 2013 *Phys. Rev.* B **87** 085308

[10] Qu H, Kanski J, Nilsson P and Karlsson U 1991 *Phys. Rev.* B **43** 9843

[11] Ebina A, Asano K and Takahashi T 1978 *Phys. Rev.* B **18** 4332

[12] Ferraz A C, Watari K and Alves J L A 1994 *Surf. Sci.* **307–309** 959

[13] Hata K, Shibata Y and Shigekawa H 2001 *Phys. Rev.* B **64** 235310

[14] Feenstra R M, Stroscio J A, Tersoff J and Fein A P 1987 *Phys. Rev. Lett.* **58** 1192

[15] Hata K, Sainoo Y and Shigekawa H 2001 *Phys. Rev. Lett.* **86** 3084

[16] Yoshida S, Kimura T, Takeuchi O, Hata K, Oigawa H, Nagamura T, Sakama H and Shigekawa H 2004 *Phys. Rev.* B **70** 235411

[17] Wijnheijmer A P, Garleff J K, Teichmann K, Wenderoth M, Loth S and Koenraad P M 2011 *Phys. Rev.* B **84** 125310

[18] Garleff J K, Wijnheijmer A P, Enden C N V D and Koenraad P M 2011 *Phys. Rev.* B **84** 075459

[19] Sinthiptharakoon K, Schofield S R, Studer P, Brázdová V, Hirjibehedin C F, Bowler D R and Curson N J 2014 *J. Phys.: Condens. Matter* **26** 012001

[20] Lee D H and Gupta J A 2010 *Science* **330** 1807

[21] Ozaki N, Okabayashi I, Kumekawa T, Nishizawa N, Kuroda S and Takita K 2005 *Appl. Phys. Lett.* **87** 192116

[22] Ozaki N, Nishizawa N, Marcet S, Kuroda S, Eryu O and Takita K 2006 *Phys. Rev. Lett.* **97** 037201

15

Giant multiferroic effects in topological GeTe-Sb$_2$Te$_3$ superlattices

Junji Tominaga[1], Alexander V Kolobov[1], Paul J Fons[1], Xiaomin Wang[1],
Yuta Saito[1], Takashi Nakano[1], Muneaki Hase[2], Shuichi Murakami[3],
Jens Herfort[4] and Yukihiko Takagaki[4]

[1] Nanoelectronics Research Institute, National Institute of Advanced Industrial Science & Technology (AIST), Tsukuba Central 4, 1-1-1 Higashi, Tsukuba 305-8562, Japan
[2] Faculty of Pure and Applied Science, University of Tsukuba, 1-1-1 Tennodai, Tsukuba 305-8573, Japan
[3] Department of Physics & TIES, Tokyo Institute of Technology, 2-12-1 Ookayama, Meguro-ku, Tokyo 152-8551, Japan
[4] Paul-Drude-Institut für Festkörperelektronik, Hausvogteiplatz 5-7, 10117 Berlin, Germany

E-mail: j-tominaga@aist.go.jp

Abstract

Multiferroics, materials in which both magnetic and electric fields can induce each other, resulting in a magnetoelectric response, have been attracting increasing attention, although the induced magnetic susceptibility and dielectric constant are usually small and have typically been reported for low temperatures. The magnetoelectric response usually depends on d-electrons of transition metals. Here we report that in $[(GeTe)_2(Sb_2Te_3)_l]_m$ superlattice films (where l and m are integers) with topological phase transition, strong magnetoelectric response may be induced at temperatures above room temperature when the external fields are applied normal to the film surface. By *ab initio* computer simulations, it is revealed that the multiferroic properties are induced due to the breaking of spatial inversion symmetry when the p-electrons of Ge atoms change their bonding geometry from octahedral to tetrahedral. Finally, we demonstrate the existence in such structures of spin memory, which paves the way for a future hybrid device combining nonvolatile phase-change memory and magnetic spin memory.

Keywords: multiferroics, magnetoresistance, topological insulator, chalcogenide superlattice, GeTe-Sb$_2$Te$_3$, phase change memory, spintronics, computer simulation

1. Introduction

Although *alternating* electric and magnetic fields are intrinsically coupled, as described by the Maxwell equations, a *static* electric (magnetic) field usually does not induce magnetization (polarization) except in a rather limited—but rapidly growing—class of materials named multiferroics. The idea that for certain classes of magnetocrystalline symmetry there may be 'a linear coupling between magnetic and electric fields in such media, which would cause, for example, a magnetization proportional to an electric field' was proposed by Landau and Lifshitz in their classical textbook on theoretical physics over 50 years ago [1]. Since the concept and theory were first proposed, various multiferroic (MF) materials have been discovered and studied, e.g., Cr$_2$O$_3$ and GaFeO$_3$ [2–5]. In multiferroics, the electric dipole, **P**, linearly responds to an external magnetic field **B** as $\Delta\mathbf{P} = \alpha\mathbf{B}$, whereas the magnetization, $\mathbf{M}$, linearly responds to an external electric field $\mathbf{E}$: $\Delta \mathbf{M} = {}^t\alpha\mathbf{E}$, where α is a 3×3 tensor and ${}^t\alpha$ is the transposed tensor of α. Multiferroic properties are closely linked to symmetry and can be characterized by their behavior under space and time inversion. Thus space inversion will reverse the direction of polarization $\mathbf{P}$, leaving the magnetization $\mathbf{M}$ invariant, whereas time reversal will change the sign of $\mathbf{M}$, preserving the sign of $\mathbf{P}$. The important role of symmetry in determining multiferroic properties of crystals makes an important parallel with topological insulators, wherein time reversal symmetry and spatial inversion symmetry also play a crucial role [6]. Although ferroelectric properties are usually determined by the off-center positions of atoms in a structure, ferromagnetic properties typically arise from the presence of transition metal d-electrons. Interestingly, the presence of d-electrons, which is essential for magnetism, reduces the tendency for off-center ferroelectric distortion [7]. Most MF materials discovered and/or engineered to date are characterized by rather small values of α.

In 2011, a chalcogenide superlattice GeTe-Sb_2Te_3 was designed, which led to as much as a 95% reduction in the switching energy of electrical non-volatile phase-change random-access memory (PC-RAM) [8, 9]. Due to the importance of interfaces, the superlattice-based phase-change memory (PCM) was named interfacial phase change memory (iPCM). It was subsequently found that iPCM possesses a giant magnetoresistance (over 2000% at room temperature under a static magnetic field ($\sim$1 kOe)) [10], whereas GeTe-Sb_2Te_3 (GST) alloys of the same average composition do not. It should also be noted that previously magnetic effects in PCM were observed only in alloys containing magnetic additives [11–13].

PC-RAM, recently commercialized by the world's largest memory makers, Samsung and Micron, is based on a phase transition between the amorphous and crystalline states in ternary GST alloys, making use of the large property contrast between the two phases. The contrast is associated with the contrasting bonding nature between the constituent atoms in the structures. When a voltage exceeding a certain value is applied to a device in the high-resistivity amorphous (RESET) state, it switches to the low-resistivity crystalline (SET) state. During the SET pulse, the GST is heated above the crystallization temperature, whereas the RESET pulse melts the material, reverting it to the amorphous phase with large entropic losses [14–16]. Pulses with typical durations of $\sim$100 ns and $\sim$500 ns are used for the RESET and SET processes, respectively. In contrast, in iPCM, which has the structure of a short-period $[(GeTe)_l(Sb_2Te_3)_m]_n$ *superlattice* (where l, m, and n are integers), where GeTe and Sb_2Te_3 share a common growth axis, the [111] direction of the rhombohedral GeTe layer and the [111] axis of the rhombohedral (A7) Sb_2Te_3 layer, being parallel to each other and normal to the substrate surface, is crystalline in both the SET and RESET states [9]. Importantly, even sputtered iPCM devices have a strong preferred [111] growth direction and

exhibit high-quality interfaces [17]. The switching mechanisms of PC-RAM and iPCM are summarized in figure 1.

In a previous work [16, 18], we examined, using *ab initio* molecular dynamics, possible structures of iPCM in the SET and RESET states. The results of *ab initio* molecular dynamics using constant temperature and constant pressure ensembles (NPT ensembles) for various studied structures are shown in figure S1. It was found that the structures with the lowest energies at the growth temperature ($\sim$520 K) consist of two buckled GeTe layers sandwiched by quintuple layers (QLs) of Sb_2Te_3. The stacking sequences within the GeTe block are (-v-Te-Ge-Te-Ge-) in the SET and (-v-Te-Ge Ge-Te-v-) in the RESET state, where v indicates interatomic distances exceeding 3.1 Å, often referred to as vacancy layers or van der Waals gaps.

The structures of the SET and RESET states are shown in the upper panel of figure 2 together with the corresponding band structures. One can see that the RESET phase has a Dirac-like density of states. This fact, combined with the calculated topological invariant $Z_2 = 1$, suggests that the RESET state is a strong topological insulator protected by spatial inversion and time reversal symmetries [19]. The SET phase, on the other hand, has a gap due to the breaking of spatial inversion symmetry, resulting in spin-split band structures. The differences between the two phases were clearly observed using high-angle annular dark field transmission electron microscopy (HAADF-TEM), as shown in figure 2. (See the supplementary material for details.) It is noted that the obtained HAADF-TEM images of real iPCM films are in good agreement with the results of earlier simulated models for the SET and RESET states [18, 19]. When one Ge layer is swapped with the adjacent Te layer by an external stimulus such as temperature or electric field, the RESET state transforms into the SET state and vice versa [18]. The change of the electric dipole moment (dielectric constant) induced by the phase transition generates large optical and electrical contrast [9].

iPCM devices have another important attribute. A number of groups have recently argued that, depending on the thicknesses of individual blocks, iPCM can be either a strong topological insulator (TI) with a single Dirac cone at the Γ point [19] or a Dirac semimetal composed of alternating layers of a narrow gap normal insulator (NI), GeTe, and a three-dimensional topological insulator (3D-TI), Sb_2Te_3 [18–21]. In addition, recent *ab initio* simulations of bulk GeTe have predicted a giant electric field-induced Rashba effect, which also originates from the p-electrons of Ge atoms [22]. These properties of iPCM offer previously unexplored possibilities for multiferroics and spintronics.

In this work, breaking either spatial inversion or time reversal symmetries or both in the topological insulating phase of the superlattices, we describe the magnetoelectric effects when external electric and magnetic fields are applied to iPCM. Subsequently we discuss the origin of magnetism in the absence of magnetic d-electrons and, finally, demonstrate the applicability of iPCM for spin-storage devices.

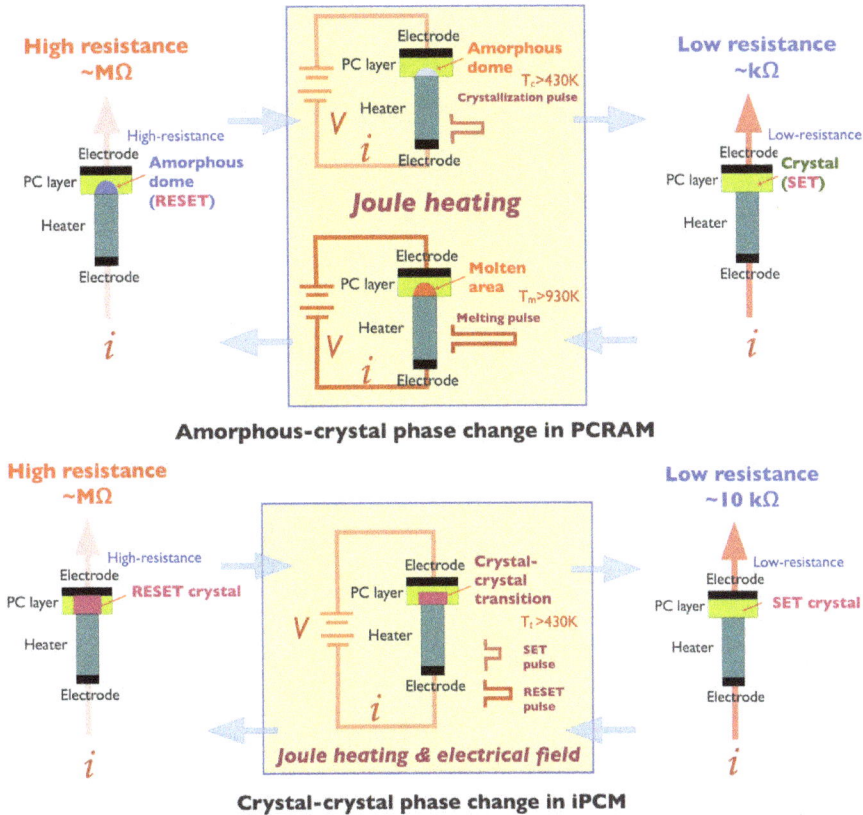

Figure 1. The switching mechanism of PC-RAM and iPCM non-volatile memories. Whereas the former utilizes an amorphous-to-crystalline phase transition, the latter is based on a crystal-to-crystal phase transition. The cell designs can be similar in both cases. The PC and iPCM layer thicknesses are typically about 40 nm, and the bottom electrode (heater) diameter is less than 100 nm. T_c, T_m, and T_t are the crystallization temperature, melting temperature, and transition temperature, respectively. V and i are the applied voltage and current. Due to the presence of a melting-free mechanism, the switching energy is reduced by 95% in the iPCM device compared with that of the PC-RAM device using a similarly designed cell platform [9, 10, 16].

2. Experimental procedures

2.1. Fabrication of iPCM devices

iPCM films were fabricated on Si wafers using a helicon-wave sputtering system that features a large (200 mm) target–substrate separation. The individual layers composing the $(GeTe)_x/(Sb_2Te_3)_y$ structure were fabricated using GeTe and Sb_2Te_3 composite targets (2-inch diameter) using an auto-mated shutter control system with a substrate temperature of about 520 K and pressures less than 0.5 Pa. The thicknesses of the GeTe and Sb_2Te_3 blocks were 0.4 nm and 1.0 nm, respectively. To ensure strong crystalline orientation, a 5 nm-thick Sb_2Te_3 layer was deposited prior to the fabrication of iPCM films. As a consequence, all the sub-layers were crystalline in the as-deposited state and exhibited a strong preferred crystallographic orientation toward the <111> direction (see [8, 9] for more details). For electrical and magnetic measurements on real devices, the iPCM film was deposited on a base device with a TiN rod electrode (70 nm

diameter) embedded in a silicon nitride layer. A 40 nm-thick TiN layer was deposited on the iPCM film to serve as an electrode.

The structure of the deposited iPCM was studied using high-angle annular dark field (HAADF) transmission electron microscopy (JEOL, JEM-ARM200F). The samples were prepared using ion milling. The images were observed using an electron beam with a diameter of 0.2 nm and an accel-erating voltage of 200 kV in combination with fast Fourier transform (FFT) analysis.

In addition, we developed an alternative device design that is more suitable for hybrid memory. In this design, the superlattice film was deposited onto a flat and cleaned (dry-etched) Si wafer, and successively a sequence of ZnS-SiO$_2$, ferromagnetic TbFeCo, and ZnS-SiO$_2$ layers, 10 nm thick each, were deposited. By exposure to a focused laser pulse ($\lambda = 405$ nm), the ZnS-SiO$_2$/TbFeCo/ZnS-SiO$_2$ layers inter-diffused, resulting in the formation of an electrical contact pillar [23] with a diameter of about 400 nm directly connected to both the ferroelectric layer and the phase-change film.

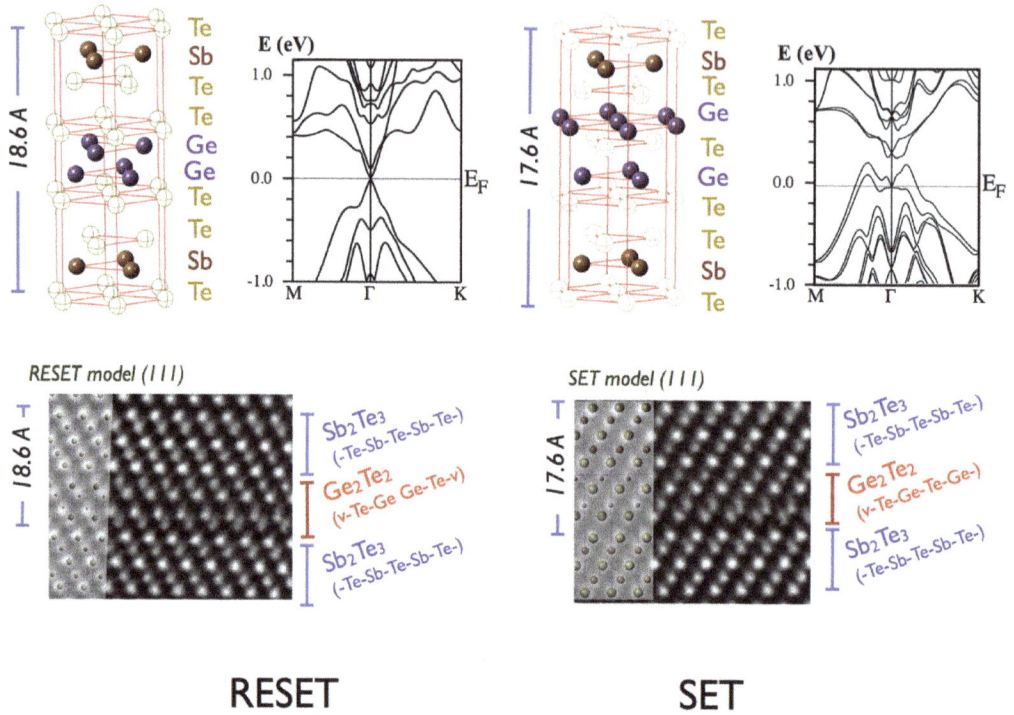

Figure 2. The upper panel shows the RESET and SET models and the corresponding bulk band structures, and HAADF TEM images of the $[(GeTe)_2/(Sb_2Te_3)_1]_{20}$ iPCM (~40 nm thick) grown on Si(111).

Finally, a TiN electrode was deposited on top. When a voltage is applied to such a device, electrons from the ferroelectric TbFeCo layer participate in the current flow. When an external magnetic field is applied to the device, spin-polarized electrons can be injected from the TbFeCo into the phase-change layer, which allows one to obtain different states when the device is SET with and without a magnetic field.

2.2. Device measurement

The Hall resistance and coefficient of the iPCM were measured maintaining a constant current of $10\,\mu A$ by a ResiTest 8400 unit (Toyo Co.) using a van der Pauw configuration. The current flowed in plane between two selected contacts among four probes, whereas the magnetic field was applied normal to the plane. Each measurement was repeated ten times with good reproducibility. As a substrate, an undoped high-resistance Si (100) wafer was used. Before the iPCM film measurement, we confirmed that the background signal from the substrate was negligible.

Resistance–voltage (R–V) curves were taken using a programmable pulse generator with a 300 ns pulse for which the voltage was increased in 50 steps. An iPCM device and a control PC-RAM device with identical architectures were compared. In the iPCM device, the threshold voltages were 0.85 V for SET and 1.5 V for RESET, whereas those in the control PC-RAM device were 1.0 V for SET and 3.5 V for RESET [10]. The iPCM switched from the SET to the RESET state at 0.20 mA, whereas the control device did not switch

until 1.1 mA. The magnetic field in the device measurements was applied using three different (0.8 kOe, 1.0 kOe, and 1.2 kOe) 1 mm-thick flat, square magnets with areal dimensions of 10 mm × 10 mm.

2.3. Simulation of band structures

The electronic band structures of iPCM films shown in figure 2 were simulated in two steps using two *ab initio* simulation codes: CASTEP and WIEN2K. iPCM models were first built and relaxed at 0 K using the plane wave code CASTEP with a GGA exchange correlation term using the Perdew–Burke–Ernzerhof 1996 (PBE) function. A $4 \times 4 \times 1$ Monkhorst–Pack grid was used for integration, and ultrasoft pseudopotentials were used with a cutoff energy of 230 eV [24, 25]. After 0 K Broyden geometrical optimization, the CASTEP-relaxed models were transferred to WIEN2K [26], which allows one to include spin–orbit coupling (SOC) effects. Wien2K is an all-electron code that uses a linearized augmented plane wave + local orbital (LAPW + lo) basis within density-functional theory using the PBE exchange correlation. The same Monkhorst–Pack grid of $4 \times 4 \times 1$ was used for integrations in the Brillouin zone, and an $R_x K_{max} = 7.0$ value was used for the plane wave component of the plane-wave basis used between augmentation spheres.

In *ab initio* molecular dynamics, on the other hand, constant temperature and constant pressure ensembles (NPT ensemble) were used at 0 GPa, using a cutoff energy of 170 eV. The smaller cutoff energy was used to increase the

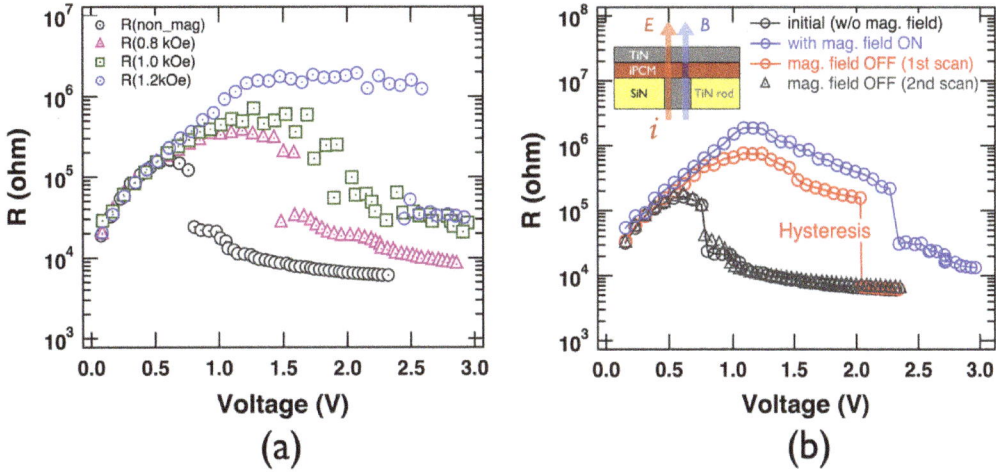

Figure 3. (a) Magnetoresistance under different magnetic field strengths 0.8, 1.0, and 1.2 kOe and (b) the associated resistance hysteresis in a typical iPCM [(GeTe)$_2$/(Sb$_2$Te$_3$)$_4$]$_8$ device (inset) at room temperature. Magnetoresistance hysteresis occurs after the electric field is turned off with a magnetic field still present. The black circles indicate the initial resistance without an applied magnetic field, the blue circles indicate resistance values with a magnetic field, the red circles indicate the first R–V curve after removal of the magnetic field, and finally the black triangles correspond to the second R–V cycle after the magnetic field was turned off. In each process, the initial and terminated states were both RESET.

simulation speed and did not have a strong effect on the result. The free energy changes were simulated at a number of different temperatures to identify the phase transition temperature between the SET and RESET states.

3. Results and discussion

3.1. Magnetoelectric response of an iPCM device

Figure 3(a) shows R–V curves of an iPCM [(GeTe)$_2$/(Sb$_2$Te$_3$)$_4$]$_8$ device when external magnetic fields of different intensities are applied normal to the interfaces. Without an external magnetic field, the transition to the SET phase occurs at the same threshold voltage as in the alloy of the same average composition (cf figure 6 of [18]) with an important difference: the resistance of iPCM increases by an order of magnitude for voltages between 0 V and 0.8 V (the threshold voltage), whereas the resistance of the composite amorphous phase in PC-RAM remains constant (or gradually decreases, which has been explained by the Frenkel–Poole mechanism [15]). Upon increasing the magnetic field to 1.2 kOe, the resistance in the RESET phase increases by two orders of magnitude between 0 V and 2.5 V and then suddenly drops to the low-resistance SET phase. The minimum magnetic field required for the magnetoelectric effect was ~0.8 kOe, and lower fields had a negligible effect (not shown). It should also be noted that the resistance curves in the range between 0 and 0.5 V almost completely overlap regardless of the magnitude of the magnetic field. Therefore, it can be surmised that the unusual resistance increase in this voltage range is due to an intrinsic magnetic field induced by the applied electric field, which is equivalent to an external magnetic field of ~0.8 kOe, after which the external magnetic field starts to have an effect

on the threshold voltage. This implies that the α of $\Delta\mathbf{M} = {}^t\alpha\mathbf{E}$ in the RESET phase has diagonal elements since both the induced magnetization and the applied electric field have the same direction normal to the surface.

Of special interest is the device behavior in subsequent cycles (high resistance state -> low resistance state -> high resistance state). If the device is in the SET state (the electric field is turned off) in the presence of a magnetic field followed by removal of the magnetic field, in the subsequent (i.e., second) cycle the R–V characteristics remain essentially unchanged (figure 3(b), red curve), namely, the threshold voltage for the second cycle is similar to the R–V curve for the first cycle in the presence of the magnetic field; i.e., the device displays a memory of the magnetic field applied during the previous cycle. In the third cycle (black curve), however, the memory of the magnetic field is erased: the R–V curve reverts to its original behavior before the application of the magnetic field. It should be noted that the difference between the R–V curves at 1.2 kOe shown in figures 3(a) and (b) is due to the fact that the magnetic field was increased in steps (0.8 to 1.0 to 1.2 kOe) in figure 3(a) and was set to 1.2 kOe from the beginning in figure 3(b). Reproducibility of this cycle was confirmed for several identical devices. It should be noted that no magnetic dopants or other impurities were included in the iPCM.

When an external electric field of 0.001 electrostatic units (a.u.) (corresponding to 0.51 V nm^{-1}) is applied normal to the iPCM film in a simulation including SOC, the Dirac cone is broken and a gap opens as shown in figure 4 (left). Because the conductivity of Sb$_2$Te$_3$ is higher than that of GeTe, the voltage drop is mainly across the GeTe block (~0.9 nm). The total voltage applied to the structure with eight repeat units (in the experiment we used eight repeats; hence we also consider eight repeats in the simulation) becomes 3.7 V. This value is four times larger than that obtained experimentally (0.85 eV),

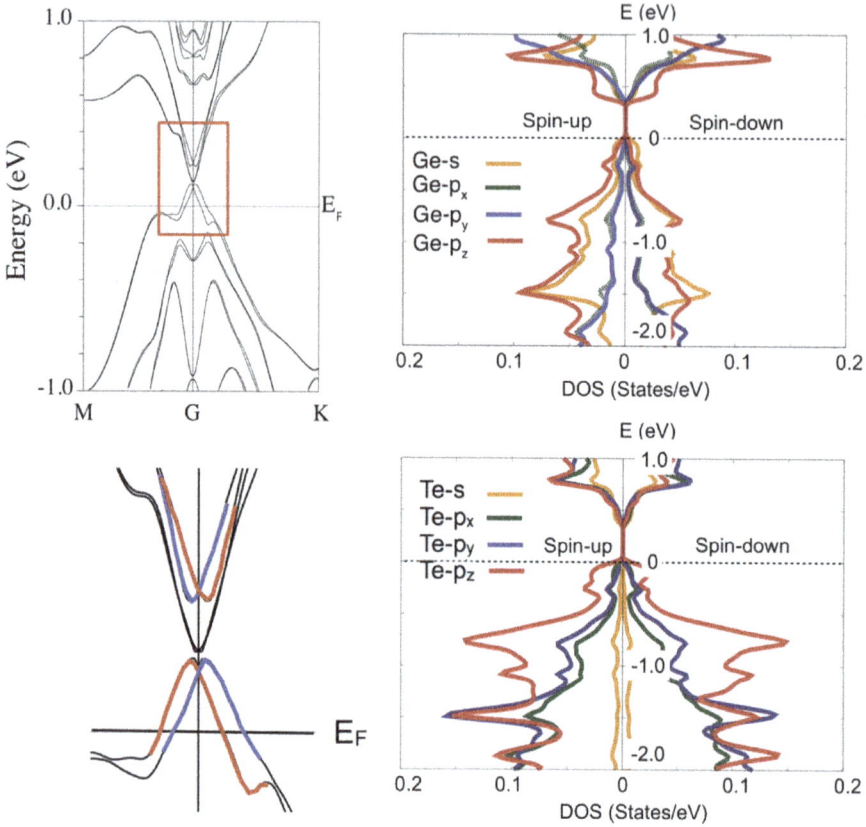

Figure 4. The band structure for the iPCM-RESET phase under an external electric field (0.51 V nm^{-1}) at 0 K (upper left), the band structure details around the Γ point (lower left), and the corresponding spin-polarized density of states for Ge (upper right) and Te (lower right) atoms.

but it should be kept in mind that even though the simulations were performed at 0 K, in the experiment the device contained a TiN heater rod, and an increase in temperature in the presence of a current is likely to facilitate the breaking of spatial inversion symmetry. In the foregoing simulation, the maximum energy difference, ΔE_{tvb}, between the spin-up and spin-down bands of 0.07 eV was obtained at $k \sim 0.06 \text{ Å}^{-1}$ (the M-point corresponds to $k = 0.76 \text{ Å}^{-1}$). In a different simulation (not shown) we found that a displacement of two Ge atoms from the stable position by 0.1 Å and 0.2 Å along the c-axis caused a splitting ΔE_{tvb} between the spin-up and spin-down bands of 0.09 eV and 0.24 eV, respectively, whereas the band gap, ΔE_{gap}, simultaneously expanded to 0.15 eV and 0.35 eV. These results support the assumption that the superlattice becomes more insulating through the breaking of spatial inversion. In the presence of the Rashba effect, the spin density of states (SDOS) for two opposite spin orientations is no longer degenerate and originates from Ge p-electrons (figure 4 (right)).

The Rashba parameter can be roughly estimated from the simulation of electric field effects, shown in figure 4 (left panel) to be about 1.2 eVÅ. This value is comparable to the values found for other materials with giant Rashba splitting such as two-dimensional Bi-Ag alloy thin layers on Ag(100) or (111) surfaces (3.6 eVÅ and 3.1 eVÅ, respectively)

[27, 28] and also to that of Pb/Ge(111) (1.5 eVÅ) [29]. Therefore, the value of 1.2 eVÅ obtained from the simulation results is a reasonable value. From the NTP ensemble simulations (supplemental figure S1, available at stacks.iop.org/STAM/16/014402/mmedia), it was found that in the iPCM, the Rashba parameter depends not only on the strength of the external electric field normal to the film but also on temperature. Since the phase transition temperature between the RESET and SET states is around 420 K, the value of the Rashba parameter is likely to be larger.

3.2. Two-dimensional transport properties of iPCM films

It is speculated that a pair of spin states exists where the spins are oriented in plane and antiparallel to each other. This is strongly supported by the band structure of the RESET phase in figure 2, which has a single Dirac cone at the Γ point. It should be noted that the unit cell of the iPCM RESET phase satisfies both spatial inversion symmetry and time reversal symmetry, although the cone is spin-degenerate. Two spin states exist in plane to maintain time reversal symmetry, i.e., $E(k, \rightarrow) = E(-k, \leftarrow)$, where k and the arrows are the electron momentum and spin directions.

We further studied transport properties using Hall effect measurements on iPCM films. In this case the electric field is

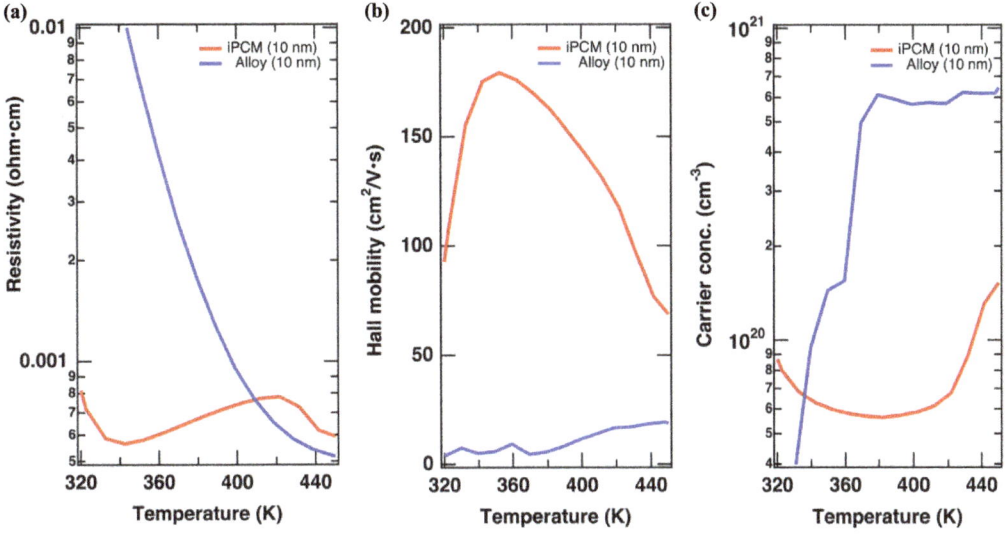

Figure 5. Temperature dependence of (a) the resistivity, (b) Hall mobility, and (c) carrier concentration of the $[(GeTe)_2(Sb_2Te_3)_1]_2$ structure (red) and the control alloy with the same composition ratio (blue). The Hall properties of (a)–(c) were measured under a 5 kOe magnetic field.

applied in plane, whereas the magnetic field (5 kOe) is applied normal to the plane. If the iPCM RESET phase does have a Dirac cone, two-dimensional metallic transport should be observed. Figure 5 shows the results obtained for an iPCM $[(GeTe)_2(Sb_2Te_3)_1]_2$ film fabricated at 520 K with a control film of GST alloy fabricated with the same film thickness and at the same temperature. After the initial drop, the resistivity (ρ_{xx}) of the RESET phase increased to $7.8 \times 10^{-4}\ \Omega$ cm at ~420 K, at which point the SET phase of the iPCM film was established. Therefore, when the electric field is applied in plane, the RESET phase of the iPCM film behaves like a metal at temperatures between 350 K and 420 K. It should be noted that in the control GST alloy film with the same average composition, the resistivity monotonically decreased as expected for a semiconductor. At the same time, the Hall mobility steeply increased, reaching a value of ~180 cm^2 V-s^{-1} at ~350 K, and then monotonically decreased, whereas the carrier concentration of 5.7×10^{19} cm^{-3} was essentially unchanged until the phase transition (apart from the initial drop). The results from the iPCM film are crucially different from the behavior of the alloy of the same average composition of our control sample and other results reported in the literature [15]. We have also checked the magnetization of the films using SQUID at 2 K and 300 K (see supplemental figure S2). As expected, the magnetization of the RESET phase was negligible (within the experimental errors) because of the spin-band degeneracy. These results support the existence of a metallic or gapless state with high mobility in the RESET phase. The change in the Hall mobility and carrier concentration with temperature is interpreted as arising from internal stresses which generate optimal conditions for the formation of a proper TI or Dirac semimetal phase, reflecting the two-dimensional nature [19, 20].

3.3. Role of p-electrons in the iPCM films

An external electric field induces, via the Rashba effect, a band opening similar to the result [22] reported for bulk GeTe, and we speculate that the magnetic properties appearing in the Hall measurement are due to the presence of thin GeTe layers. As shown in figure 4, under an applied electric field the spin-polarized electron densities of states in Ge and Te are different for spin-up and spin-down states. For this reason, when an electric field is applied and current flows, the iPCM RESET structure can have a magnetic moment. The magnetic moment disappears when the electric field is turned off and the structure reverts to the original symmetrical structure.

Ge atoms in GeTe, such as those present in iPCM, were shown to possess lone-pair electrons [30], a situation that is not uncommon for elements containing paired s-electrons that formally do not participate in nominal σ-electron (bonding) formed using p-electron orbitals [31] but become stereo-chemically active when the bonding angles deviate from 90°. Such lone-pair electrons have been considered as the origin of ferroelectricity in GeTe and related materials [32, 33]. Real GeTe crystals are characterized by a high concentration of Ge vacancies [34] (which have a very low formation energy [35]), generating a change in the density of states. In particular, Ge atoms nearest the vacancy may have lower electron density on the lone-pair orbital, in the limit changing it to an unpaired p-electron (supplemental figure S3). In the RESET phase, where the two GeTe buckled layers have an anti-ferroelectric arrangement of dipole moments with the adjacent Ge planes, interaction of the unpaired electrons from the neighboring layers results in a net zero spin moment.

However, the situation with the iPCM SET phase is different. Since the GeTe block loses spatial inversion symmetry, spins located at Ge atoms in different layers do not

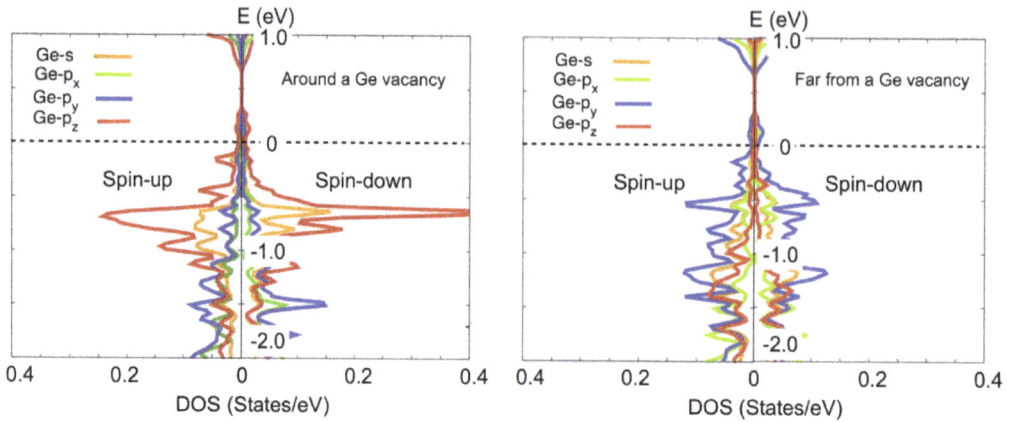

Figure 6. Spin-polarized density of states for a Ge atom near a Ge-site vacancy (left) compared with a Ge atom located away from the vacancy (right) in the iPCM-SET phase.

cancel each other. Indeed, as shown in figure 6 (left), Ge p-electrons around the vacancy exhibit a large asymmetry between the spin-up and spin-down configurations, the situation being significantly different from Ge atoms located away from vacancies (figure 6 right) and similar to the case of a graphene ribbon where zigzag edges becomes ferromagnetic due to unpaired electrons localized at the edge C atoms [36, 37]. The magnetization of the SET phase when the spatial inversion was broken was recently confirmed experimentally using magneto-optical Kerr rotation measurement [38]. It was reported that the Kerr rotation was observed only in the high-temperature SET phase (>420 K) with unusual mirror-symmetric curves with a relatively large angle (~0.3°) [38].

It should also be noted that the magnetoelectric response of iPCM is unusually strong. We believe this is caused by the rather special origin of magnetism in these structures. Although typically ferromagnetic properties are due to the presence of transition metal d-electrons that tend to reduce the off-centering of atoms like Bi or Pb that are responsible for ferroelectricity, in iPCM, the ferromagnetic properties arise from unpaired electrons located on Ge atoms along with the ferroelectric properties caused by off-centering of the same Ge atoms; i.e., *both the ferroelectricity and ferromagnetism in iPCM are due to the same atomic species and p-electrons*.

3.4. Proposal of topological switching random access memory (TRAM or TSRAM)

To implement the observed spin storage practically, we developed a special device design, in which there is an additional ferromagnetic TbFeCo layer that can act as a spin-injection source. When voltage is applied to this structure and current flows under an applied magnetic field, spins in the TbFeCo layer, aligned by the magnetic field, are injected into the superlattice through the conducting pillar.

Figure 7 shows I–V curves for this structure (shown in the inset) SET either with or without a magnetic field. The curves shown clearly demonstrate a magnetoresistance response depending on the strength of the magnetic field. In

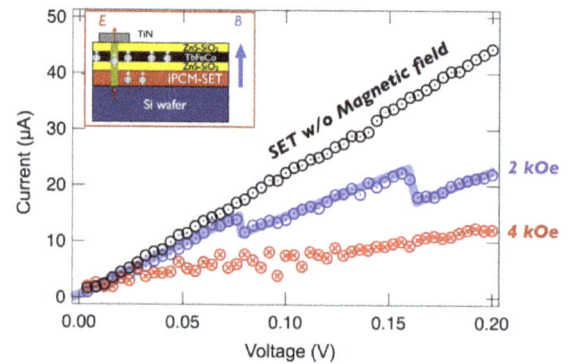

Figure 7. I–V curves from the iPCM-SET phase with and without a magnetic field (2 kOe and 4 kOe) for the spin-storage device (inset) at room temperature, using iPCM $[(GeTe)_2(Sb_2Te_3)_4]_8$.

particular, at intermediate field strength (2 kOe), an I–V curve with several steps was observed, which supports practical applicability for multilevel magnetoresistance memory devices.

4. Conclusions

In this work, we described the important role of the topological phase transition in $[(GeTe)_2(Sb_2Te_3)_l]_m$ superlattices due to the breaking of the spatial inversion and time reversal symmetries when external electric and magnetic fields are applied. Through the combined use of experiments and *ab initio* simulations, we demonstrated that superlattices with appropriately chosen thicknesses of individual layers are high-temperature multiferroic materials. We argue that the giant magnetoelectric response of iPCM is due to the fact that, different from most other cases, both ferroelectric and ferromagnetic properties are associated with the same atomic species, namely Ge atoms. The demonstrated spin storage, TRAM, in iPCM-based devices opens up new avenues for use by future nonvolatile storage memory in the form of new hybrid devices combining the attributes of PCM, MRAM, and topological insulators.

Acknowledgments

Part of the work was supported by the FIRST program initiated by the Council for Science and Technology Policy (CSTP). We thank Ms Reiko Kondou for her assistance in the fabrication of iPCMs.

References

[1] Landau L D and Lifshitz E M 1959 *Statistical Physics* (London: Pergamon)
[2] Ramesh R and Spaldin N A 2007 *Nat. Mater.* **6** 21
[3] Aizu K 1969 *J. Phys. Soc. Jpn.* **27** 387
[4] Folen V J, Rado G T and Stalder E W 1961 *Phys. Rev. Lett.* **6** 607
[5] Arima T *et al* 2004 *Phys. Rev. B* **70** 064426
[6] Qi X L and Zhang S C 2011 *Rev. Mod. Phys.* **83** 1057
[7] Nichola A H 2000 *J. Phys. Chem. B* **104** 6694
[8] Tominaga J, Kolobov A V, Simpson R and Fons P 2009 *E/PCOS: Proc. of the European Symp. on Phase Change and Ovonic Science (Aachen, Germany)* pp 148
[9] Simpson R E, Fons P, Kolobov A V, Fukaya T, Krbal M, Yagi T and Tominaga J 2011 *Nat. Nano.* **6** 501
[10] Tominaga J, Simpson R, Fons P and Kolobov A V 2011 *Appl. Phys. Lett.* **99** 152105
[11] Song W D, Shi L, Miao X S and Chong C T 2008 *Adv. Mater.* **20** 2394
[12] Ding D, Bai K, Song W D, Shi L P, Zhao R, Ji R, Sullivan M and Wu P 2011 *Phys. Rev. B* **84** 214416
[13] Skelton J M and Elliott S R 2013 *J. Phys.: Condens. Matter* **25** 205801
[14] Wuttig M and Yamada N 2007 *Nat. Mater.* **6** 824
[15] Raoux S and Wuttig M (ed) 2009 *Phase Change Materials* (Heidelberg: Springer) 175

[16] Tominaga J, Wang X, Kolobov A V and Fons P 2012 *Phys. Status Solidi* B **249** 1932
[17] Simpson R E, Fons P, Kolobov A V, Krbal M and Tominaga J 2012 *Appl. Phys. Lett.* **100** 021911
[18] Tominaga J, Kolobov A V, Fons P, Murakami S and Nakano T 2014 *Adv. Mater. Interfaces* **1** 1300027
[19] Sa B, Zhou J, Sun Z, Tominaga J and Ahuja R 2012 *Phys. Rev. Lett.* **109** 096802
[20] Sa B, Zhou J, Song Z, Sun Z and Ahuja R 2011 *Phys. Rev. B* **84** 085130
[21] Kim J, Kim J, Kim K S and Jhi S H 2012 *Phys. Rev. Lett.* **109** 146601
[22] Sante D D, Barone P, Bertacco R and Picozzi S 2013 *Adv. Mater.* **25** 509
[23] Kim J, Akinaga H, Atoda N and Tominaga J 2002 *Appl. Phys. Lett.* **80** 2764
[24] Martin R M 2004 *Electronic Structure—Basic Theory and Practical Methods* (Cambridge: Cambridge University Press)
[25] Vanderbilt D 1990 *Phys. Rev. B* **41** 7892
[26] Schwarz K and Blaha P 2003 *Comp. Mater. Sci.* **28** 259
[27] Nakagawa T, Ohgami O, Saito Y, Okuyama H, Nishijima M and Aruga T 2007 *Phys. Rev. B* **75** 155409
[28] Ast C R, Henk J, Ernst A, Moreschini L, Falub M C, Pacilé D, Bruno P, Kern K and Grioni M 2007 *Phys. Rev. Lett.* **98** 186807
[29] Lin C H, Chang T R, Liu R Y, Cheng C M, Tsuei K D, Jeng H T, Mou C Y, Matsuda I and Tang S J 2014 *New J. Phys.* **16** 045003
[30] Kolobov A V, Fons P and Tominaga J 2013 *Phys. Rev. B* **87** 155204
[31] Masterton W L, Hurley C N and Neth E J 2009 *Chemistry: Principles and Reactions* (Belmont: Wadsworth)
[32] Waghmare U, Spaldin N, Kandpal H and Seshadri R 2003 *Phys. Rev. B* **67** 125111
[33] Picozzi S 2014 *Front. Physics* **2** 10
[34] Kolobov A V, Tominaga J, Fons P and Uruga T 2003 *Appl. Phys. Lett.* **82** 382
[35] Edwards A H, Pineda A C, Schultz P A, Martin M G, Thompson A P, Hjalmarson H P and Umrigar C J 2006 *Phys. Rev. B* **73** 045210
[36] Fujita M, Wakabayashi K, Nakada K and Kusakabe K 1996 *J. Phys. Soc. Jpn.* **65** 1920
[37] Hou D, Wei J and Xie S 2011 *Phys. Chem.* **13** 13202
[38] Bang D *et al* 2014 *Sci. Rep.* **4** 5727

16

Magnetic properties of solid solutions between $BiCrO_3$ and $BiGaO_3$ with perovskite structures

Alexei A Belik

International Center for Materials Nanoarchitectonics (WPI-MANA), National Institute for Materials Science (NIMS), 1-1 Namiki, Tsukuba, Ibaraki 305-0044, Japan

E-mail: Alexei.BELIK@nims.go.jp

Abstract

Magnetic properties of $BiCr_{1-x}Ga_xO_3$ perovskite-type solid solutions are reported, and a magnetic phase diagram is established. As-synthesized $BiCrO_3$ and $BiCr_{0.9}Ga_{0.1}O_3$ crystallize in a monoclinic (m) C2/c structure. The Néel temperature (T_N) decreases from 111 K in $BiCrO_3$ to 98 K in $BiCr_{0.9}Ga_{0.1}O_3$, and spin-reorientation transition temperature increases from 72 K in $BiCrO_3$ to 83 K in $BiCr_{0.9}Ga_{0.1}O_3$. o-$BiCr_{0.9}Ga_{0.1}O_3$ with a $PbZrO_3$-type orthorhombic structure is obtained by heating m-$BiCr_{0.9}Ga_{0.1}O_3$ up to 573 K in air; it shows similar magnetic properties with those of m-$BiCr_{0.9}Ga_{0.1}O_3$. T_N of $BiCr_{0.8}Ga_{0.2}O_3$ is 81 K, and T_N of $BiCr_{0.7}Ga_{0.3}O_3$ is 63 K. Samples with $x = 0.4, 0.5, 0.6$ and 0.7 crystallize in a polar R3c structure. Long-range antiferromagnetic order with weak ferromagnetism is observed below $T_N = 56$ K in $BiCr_{0.6}Ga_{0.4}O_3$, $T_N = 36$ K in $BiCr_{0.5}Ga_{0.5}O_3$ and $T_N = 18$ K in $BiCr_{0.4}Ga_{0.6}O_3$. $BiCr_{0.3}Ga_{0.7}O_3$ shows a paramagnetic behaviour because the Cr concentration is below the percolation threshold of 31%.

Keywords: perovskites, multiferroics, high-pressure, $BiCrO_3$, magnetic properties

1. Introduction

Bi-containing perovskites have received a lot of attention as multiferroic materials and lead-free ferroelectrics [1–4]. The stereochemically active $6s^2$ lone pair of a Bi^{3+} ion plays an important role in producing polar distortions in Bi-containing perovskites. They form a basis for materials with super-tetragonality and a huge spontaneous polarization (P_S) observed, for example, in $BiCoO_3$ [4], Bi_2ZnTiO_6 [5], Bi_2ZnVO_6 [6], $BiCo_{0.3}Fe_{0.7}O_3$ [7], and highly strained $BiFeO_3$ thin films

[3, 8]. Bulk $BiFeO_3$ [3] and $BiAlO_3$ [4] also crystallize in a polar structure with space group R3c and large P_S.

On the other hand, $BiCrO_3$ [9, 10] crystallizes in a centrosymmetric structure with space group C2/c [4, 11–13]. There is a structural phase transition to a $GdFeO_3$-type Pnma structure from 420 K in $BiCrO_3$ [9–12]. It is interesting that during the Pnma-to-C2/c phase transition, twin domains with the size of less than 10 nm are formed [12, 14]. First-principle calculations could not explain the experimental C2/c structure and predicted the $GdFeO_3$-type Pnma structure as the ground-state structure [15, 16]. The Néel temperature (T_N) of $BiCrO_3$ is 109–111 K [9–13, 17], and there is a spin-reorientation transition below 75–80 K in $BiCrO_3$ [12, 13, 17].

Among $BiMO_3$ (M = 3d transition metals) compounds, $BiCrO_3$ is one of the least studied compounds, especially its solid solutions with isovalent substitutions. From the $BiCrO_3$-

rich side, just a few selected compositions have been investigated in $Bi_{1-x}Y_xCrO_3$ ($x = 0.01$, 0.05, 0.2 and 0.5 [13] and 0.1 [18]), $BiCr_{1-x}Fe_xO_3$ ($x = 0.5$ [19]) and $BiCr_{1-x}Mn_xO_3$ ($x = 0.5$ [20]) bulk systems. We have recently identified $BiGaO_3$-based perovskites, $BiM_{1-x}Ga_xO_3$ ($M = Cr$, Mn and Fe), as a large family of polar materials, which includes phases with (pseudo) super-tetragonality and R3c symmetry [21]. In particular, the R3c polar phase was found at $0.4 \leqslant x \leqslant 0.7$ in $BiCr_{1-x}Ga_xO_3$ solid solutions even though the end members, $BiCrO_3$ [4, 11] and $BiGaO_3$ [4], crystallize in centrosymmetric crystal structures. In this work, we report on detailed magnetic properties of the $BiCr_{1-x}Ga_xO_3$ solid solutions.

2. Experimental section

$BiCr_{1-x}Ga_xO_3$ with $x = 0$, 0.1, 0.2, 0.3, 0.4, 0.5, 0.6 and 0.7 were prepared from stoichiometric mixtures of Bi_2O_3 (99.9999%), Cr_2O_3 (99.99%) and Ga_2O_3 (99.99%). The mixtures were reground under acetone several times, placed in Pt capsules, dried at 573 K for several days and finally treated in a belt-type high-pressure apparatus at 6 GPa and 1700 K for 2 h (heating rate to the desired temperature was 10–15 min). After heat treatment, the samples were quenched to room temperature (RT), and the pressure was slowly released. Samples with other x values were mixtures of two perovskite phases (R3c and Cm phases for $x = 0.8$) and perovskite (Cm) and pyroxene phases for $x = 0.9$ [21]; therefore, their magnetic properties were not studied.

X-ray powder diffraction (XRPD) data were collected at RT on a RIGAKU Ultima III diffractometer using CuK_α radiation (2θ range of 10°–120°, a step width of 0.02°, and a counting time of 2–14 s/step). The samples contained small amounts of Cr_2O_3 and $Bi_2O_2CO_3$ impurities. The formation of $Bi_2O_2CO_3$ impurity was observed in many other works [13], and it is usually attributed to the diffusion of carbon from carbon heaters through cracks in capsules. Cr_2O_3 impurity could remain because a small amount of Bi_2O_3 was removed as $Bi_2O_2CO_3$ from stoichiometric mixtures.

Magnetic susceptibilities ($\chi = M/H$) were measured using pellets on SQUID magnetometers (Quantum Design, MPMS XL and 1 T) between 2–5 and 300–400 K in different applied fields. Samples were rapidly inserted into magnetometers kept at 10 K and having a zero magnetic field; then, temperature was set to 2 or 5 K; at 2 or 5 K, measurement magnetic fields were applied; and finally measurements were performed on heating up to 300, 350, or 400 K. This procedure gave zero-field-cooled (ZFC) curves. After ZFC measurements, samples were measured on cooling resulting in field-cooled (FC) curves. Isothermal magnetization measurements were performed between −50 and 50 kOe or between −10 and 10 kOe at different temperatures. Frequency dependent ac susceptibility measurements at a zero static magnetic field were performed with a Quantum Design MPMS 1 T instrument from 150 to 2 K at frequencies (f) of 2, 110, and 300 Hz and an applied oscillating magnetic field (H_{ac}) of 5 Oe. Specific heat, C_p, at a zero magnetic field was recorded between 2 and 300 K on cooling by a pulse relaxation method using a commercial calorimeter (Quantum Design PPMS).

Differential scanning calorimetry (DSC) curves were recorded on a Mettler Toledo DSC1 STAR^e system at a heating/cooling rate of 10 K min^{-1} from 290 to 573 K in open aluminium capsules; three cycles were performed to check the reproducibility.

3. Results and discussion

As-synthesized $BiCrO_3$ and m-$BiCr_{0.9}Ga_{0.1}O_3$ crystallize in the monoclinic (m) C2/c structure. The lattice parameters refined by the Rietveld method are $a = 9.4786(4)$ Å, $b = 5.4852(2)$ Å, $c = 9.5824(4)$ Å and $\beta = 108.587(3)°$ for $BiCrO_3$ and $a = 9.4762(4)$ Å, $b = 5.4899(2)$ Å, $c = 9.5791(4)$ Å and $\beta = 108.571(3)°$ for m-$BiCr_{0.9}Ga_{0.1}O_3$. However, both $BiCrO_3$ and $BiCr_{0.9}Ga_{0.1}O_3$ show noticeable anisotropic broadening (AB) of some reflections, the appearance of continuous diffuse scattering (DS) between some reflections, and shifts of some reflections (RS) from their ideal/expected positions (figure 1). Those features are most probably originated from local disorder, the presence of nanodomains [12, 14] and high concentration of domain boundaries and defects; they make the precise Rietveld analysis impossible and result in large variations in refined lattice parameters depending on models used. Similar features are observed in as-synthesized m-$BiCr_{0.8}Ga_{0.2}O_3$ and m-$BiCr_{0.7}Ga_{0.3}O_3$ (figures S1–S3 of the electronic supporting information (ESI)). If those features are modelled as a second perovskite phase (a $PbZrO_3$-related phase with space group Pnma [21]) in the Rietveld analysis, the weight fraction of the second phase is estimated to be about 20% in $BiCrO_3$, 20% in m-$BiCr_{0.9}Ga_{0.1}O_3$, 30% in m-$BiCr_{0.8}Ga_{0.2}O_3$ and 50% in m-$BiCr_{0.7}Ga_{0.3}O_3$ [21]. Electron microscopy and electron diffraction studies showed that the AB, DS and RS features are intrinsic for $BiCrO_3$ and do not originate from the presence of other perovskite phases [12, 14]. They can also be considered as intrinsic in m-$BiCr_{0.9}Ga_{0.1}O_3$ based on the above numbers. However, for unambiguous interpretation of those features in m-$BiCr_{0.8}Ga_{0.2}O_3$ and m-$BiCr_{0.7}Ga_{0.3}O_3$ detailed high-resolution electron microscopy studies are needed.

m-$BiCr_{0.9}Ga_{0.1}O_3$ exhibits a structural phase transition and shows a peak on the DSC curve at about 500 K on heating, and at 480 K on cooling. After heating as-synthesized m-$BiCr_{0.9}Ga_{0.1}O_3$ up to 573 K, its XRD pattern changes. All fundamental perovskite reflections remain the same; however, weak superstructure reflections change. Reflections can be indexed in the $PbZrO_3$-related structure [21] with space group Pnma and lattice parameters of $a = 5.4892(4)$ Å, $b = 15.4761(9)$ Å and $c = 11.1269(7)$ Å (figure 3); this sample will be called o-$BiCr_{0.9}Ga_{0.1}O_3$. Irreversible transformations of high-pressure phases were also found, for example, in $BiFe_{0.5}Sc_{0.5}O_3$ [22]. The origin of the irreversible behaviour is the existence of competing phases.

$BiCrO_3$ has the Néel temperature (T_N) of 111 K (determined/defined here by peak positions on the FC $d\chi/dT$ versus T curves at 100 Oe). Below T_N, the magnetic moments of

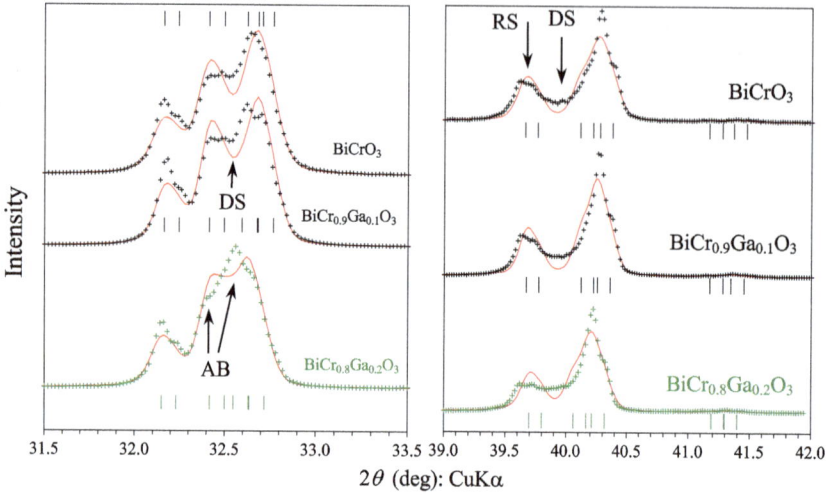

Figure 1. Portions of experimental (crosses) and calculated (lines) x-ray powder diffraction patterns of as-synthesized $BiCrO_3$, m-$BiCr_{0.9}Ga_{0.1}O_3$ and m-$BiCr_{0.8}Ga_{0.2}O_3$ measured with the $CuK\alpha$ radiation at room temperature. Possible Bragg positions of the C2/c structure are indicated by tick marks. 'AB' means reflections with anisotropic broadening, 'DS' means diffuse scattering between some reflections, and 'RS' means reflection shifts from their ideal expected positions.

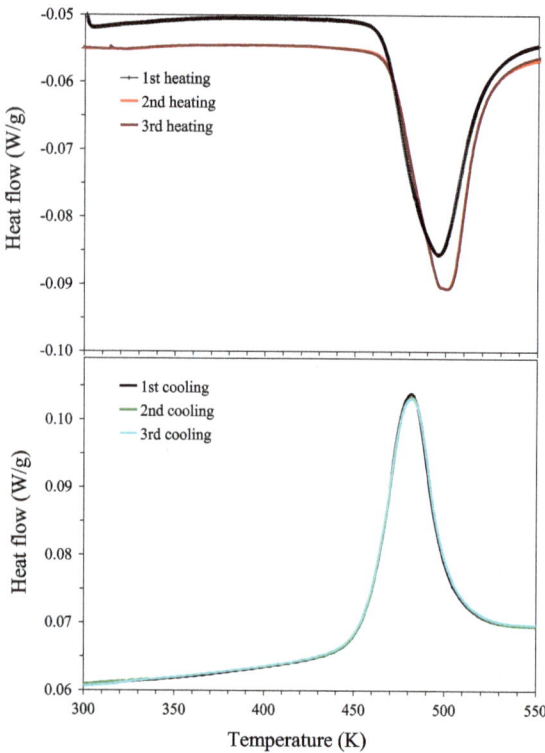

Figure 2. Differential scanning calorimetry curves of $BiCr_{0.9}Ga_{0.1}O_3$ measured at a heating/cooling rate of $10\,K\,min^{-1}$ between 290 and 573 K. Results of three heating–cooling cycles are shown.

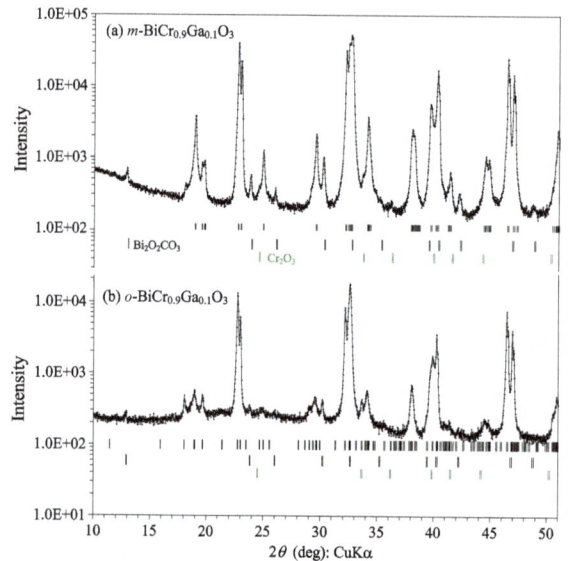

Figure 3. Portions of experimental x-ray powder diffraction patterns of (a) m-$BiCr_{0.9}Ga_{0.1}O_3$ and (b) o-$BiCr_{0.9}Ga_{0.1}O_3$ measured with the $CuK\alpha$ radiation at room temperature in the logarithmic scale to emphasize the difference in weak reflections. Possible Bragg positions are indicated by tick marks for the main perovskite phases and $Bi_2O_2CO_3$ and Cr_2O_3 impurities.

Cr^{3+} ions in $BiCrO_3$ are aligned along the a axis [14] in the G-type antiferromagnetic (AFM) structure, where all Cr–O–Cr interactions are AFM. $BiCrO_3$ has a spin-reorientation transition (T_{SR}) at 72 K (also defined by peak positions on the FC χ^{-1} versus T curve (figure 4(c)); they were suggested to

$d\chi/dT$ versus T curves at 100 Oe), where Cr^{3+} spins start to rotate away from the a axis in the (a, c) plane [13], but keeping the G-type AFM arrangement. Characteristic anomalies at T_N and T_{SR} can be clearly seen on the χ versus T, $d\chi/dT$ versus T and χ' versus T curves of $BiCrO_3$ and m-$BiCr_{0.9}Ga_{0.1}O_3$ (figures 4(a), (b) and 5). We note that small anomalies are observed at 165 K in $BiCrO_3$ on the 100 Oe FC χ^{-1} versus T curve (figure 4(c)); they were suggested to

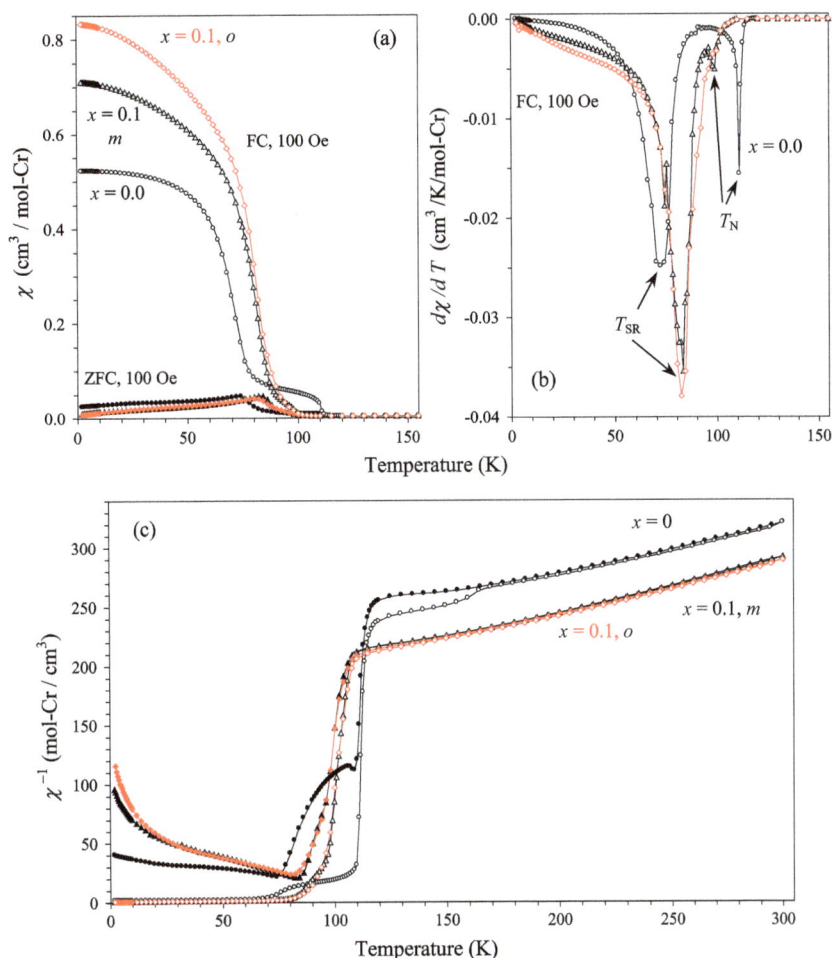

Figure 4. Magnetic properties of $BiCr_{1-x}Ga_xO_3$ ($x=0$, m—0.1 and o—0.1). (a) ZFC (filled symbols) and FC (empty symbols) dc magnetic susceptibility curves measured at 100 Oe. (b) FC $d\chi/dT$ versus T curves at 100 Oe; peaks define phase transition temperatures. (c) ZFC and FC χ^{-1} versus T curves measured at 100 Oe.

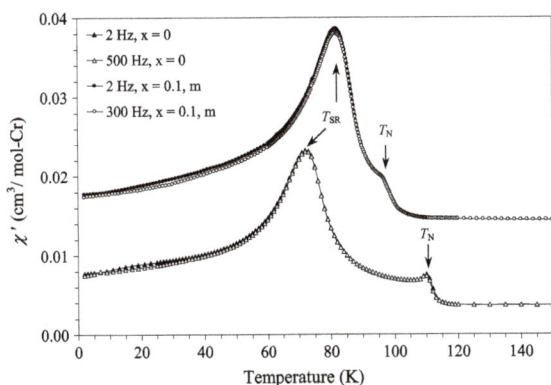

Figure 5. Real parts of the ac susceptibility curves of $BiCr_{1-x}Ga_xO_3$ ($x=0$ [17] and m—0.1). T_N is the Néel temperature, and T_{SR} is temperature of a spin-reorientation transition. The data for m-$BiCr_{0.9}Ga_{0.1}O_3$ are shifted by +0.01 cm^3/mol-Cr for the clarity.

originate from a very small amount of the $GdFeO_3$-type Pnma modification of BiCrO3 [4, 18]. We note that those anomalies at 165 K with different magnitudes were observed in all checked BiCrO3 samples (about a dozen of different samples (figure S17 of ESI), even synthesized by different groups); and those anomalies cannot be eliminated by further annealing and very slow cooling [18] suggesting that they are 'intrinsic' for bulk BiCrO3 samples. As-synthesized m-$BiCr_{0.9}Ga_{0.1}O_3$ shows very similar magnetic behaviour with that of BiCrO3, but with $T_N=98$ K and $T_{SR}=83$ K. No additional magnetic anomalies above T_N are found in m-$BiCr_{0.9}Ga_{0.1}O_3$ in comparison with BiCrO3; it can be related with the higher temperature of the C2/c-to-Pnma transition in $BiCr_{0.9}Ga_{0.1}O_3$ that results in a complete transformation. Magnetic properties of m-$BiCr_{0.9}Ga_{0.1}O_3$ and o-$BiCr_{0.9}Ga_{0.1}O_3$ are very similar with each other (figures 4 and S13 of ESI). Inverse magnetic susceptibilities of BiCrO3, m-$BiCr_{0.9}Ga_{0.1}O_3$ and o-$BiCr_{0.9}Ga_{0.1}O_3$ are given on figure 4(c). They show a noticeable deviation from the Curie–Weiss

Table 1. Temperatures of intrinsic magnetic transitions and results of the Curie–Weiss fits for as-synthesized $BiCr_{1-x}Ga_xO_3$.

Sample	T_N (K)	T_{SR} (K)	μ_{eff} per Cr^{3+}	θ
$BiCrO_3$	111	72	$4.161(14)\mu_B$	$-351(5)$ K
m-$BiCr_{0.9}Ga_{0.1}O_3$	98	83	$4.069(10)\mu_B$	$-282(3)$ K
m-$BiCr_{0.8}Ga_{0.2}O_3$	81		$3.965(4)\mu_B$	$-239(1)$ K
m-$BiCr_{0.7}Ga_{0.3}O_3$	63		$4.116(10)\mu_B$	$-239(3)$ K
$BiCr_{0.6}Ga_{0.4}O_3$	56		$3.839(3)\mu_B$	$-148.6(6)$ K
$BiCr_{0.5}Ga_{0.5}O_3$	36		$3.884(1)\mu_B$	$-119.8(2)$ K
$BiCr_{0.4}Ga_{0.6}O_3$	18		$3.889(2)\mu_B$	$-96.3(3)$ K
$BiCr_{0.3}Ga_{0.7}O_3$	No		$3.828(3)\mu_B$	$-73.3(5)$ K

μ_{eff} is an effective magnetic moment, θ is the Curie–Weiss temperature. The calculated μ_{eff} for Cr^{3+} $(S = 3/2)$ is $3.87\mu_B$.
For the Curie–Weiss fits, the FC curves at 10 kOe are used, and the data are corrected for diamagnetic contributions from sample holders and core diamagnetism. The Curie–Weiss fits are performed between 250 and 400 K for $x = 0$ and 0.1, 250 and 340 K for $x = 0.2$ and 0.3 and 150 and 400 K for $x = 0.4$–0.7 (figure S9 of the ESI).

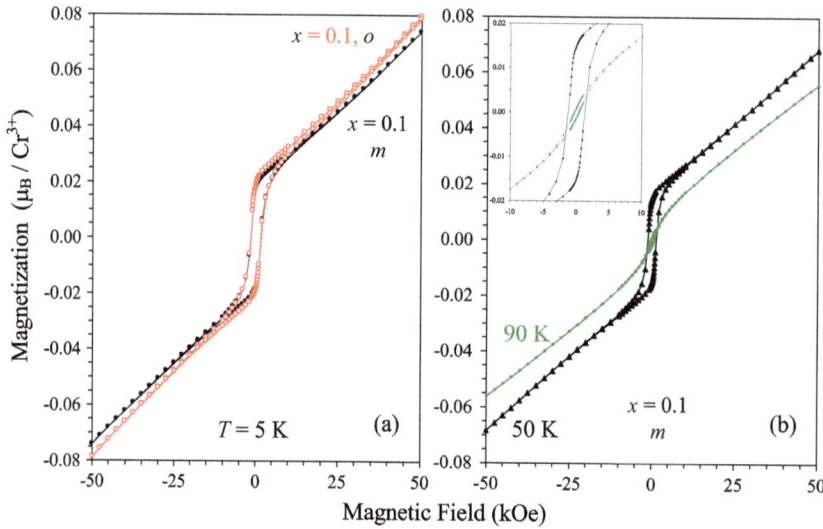

Figure 6. (a) Isothermal magnetization curves of m-$BiCr_{0.9}Ga_{0.1}O_3$ and o-$BiCr_{0.9}Ga_{0.1}O_3$ at 5 K. (b) Isothermal magnetization curves of m-$BiCr_{0.9}Ga_{0.1}O_3$ at 50 and 90 K. Insert shows details near the origin.

behaviour far above T_N; this is why the Curie–Weiss fits are performed above 250 K for those samples (table 1). This fact can also explain why the effective magnetic moments are slightly larger than expected ones.

Isothermal magnetization curves of $BiCr_{0.9}Ga_{0.1}O_3$ are given on figure 6. They show that a very weak ferromagnetic moment is developed below T_N in m-$BiCr_{0.9}Ga_{0.1}O_3$; below T_{SR}, the hysteresis loop becomes more defined. The magnetization reaches $0.074\mu_B/Cr$ at 5 K and 50 kOe; at 5 K, the remnant magnetization is $0.0177\mu_B/Cr$, and the coercive field is about 1.8 kOe in m-$BiCr_{0.9}Ga_{0.1}O_3$. Very similar M versus H curves were obtained in $BiCrO_3$ [17]. Spin canting angles in $BiCrO_3$ could not be determined from neutron diffraction because of their small values [12, 13].

Detailed magnetic properties (χ versus T, χ^{-1} versus T, $d\chi/dT$ versus T, χ' versus T, χ'' versus T, M versus H and C_p/T

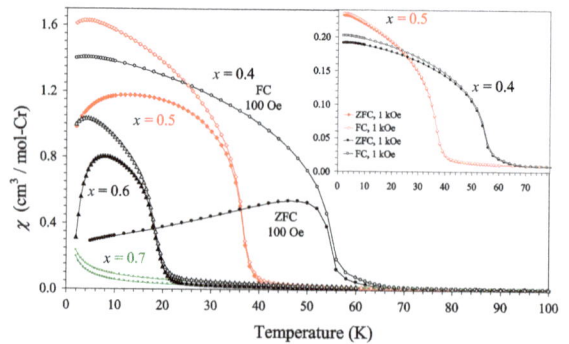

Figure 7. ZFC (filled symbols) and FC (empty symbols) dc magnetic susceptibility curves of $BiCr_{1-x}Ga_xO_3$ (R3c) ($x = 0.4$, 0.5, 0.6 and 0.7) measured at 100 Oe. The insert gives the ZFC and FC χ versus T curves for $x = 0.4$ and 0.5 measured at 1 kOe.

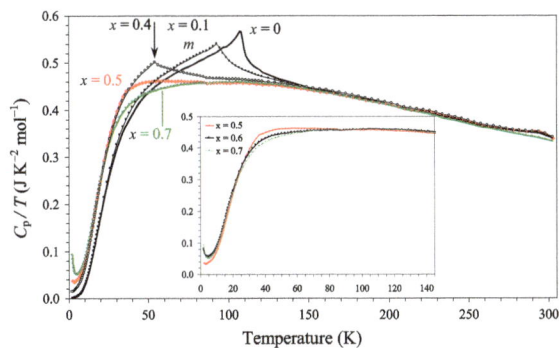

Figure 8. Specific heat data of BiCr$_{1-x}$Ga$_x$O$_3$ at 0 Oe plotted as C_p/T versus T. The vertical arrow shows the Néel temperature (T_N) of BiCr$_{0.6}$Ga$_{0.4}$O$_3$. The insert gives a fragment of the C_p/T versus T curves of BiCr$_{0.5}$Ga$_{0.5}$O$_3$, BiCr$_{0.4}$Ga$_{0.6}$O$_3$ and BiCr$_{0.3}$Ga$_{0.7}$O$_3$.

versus T) of as-synthesized m-BiCr$_{0.8}$Ga$_{0.2}$O$_3$ and m-BiCr$_{0.7}$Ga$_{0.3}$O$_3$ are given in figures S8–S11 of the ESI. T_N is found to be 81 K in BiCr$_{0.8}$Ga$_{0.2}$O$_3$ and 63 K in BiCr$_{0.7}$Ga$_{0.3}$O$_3$ from dχ/dT versus T and C_p/T versus T curves. T_N remains the same (figure S13 of ESI) for the as-synthesized samples and samples after DSC experiments (up to 623 K for BiCr$_{0.8}$Ga$_{0.2}$O$_3$ and 773 K for BiCr$_{0.7}$Ga$_{0.3}$O$_3$ [21]). Magnetic entropy is almost independent of x for $0.0 \leqslant x$

$\leqslant 0.3$ and varies between 5.2 and 5.5 J K^{-1}mol^{-1} (figure S14 of the ESI).

The χ versus T curves of BiCr$_{1-x}$Ga$_x$O$_3$ with $x=0.4$, 0.5, 0.6 and 0.7 having the R3c polar symmetry are shown on figure 7, and the parameters of the Curie–Weiss fits are summarized in table 1 (the fits are given in figure S9 of the ESI). Effective magnetic moments per a Cr^{3+} ion are close to the expected value for those samples. The temperature of magnetic transitions in BiCr$_{1-x}$Ga$_x$O$_3$ monotonically decreases with increasing x indicating that the magnetic transitions are intrinsic. The ZFC and FC curves of BiCr$_{0.6}$Ga$_{0.4}$O$_3$ are typical for canted antiferromagnets; in particular, the FC curves demonstrate the saturation behaviour. In addition, BiCr$_{0.6}$Ga$_{0.4}$O$_3$ shows a weak specific heat anomaly at 55 K (figure 8). These features indicate that there is a long-range magnetic ordering in BiCr$_{0.6}$Ga$_{0.4}$O$_3$. The χ' versus T and χ'' versus T curves of BiCr$_{0.6}$Ga$_{0.4}$O$_3$ (figure 9(a)) show sharp and frequency-independent peaks at T_N with additional very broad anomalies near 22 K (on the χ'' versus T curves).

No characteristic λ-type anomaly is observed on specific heat of BiCr$_{0.5}$Ga$_{0.5}$O$_3$. But we observe an excess of magnetic specific heat between about 20 and 80 K in BiCr$_{0.5}$Ga$_{0.5}$O$_3$ in comparison with BiCr$_{0.3}$Ga$_{0.7}$O$_3$ (figures 8 and S15 of ESI), which comes from magnetic interactions between Cr^{3+} ions. A low temperature, the tail on the specific heat increases with increasing x in BiCr$_{1-x}$Ga$_x$O$_3$ probably from short-range

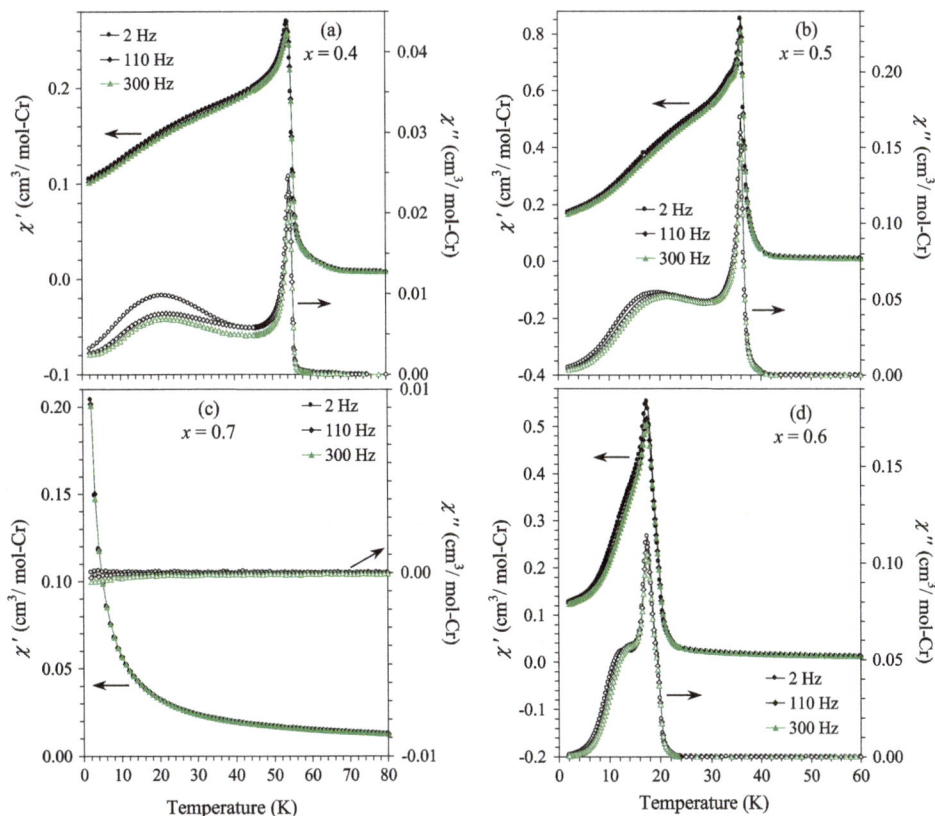

Figure 9. Real (χ' versus T) and imaginary (χ'' versus T) parts of the ac susceptibility curves of BiCr$_{1-x}$Ga$_x$O$_3$ (R3c) with $x=$ (a) 0.4, (b) 0.5, (c) 0.7 and (d) 0.6.

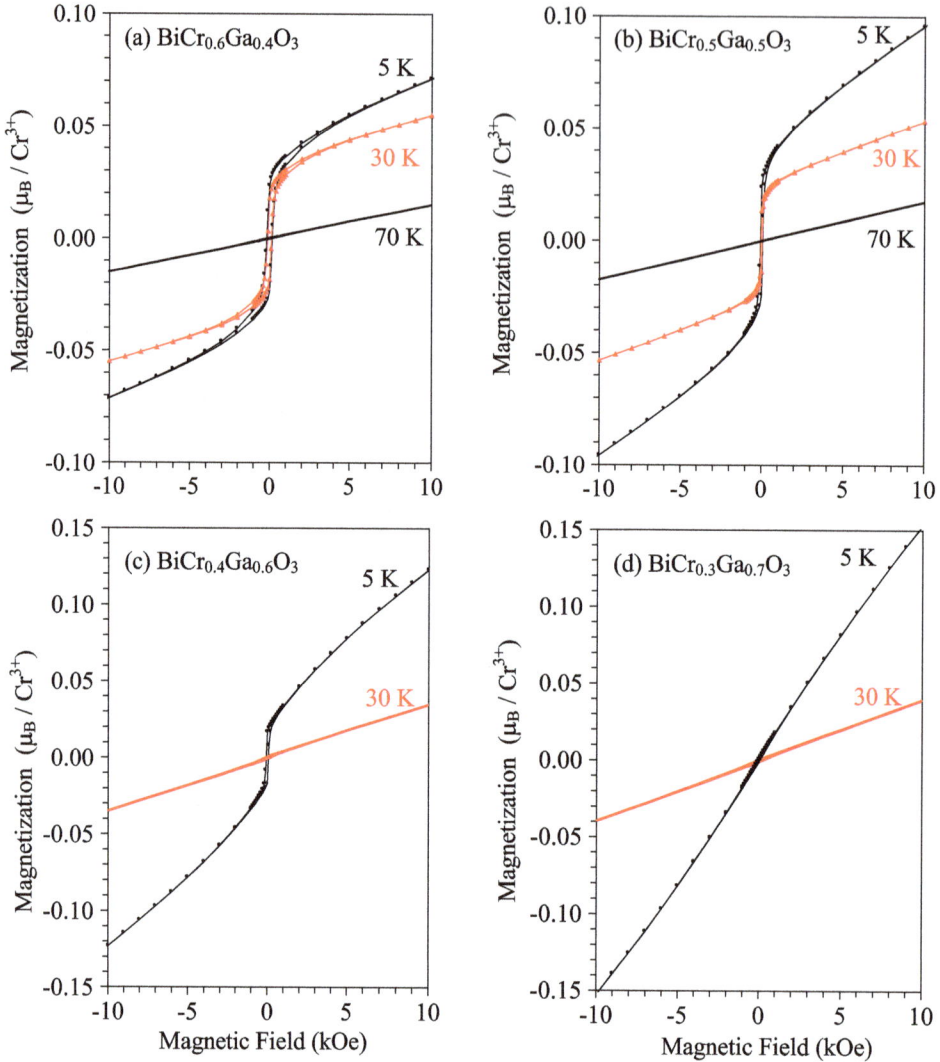

Figure 10. Isothermal magnetization curves of (a) $BiCr_{0.6}Ga_{0.4}O_3$ (R3c) at 5, 30 and 70 K, (b) $BiCr_{0.5}Ga_{0.5}O_3$ (R3c) at 5, 30 and 70 K, (c) $BiCr_{0.4}Ga_{0.6}O_3$ (R3c) at 5 and 30 K and (d) $BiCr_{0.3}Ga_{0.7}O_3$ (R3c) at 5 and 30 K.

magnetic interactions. Despite the absence of specific heat anomalies, magnetic properties of $BiCr_{0.5}Ga_{0.5}O_3$ and $BiCr_{0.4}Ga_{0.6}O_3$ are very similar with those of $BiCr_{0.6}Ga_{0.4}O_3$, but with transitions at lower temperatures of 36 and 18 K, respectively. In particular, the χ' versus T and χ'' versus T curves of $BiCr_{1-x}Ga_xO_3$ with $x=0.6$, 0.5 and 0.4 are remarkably similar to each other (figure 9) demonstrating sharp and frequency-independent anomalies at T_N. Therefore, we can assume that $BiCr_{0.5}Ga_{0.5}O_3$ and $BiCr_{0.4}Ga_{0.6}O_3$ also have long-range magnetic ordering at $T_N=36$ and 18 K, respectively. Because of the highly diluted magnetic sub-lattice, the entropy change associated with the transitions is very small resulting in the absence of specific heat anomalies. Note that long-range magnetic order was found in $BiFe_{0.5}Sc_{0.5}O_3$ [22] and $LaMn_{0.4}Ga_{0.6}O_3$ [23] by neutron diffraction, thus, confirming that long-range magnetic

ordering can occur in samples with highly diluted magnetic sublattices.

Magnetic properties of $BiCr_{0.3}Ga_{0.7}O_3$ are principally different (figures 7 and 9(c)). $BiCr_{0.3}Ga_{0.7}O_3$ demonstrates basically paramagnetic behaviour as can be clearly seen from the χ' versus T curve and the absence of any anomalies on the χ'' versus T curves (figure 9(c)). $BiCr_{0.3}Ga_{0.7}O_3$ still has a large Curie–Weiss temperature of about -73 K (table 1) indicating strong short-range AFM coupling between Cr^{3+} ions, but there should be no long-range ordering because $x=0.7$ is above the percolation threshold of $x=0.69$ for per-ovskite structures [24]; below the 31% concentration, a dopant cannot be linked in a continuous path throughout a crystal. We note that $BiCr_{0.3}Ga_{0.7}O_3$ and other samples with $x=0.4$–0.6 show a small divergence between the ZFC and FC curves below about 90 K; this feature most probably

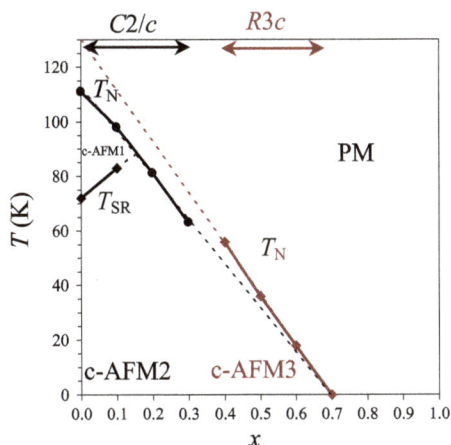

Figure 11. Magnetic phase diagram of $BiCr_{1-x}Ga_xO_3$. T_N is the Néel temperature, T_{SR} is temperature of a spin-reorientation transition, PM is a paramagnetic phase and c-AFM is a canted antiferromagnetic phase.

originates from trace amounts of a magnetic impurity (figures S5 and S6 of the ESI).

Isothermal magnetization curves of $BiCr_{1-x}Ga_xO_3$ with $x = 0.4$, 0.5, 0.6 and 0.7 are given in figure 10. They show that a weak ferromagnetic moment is developed below T_N in $BiCr_{1-x}Ga_xO_3$ with $x = 0.4$, 0.5 and 0.6. In $BiCr_{0.6}Ga_{0.4}O_3$, the magnetization reaches $0.17\mu_B$/Cr at 5 K and 50 kOe; at 5 K, the remnant magnetization is $0.023\mu_B$/Cr, and the coercitive field is about 180 Oe (figure S11 of the ESI).

The resulting magnetic phase diagram of $BiCr_{1-x}Ga_xO_3$ is given in figure 11. The T_N gradually decreases with increasing x, as one would expect because of the dilution of the magnetic sublattice by a nonmagnetic ion, while T_{SR} increases. By the extrapolation, the T_N and T_{SR} should merge near $x = 0.15$. There is almost linear dependence of T_N on x in the compositional ranges of $0.0 \leqslant x \leqslant 0.3$ and $0.4 \leqslant x \leqslant 0.7$. In both ranges, T_N vanishes near $x = 0.7$ by the extrapolation, that is, near the percolation threshold. From the extrapolation, we can also estimate that T_N should be about 130 K for a hypothetical R3c phase of $BiCrO_3$, which was studied theoretically in some papers [16, 25]; the theoretically estimated T_N is about 80–120 K [25]. The G-type AFM structure should realize in $BiCrO_3$-based perovskites as predicted in many theoretical papers [15, 16, 25] and found experimentally [11–13]. However, spin canting mechanisms might be different depending on the symmetry; this is why different regions are marked as c-AFM1, c-AFM2 and c-AFM3 in figure 11. Spin canting is allowed by the symmetry in the C2/c and R3c structures and G-type magnetic arrangements.

In conclusion, we investigated magnetic properties of $BiCr_{1-x}Ga_xO_3$ solid solutions. AFM order with weak ferromagnetism is observed below $T_N = 54$, 36 and 18 K in the samples with $x = 0.4$, 0.5 and 0.6, respectively, having a polar R3c structure. T_N decreases from 111 K in $BiCrO_3$ to 98 K in $BiCr_{0.9}Ga_{0.1}O_3$, and the spin-reorientation transition temperature increases from 72 K in $BiCrO_3$ to 83 K in

$BiCr_{0.9}Ga_{0.1}O_3$, having C2/c symmetry. A magnetic phase diagram of $BiCr_{1-x}Ga_xO_3$ is constructed.

Acknowledgments

This work was supported by World Premier International Research Center Initiative (WPI Initiative, MEXT, Japan), the Japan Society for the Promotion of Science (JSPS) through its 'Funding Program for World-Leading Innovative R&D on Science and Technology (FIRST Program)'.

References

[1] Izyumskaya N, Ya Alivov and Morkoc H 2009 *Crit. Rev. Solid State Mater. Sci.* **34** 89
[2] Rodel J, Jo W, Seifert K T P, Anton E M, Granzow T and Damjanovic D 2009 *J. Am. Ceram. Soc.* **92** 1153
[3] Catalan G and Scott J F 2009 *Adv. Mater.* **21** 2463
[4] Belik A A 2012 *J. Solid State Chem.* **195** 32
[5] Suchomel M R, Fogg A M, Allix M, Niu H J, Claridge J B and Rosseinsky M J 2006 *Chem. Mater.* **18** 4987
[6] Yu R, Hojo H, Oka K, Watanuki T, Machida A, Shimizu K, Nakano K and Azuma M 2015 *Chem. Mater.* **27** 2012
[7] Oka K, Koyama T, Ozaaki T, Mori S, Shimakawa Y and Azuma M 2012 *Angew. Chem., Int. Edn* **51** 7977
[8] Zhang J X *et al* 2011 *Phys. Rev. Lett.* **107** 147602
[9] Sugawara F, Iiida S, Syono Y and Akimoto S 1968 *J. Phys. Soc. Japan* **25** 1553
[10] Niitaka S, Azuma M, Takano M, Nishibori E, Takata M and Sakata M 2004 *Solid State Ion.* **172** 557
[11] Belik A A, Iikubo S, Kodama K, Igawa N, Shamoto S and Takayama-Muromachi E 2008 *Chem. Mater.* **20** 3765
[12] Darie C, Goujon C, Bacia M, Klein H, Toulemonde P, Bordet P and Suard E 2010 *Solid State Sci.* **12** 660
[13] Colin C V, Perez A G, Bordet P, Goujon C and Darie C 2012 *Phys. Rev. B* **85** 224103
[14] Goujon C, Darie C, Bacia M, Klein H, Ortega L and Bordet P 2008 *J. Phys.: Conf. Ser.* **121** 022009
[15] Ding J, Yao Y G and Kleinman L 2011 *J. Appl. Phys.* **109** 103905
[16] Ding J, Wen L, Kang X, Li H and Zhang J 2014 *Comput. Mater. Sci.* **96** 219
[17] Belik A A, Tsujii N, Suzuki H and Takayama-Muromachi E 2007 *Inorg. Chem.* **46** 8746
[18] Belik A A and Takayama-Muromachi E 2009 *J. Phys.: Conf. Ser.* **165** 012035
[19] Suchomel M R, Thomas C I, Allix M, Rosseinsky M J, Fogg A M and Thomas M F 2007 *Appl. Phys. Lett.* **90** 112909
[20] Mandal P, Iyo A, Tanaka Y, Sundaresan A and Rao C N R 2010 *J. Mater. Chem.* **20** 1646
[21] Belik A A, Rusakov D A, Furubayashi T and Takayama-Muromachi E 2012 *Chem. Mater.* **24** 3056
[22] Khalyavin D D, Salak A N, Olekhnovich N M, Pushkarev A V, Radyush Yu V, Manuel P, Raevski I P, Zheludkevich M L and Ferreira M G S 2014 *Phys. Rev. B* **89** 174414
[23] Blasco J, Garcia J, Campo J, Sanchez M C and Subias G 2002 *Phys. Rev. B* **66** 174431
[24] Arevalo-Lopez A M and Alario-Franco M A 2011 *Inorg. Chem.* **50** 7136
[25] Baettig P, Ederer C and Spaldin N A 2005 *Phys. Rev. B* **72** 214105

17

High-sensitivity piezoelectric perovskites for magnetoelectric composites

Harvey Amorín[1], **Miguel Algueró**[1], **Rubén Del Campo**[1], **Eladio Vila**[1],
Pablo Ramos[2], **Mickael Dollé**[3], **Yonny Romaguera-Barcelay**[4],
Javier Pérez De La Cruz[4] **and Alicia Castro**[1]

[1] Instituto de Ciencia de Materiales de Madrid, CSIC, Cantoblanco, 28049 Madrid, Spain
[2] Universidad de Alcalá, 28871 Alcalá de Henares, Spain
[3] Département de Chimie, Université de Montréal C.P. 6128, succursale Centre-Ville Montréal, QC, H3C 3J7, Canada
[4] IFIMUP and IN-Institute of Nanoscience and Nanotechnology, Faculdade de Ciências da Universidade do Porto, Rua do Campo Alegre, 687, 4169-007 Porto, Portugal

E-mail: hamorin@icmm.csic.es

Abstract

A highly topical set of perovskite oxides are high-sensitivity piezoelectric ones, among which Pb $(Zr,Ti)O_3$ at the morphotropic phase boundary (MPB) between ferroelectric rhombohedral and tetragonal polymorphic phases is reckoned a case study. Piezoelectric ceramics are used in a wide range of mature, electromechanical transduction technologies like piezoelectric sensors, actuators and ultrasound generation, to name only a few examples, and more recently for demonstrating novel applications like magnetoelectric composites. In this case, piezoelectric perovskites are combined with magnetostrictive materials to provide magnetoelectricity as a product property of the piezoelectricity and piezomagnetism of the component phases. Interfaces play a key issue, for they control the mechanical coupling between the piezoresponsive phases. We present here main results of our investigation on the suitability of the high sensitivity MPB piezoelectric perovskite $BiScO_3$–$PbTiO_3$ in combination with ferrimagnetic spinel oxides for magnetoelectric composites. Emphasis has been put on the processing at low temperature to control reactions and interdiffusion between the two oxides. The role of the grain size effects is extensively addressed.

Keywords: perovskite oxides, piezoelectrics, magnetoelectrics, ceramic composites

1. Introduction

Piezoelectric materials develop a linear electrical polarization in response to a mechanical stress: the direct piezoelectric effect, and linearly deform under the application of an electric field: the converse piezoelectric effect [1]. They are thus also electromechanical transducers, and show enhanced conversion efficiency at mechanical resonance frequencies. Piezoelectric ceramics are a mature and ubiquitous technology, which is the basis of a wide range of applications, such as acceleration sensors, flow meters, actuators (fuel injectors, positioning systems, etc), smart systems (active control of vibrations and structures, adaptive optics, etc), ultrasound transducers (either for non destructive testing or for medical imaging), underwater acoustics, and many others [2–4]. Currently, they are also being considered for novel applications like energy harvesting and magnetoelectric transduction [5, 6].

High-sensitivity piezoelectrics are mostly perovskite ABO_3 oxides. The current state of the art material is $Pb(Zr,Ti)O_3$ (PZT) at the morphotropic phase boundary (MPB) between the ferroelectric rhombohedral $R3m$ and tetragonal $P4mm$ polymorphs [1–4]. It is acknowledged that ferro-electrics at phase instability regions and the presence of monoclinic phases between the rhombohedral and tetragonal ones are key to high piezoelectric response [7, 8]. Domain dynamics also play a major role, and indeed, large piezo-electric coefficients can be achieved by enhancing the ferro-electric/ferroelastic domain walls mobility [9]. A main line of research at the cutting edge of the field is the design and synthesis of novel perovskite systems with ferroelectric MPBs and high piezoelectric response that can replace PZT, either lead-free, necessary for environmentally friendly pie-zoelectric elements [10], or high Curie temperature T_C to enable electromechanical transduction in harsh environ-ments [11].

Bi-containing perovskites are playing a major role in both topics, though advances are hindered by thermodynamic instability associated with a low tolerance factor, which causes a number of very promising systems to be obtained only by high-pressure synthesis [12]. Mechanochemical activation is being considered an alternative route to obtain low-tolerance factor perovskites that cannot be obtained by conventional synthesis routes, and indeed we have succeeded in mechanosynthesizing a number of examples [13–15]. The solid solution of $(1-x)BiScO_3-(x)PbTiO_3$ is receiving a lot of attention as an alternative system to PZT for the next gen-eration of high-temperature and high-sensitivity piezoelectric devices. This system exhibits large piezoelectric coefficient d_{33} of ~450 pC N^{-1} and high T_C of ~450 °C [16] (this is 100 °C above that of PZT), and better properties on grain size down-scaling [17–19]. As expected, best properties are found around a MPB (for $x \sim 0.64$) analogous to that of PZT [20].

Perovskite MPB piezoelectrics are being considered in combination with magnetostrictive materials for the devel-opment of magnetoelectric composites, in which magnetoe-lectricity is obtained as a product tensor property of the piezoelectricity and effective piezomagnetism of the two ferroic phases [21–23]. Direct and converse effects also result: the development of an electric polarization $\mathbf{P}$ propor-tional to an applied magnetic field $\mathbf{H}$ and of magnetization $\mathbf{M}$ in response to an applied electric field $\mathbf{E}$, respectively. This opens the possibility of enabling a range of novel disruptive technologies, such as: high-sensitivity magnetic field sensors with room-temperature (RT) operation, microgenerators for remote powering of wireless bio-implanted devices from ambient or directed magnetic fields, or dual electric-field magnetic-field tunable microwave devices such as filters, resonators and phase shifters [24–26].

Perovskite PZT is the usual choice for the piezoelectric phase, while magnetostrictive phases could be either metallic alloys or magnetic oxides [21–23]. These latter materials are not actual piezomagnetics, but effective responses can be obtained under bias magnetic field. A range of two-phase composites with 0–3 and 2–2 connectivities have been demonstrated, among which laminate composites fabricated by the epoxy-bonding method with modified PZT and giant magnetostrictive metal alloys like tefenol-D and metglas stand out [27]. However, this method presents a challenge for mass fabrication, and cofiring techniques to obtain direct interfaces would be preferred. In this case, research focuses in all-oxide ceramic composites obtained by co-sintering at high temperatures, among which ferrimagnetic spinels and ferro-electric perovskite materials play a major role, with magne-toelectric coefficients of several orders of magnitudes above those shown by single-phase multiferroics [28–30].

Nevertheless, the magnetoelectric effects so far observed in high-temperature cofired bulk ceramic composites are still much lower than those theoretically predicted, due mainly to the poor mechanical coupling between the phases and defects concentrated at the interfaces [22]. The novel functionality in these two-phase composite oxides is a strain-mediated effect, so a major issue is to optimize strain continuity across the piezoelectric–magnetic interfaces, and a main line of research is controlling diffusion phenomena across and chemical reac-tions at the interfaces during high-temperature preparation.

In this context, we are investigating novel methods for the preparation of laminate ceramic composites with high-quality interfaces, by using nanopowders of the ferroic oxides obtained by mechanosynthesis and wet-chemistry routes, and cofiring them by spark plasma sintering. We present here results on the suitability of using the high-sensitivity piezo-electric perovskite $BiScO_3-PbTiO_3$ in combination with the ferrimagnetic spinel $NiFe_2O_4$ for the preparation of multilayer magnetoelectric composites made of alternating piezoelectric and magnetostrictive layers obtained by tape-casting. Emphasis has been put on controlling chemical reactions and interdiffusion across the interfaces to obtain high magneto-electric response. Microstructural and interface features are described, and the main issues relevant to functionality are discussed.

2. Experimental procedures

Perovskite phase nanocrystalline powders of $0.36BiScO_3-0.64PbTiO_3$ (BSPT) were obtained by mechanochemical treatment of stoichiometric mixtures of analytical grade Bi_2O_3, Sc_2O_3, PbO and TiO_2, with a Pulverisette 6 (Fritsch) planetary mill. The mechanosynthesis of the perovskite as single phase was successfully achieved after only 20 h of milling, as shown in figure 1(a). An average particle size of 23 nm with standard deviation (SD) of 11 nm resulted, as measured by transmission electron microscopy (TEM) [31]. For the mechanochemical synthesis of the $NiFe_2O_4$, analytical grade NiO and Fe_2O_3 were used as starting reagents. The spinel phase could be completely isolated after a thermal treatment at 600 °C/2 h, as shown in figure 1(b). An average particle size of 30 nm with SD of 13 nm resulted. Details of the procedures and of the mechanisms taking place during the mechanosynthesis of these two systems, the perovskite and the spinel, can be found elsewhere [13, 31].

A wet-chemistry route was also employed for the synthesis of reactive precursors of the ferrite $NiFe_2O_4$, as

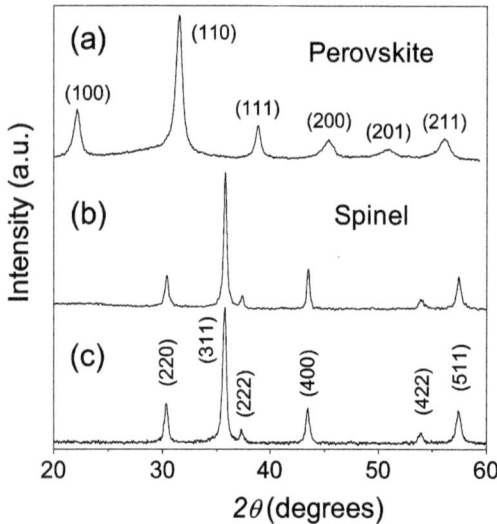

Figure 1. XRD patterns corresponding to (a) the $0.36BiScO_3$–$0.64PbTiO_3$ perovskite phase; and those for the $NiFe_2O_4$ spinel phases obtained by (b) mechanochemical activation and (c) wet-chemistry, after annealing at 600 °C.

described elsewhere [32]. Reagent grade 0.06 mole of Fe$(NO_3)_3 \cdot 9H_2O$ and 0.03 mole of $Ni(NO_3)_2 \cdot 6H_2O$ were dissolved in 300 ml of distilled water with 3 ml of HNO_3 (65% concentration). Precursor materials were obtained by precipitation of Fe^{3+} and Ni^{2+} cations with dropwise addition (2 ml min^{-1}) of 1 M solution of n-butylamine at RT, under vigorous stirring, up to pH = 10. Then, these suspensions were filtered, washed with distilled water, and dried at 80 °C. The solid precursors were annealed at 600 °C for removing water, nitrates, and organic ions, which allows the synthesis of the spinel as single phase, as shown in figure 1(c). In this case, an average particle size of 35 nm with SD of 16 nm resulted.

Tape casting was used to obtain laminate composites of alternating magnetostrictive and piezoelectric thick films. Ethanol-based slurries with 20 vol% of powders were developed to avoid the use of highly toxic toluene-based solvents [33]. The binder system was optimized to ensure similar organic volume fractions in the perovskite and spinel phases, to prevent delamination due to uneven shrinkage during organics burn out process. After drying, tapes with thickness of 35 to 50 μm resulted, from which multilayer composites were prepared by lamination under uniaxial pressure of 10 MPa. Organics were burned out at 500 °C/2 h, with slow heating/cooling rates of 0.5 °C min^{-1} to avoid delamination. Ceramic composites were prepared by spark plasma sintering (SPS, Dr Sinter 2080 Sumitomo apparatus) under uniaxial pressure of 100 MPa and sintering temperature between 700 and 1000 °C with heating rate of 375 °C min^{-1}. The soaking time at the final temperatures and pressure was 10 min. Density was measured by Archimedes's method in distilled water after polishing.

X-ray diffraction (XRD) was carried out with a Bruker D8 Advance diffractometer (CuKα radiation) in the range

from 5 to 70° (2θ), in steps of 0.05° and counting time of 1.5 s/step. Ceramic microstructures were characterized by field emission scanning electron microscopy, FE-SEM (FEI Nova NanoSEM 230 microscope) on cross-sections perpendicular to the casting plane. Samples were prepared by polishing with Al_2O_3 suspensions down to 0.1 μm. The Feret diameters of more than 300 grains were measured from SEM images to obtain reliable size distributions. The average size and SD were obtained with statistical analysis by using probability plots. Linear fits with correlation factor above 0.99 were obtained.

Electrical characterization was carried out on ceramic discs, on which Ag electrodes were painted and annealed at 700 °C/1 h. The temperature dependence of the dielectric permittivity and losses were dynamically measured under heating at 1.5 °C min^{-1}, with an HP4284A precision LCR Meter in the frequency range 100 Hz–1 MHz. Polarization–electric field (P–E) ferroelectric hysteresis loops were obtained by current integration. High voltage sine waves (0.1 Hz) were applied by the combination of a synthesizer/function generator (HP3325B) and a high-voltage amplifier (Trek Model 10/40A), while charge was measured with a homebuilt charge to voltage converter. Loops are presented after subtracting linear polarization and conduction contributions from the current response, assuming a resistance and a capacitance in parallel. The d_{33} piezoelectric coefficient was measured with a Berlincourt-type piezometer.

Magnetic measurements were carried out with a Quantum Design MPMS-5S SQUID magnetometer. Isothermal magnetization–magnetic field (M–H) curves were recorded at RT with magnetic fields of ±3 kOe. The characterization of the magnetoelectric response was carried out with a system (Serviciencia SL) consisting of a combination of two Helmholtz coils: a high power coil and a high frequency one, designed to independently provide a dc magnetic field up to 1 kOe (to magnetize the material), and an ac magnetic field up to 10 Oe at 10 kHz (the stimulus). Magnetoelectric output voltages were monitored with a lock-in amplifier. The thickness of the piezoelectric element was used to compute the magnetoelectric coefficient, which allows comparing the response of trilayers and multilayers.

3. Results and discussion

The magnetoelectric response of laminated composites is determined by three main issues [26]: (i) the inherent properties of the constituent phases (e.g., dielectric permittivity and losses, elastic stiffness, piezoelectric and effective piezomagnetic coefficients, etc); (ii) the quality of the interfaces and geometry of the composites (e.g., pores, defects and stress along the interfaces, volume to thickness ratio of the layers, etc); and (iii) the operation mode (i.e., orientation of the applied magnetic field with respect to the alternating layers). For the latter, it is well-known that measuring the magnetoelectric effect in a longitudinal–transverse (L–T) field mode, in which the applied magnetic field is oriented parallel to the alternating layers whereas the sample is poled and the

16 (2015) 016001

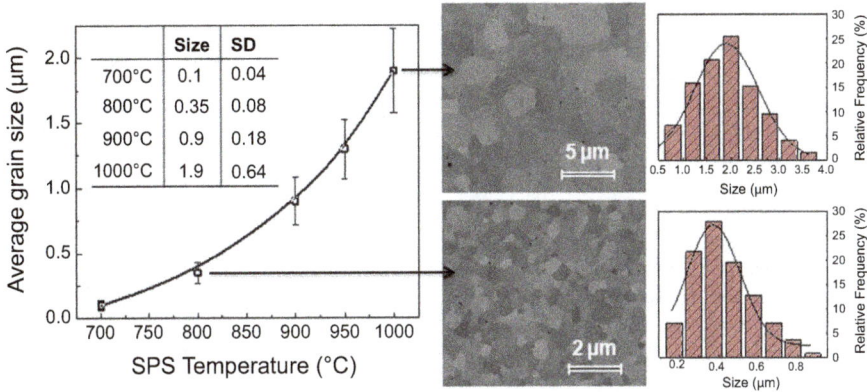

Figure 2. Evolution of the average grain size for the perovskite BSPT with the SPS temperature, average sizes and standard deviation (SD) are given. SEM images of polished surfaces of two representative samples along with their size distributions.

output voltage measured perpendicular to them, results in a much larger signal [22, 27]. On the other hand, the control of the functional properties of the constituent phases near the interfaces is as important as the effectiveness of the elastic coupling and strain continuity through them.

The processing method is key to obtain tailored interfaces in composites, and this usually requires low-temperature cofiring techniques, which also limits grain growth [28–30]. The unique features of the SPS makes possible to obtain fully dense materials at relatively low temperatures with controlled grain growth [34]. Besides, small grains at the interfaces would be advantageous for diminishing the thermal expansion mismatch between the dissimilar ceramic phases, thus reducing stress effects that may generate cracks and delamination. In such cases, however, it would be necessary to consider the effects of the grain size reduction on the functional properties of the perovskite phase, which may be an issue for obtaining large magnetoelectric response in low-temperature cofiring composites.

Grain size effects were studied for the functional properties of perovskite BSPT ceramics with grain sizes in the range obtained for the composites. Samples were processed by SPS under tailored conditions of nanopowders obtained by mechanosynthesis. Figure 2 shows the evolution of the average grain size with the SPS temperature, whose exponential increment suggests a single mechanism of grain growth. Error bars represent the SD obtained from the size distributions for each data point. SEM images of polished surfaces of two representative samples are also shown, along with the size distributions resulting, to illustrate the high-density and quality of the materials prepared. In these images, individual grains are distinguished by a gray scale map based on the electron backscattering diffraction pattern, in which gray contrasts correspond to changes in crystal orientations [35].

Figure 3(a) illustrates the grain size effect on the ferroelectric switching of these ceramics. Square hysteresis loops close to saturation of the polarization are obtained for ceramics with grain sizes about and above $1\,\mu m$. Remnant polarization P_r slowly decreases from 40 to $36\,\mu C\,cm^{-2}$ when

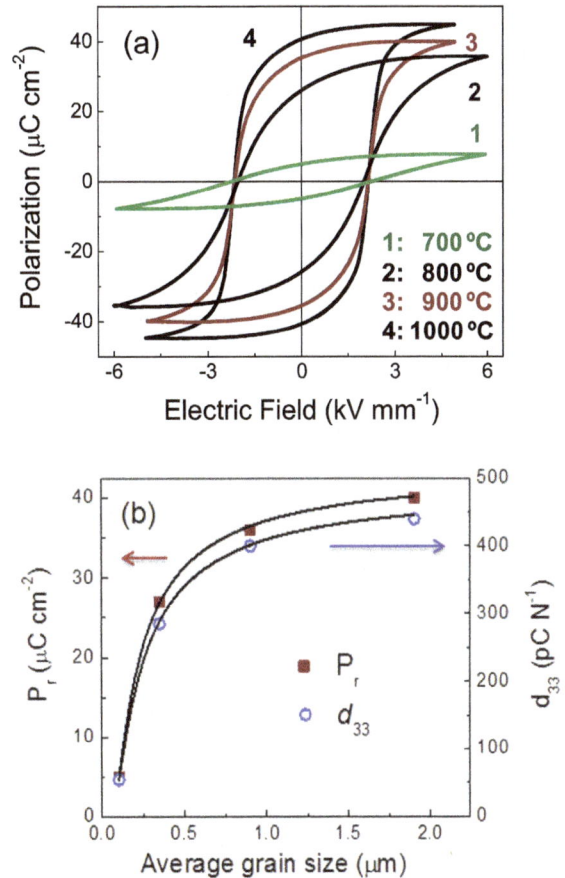

Figure 3. (a) Ferroelectric hysteresis loops for $BiScO_3$–$PbTiO_3$ ceramics obtained at different SPS temperatures, and (b) evolution of the remnant polarization (P_r) and d_{33} piezoelectric coefficient with increasing grain size.

entering the submicron range, and then drops down to $27\,\mu C\,cm^{-2}$ as the grain size was reduced to $0.35\,\mu m$, and only $5\,\mu C\,cm^{-2}$ was achieved for $0.1\,\mu m$. Size effects were also observed in the piezoelectric charge coefficient d_{33}

(figure 3(b)), which slowly decreases from 440 to 400 pC N^{-1} when entering the submicron range, and suddenly drops down to 285 and 55 pC N^{-1} for 0.35 and 0.10 μm, respectively.

It has been recently shown that the grain size dependence of P_r (and also d_{33}) actually reflects the increase of the effective coercivity as the grain size decreases rather than a real decrease of the spontaneous polarization [36]. Higher electric fields would be required to approach saturation, and thus achieve an efficient poling (and a higher d_{33}) of the material. However, infinitely high fields cannot be applied because electrical breakdown takes place (e.g., above 6 kV mm^{-1}). The increase of the effective coercivity resulting in the loss of functionality was associated with the presence of grain boundaries with reduced permittivity. The decrease of d_{33} with grain size thus reflects how one moves away from the saturation condition as the effective E_c increases with decreasing grain size, and an efficient poling of the material cannot be achieved. Therefore, small grain sizes in the perovskite component of laminated composites would be detrimental for the magnetoelectric response.

Perovskite/spinel/perovskite trilayers with 1 mm thickness each element were prepared, as they are especially suitable to study chemical reactions and interdiffusion at the interfaces, and for addressing compatibility issues during SPS of the component phases. The different shrinkage behaviour of these dissimilar phases may lead to cracking and delamination, and thus result in poor properties [29]. Results in terms of microstructure and properties are the same if the perovskite is the one sandwiched between spinel phases in the trilayers.

Figure 4 shows representative SEM images of polished cross-sections of trilayers prepared at 800, 900 and 1000 °C, showing the interface between the perovskite and spinel phases. Chemical reaction and interdiffusion phenomena at the interfaces were not detected, and nor was the occurrence of abnormal grain growth in the perovskite side, as previously reported for composites obtained by hot-pressing or conventional sintering [29, 31]. These abnormal grains were attributed to diffusion of ions from the spinel phase into the perovskite with diffusion lengths of 20 to 30 μm, and thus should result in poorer piezoelectric properties.

A common feature of the samples at 800 and 900 °C is the high-quality of the interfaces in terms of the absence of cracks or delamination, where differences between them are basically in the grain sizes. Fairly dense microstructures with very few isolated pores were obtained. The composite prepared at 800 °C (figure 4(a)) shows negligible porosity and very small grain sizes approaching the nanoscale, for both phases. As expected, grain size increases with the SPS temperature. The composite prepared at 900 °C (figure 4(b)) shows submicron-sized grains in both phases, with a high-quality interface; whereas the sample prepared at 1000 °C (figure 4(c)) shows micron-sized grains, with the perovskite phase denser than the ferrite one. However, this sample showed a bimodal and degraded microstructure at the ferrite side, in which cracks and large voids can be observed. The densification of this trilayer decreased to less than 92% of the theoretical density (due basically to the porosity of the spinel

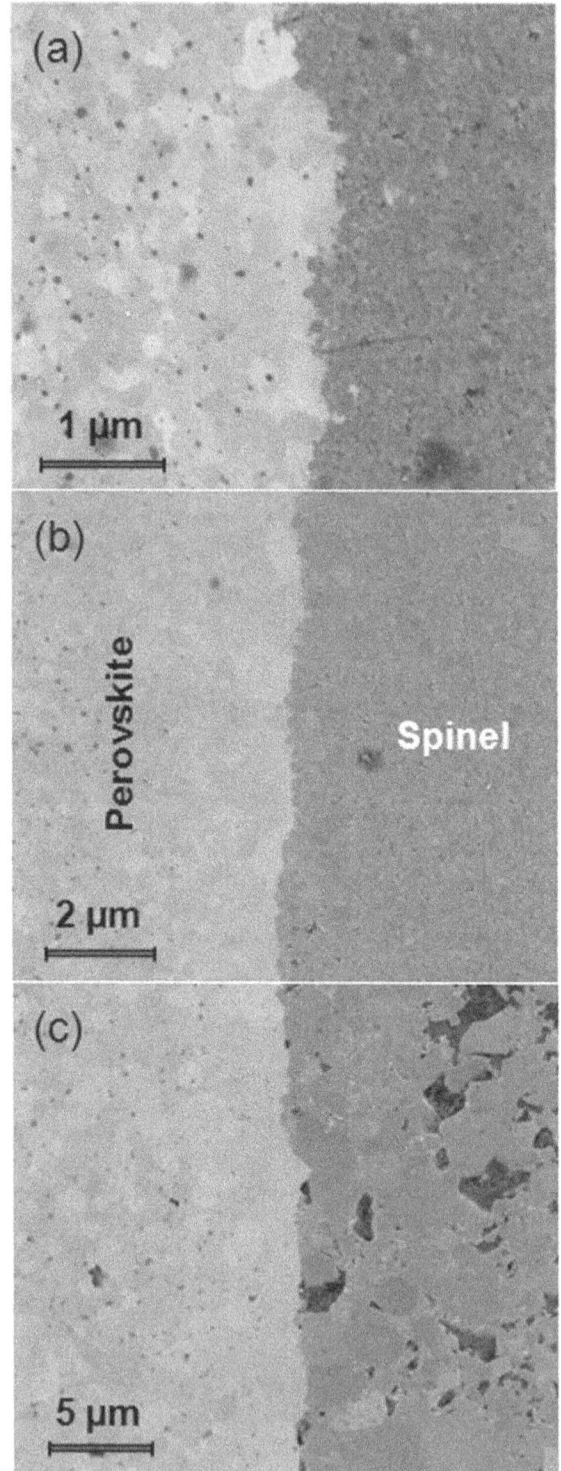

Figure 4. Representative SEM images of polished cross-sections at perovskite/spinel interfaces in trilayers prepared by SPS at (a) 800, (b) 900 and (c) 1000 °C, using the ferrite obtained by mechanochemical activation.

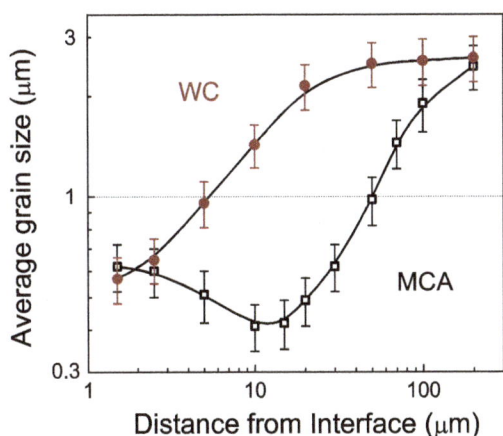

Figure 5. Evolution of the perovskite average grain size with the distance from the interface for trilayers prepared by SPS at 900 °C, using ferrites obtained by wet-chemistry (WC) and mechanochemical activation (MCA).

phase), as compared to that about 99% for the other two samples (densification was calculated taking into account the characteristics of the trilayers). The degradation of the spinel when the sintering temperature reaches 1000 °C is the result of an abnormal grain growth.

Nevertheless, none of the trilayers of figure 4 showed a homogeneous microstructure over the whole perovskite phase, and instead grain size gradients were obtained from the interface to the inner BSPT matrix, which spreads over a region that extends up to 200 μm. Figure 5 shows the evolutions of the perovskite average grain size with distance from the interface for the trilayer prepared at 900 °C (curve with open dots). The plot is given in log–log scale for a better visualization of the data evolution, and the SD calculated from the size distributions are indicated as the error bars. The grain size shows a peculiar evolution with distance for this trilayer prepared with the spinel obtained by mechanochemical synthesis, for which a minimum average size of ~0.4 μm was achieved at about 10 to 20 μm from the interface, and then evolves up to about 2.5 μm (at 200 μm away the interface).

The picture changes significantly when the spinel powder obtained by the wet-chemistry route is used whereas the perovskite is still mechanosynthesized. In this case, the perovskite grain size increases continuously with distance and average sizes in the micron range are achieved at only 5 μm from the interface (curve with closed dots in figure 5). Notably, both trilayers have basically the same average sizes at the interface and far away thereof, but its evolution with distance from the interface greatly differs. This is the result of the different sinterability of the phases involved, i.e., the delay in achieving high densification before the thermally activated processes for grain growth are triggered. A deeper insight into the underlying mechanisms can be obtained by studying the *in situ* shrinkage curves recorded during the SPS experiments, and the role of the different parameters on the microstructural features. A direct influence on the

magnetoelectric response of multilayers obtained from these materials can be anticipated, attending to the functionality expected for these interfaces.

Figure 6(a) shows a representative SEM micrograph of the cross-section of a multilayer structure made of alternating perovskite (light stripes) and spinel (dark stripes) phases by SPS at 900 °C. The approach is demonstrated suitable to obtain highly-dense microstructures with high-quality interfaces, avoiding thermal expansion mismatch between the two ceramics that usually results in cracks and delamination. The peculiar evolution of the perovskite grain size when the spinel is obtained by mechanosynthesis in trilayers does not occur similarly in multilayers, due to the small thickness of the layers. Note in figure 5 that similar grain sizes are obtained up to 30 μm from the interface. In the multilayers, submicron grains are obtained across the whole perovskite layers. The different interfaces, basically in the perovskite grain size, of the multilayers prepared with the ferrites obtained by mechanosynthesis and wet-chemistry are shown in figures 6(b) and (c), respectively. Fairly dense microstructures were obtained in both cases at any distance from the interface. Improved properties would be anticipated for the latter, attending to the larger grain size near the interfaces.

Figure 7(a) shows the relative dielectric permittivity K' and losses (tan δ) as a function of temperature at several frequencies for the trilayer prepared at 900 °C with the ferrite obtained by wet-chemistry. K' has been computed with the thickness of the piezoelectric element for comparison with the multilayers. Both K' and tan δ curves hold a strong resemblance with the typical ones of BSPT ceramics [36], and indicate a very small influence of the spinel in the dielectric response of the trilayer. The anomaly associated with the ferroelectric transition is clearly observed at $T_C \sim 470$ °C. The pronounced dispersion with frequency in K' around the ferroelectric transition also resembles the previously reported data for BSPT ceramics, which was described as a grain boundary effect. Indeed, simulation of the electrical response for the grain bulk and boundary contributions shows the low-frequency data to be the closest one to the bulk response [37].

The dielectric response changes significantly for the multilayer composites. Figures 7(b) and (c) show the K' and tan δ curves for multilayers prepared at 900 °C, using the ferrites obtained by wet-chemistry and mechanochemical activation, respectively. In both cases, the dielectric dispersion increases by a large extent, i.e., the maximum K' decreases about 5 times on increasing frequency and shifts to higher temperatures for the highest frequencies. Note also the increase of the tan δ in more than an order of magnitude. This is not a grain boundary effect of the perovskite as described in trilayers, and neither a ferroelectric nor a relaxor-type phenomenon. This is most probably associated with a Maxwell–Wagner relaxation process that overlaps with the ferroelectric transition. This effect accounts for charge accumulation at the interfaces between the perovskite and the spinel components of the multilayers on the basis of the difference in the characteristic relaxation times in these two phases. Several relaxations would be expected by the presence of different

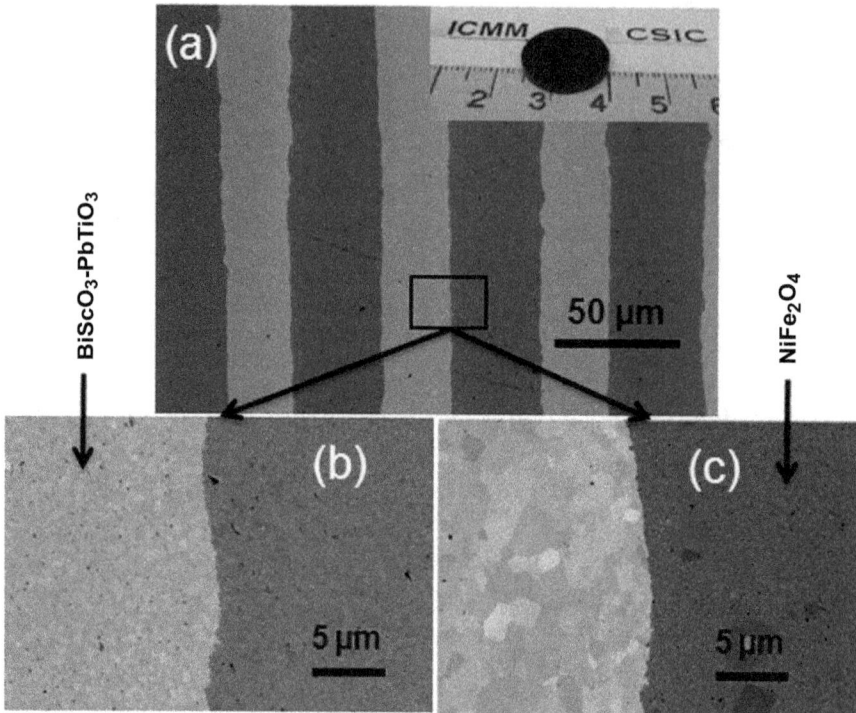

Figure 6. (a) Representative SEM micrograph of the cross-section of a multilayer structure made of alternating perovskite (light stripes) and spinel (dark stripes) phases by SPS at 900 °C. Images of the interfaces of multilayers prepared with ferrites obtained by (b) mechanochemical activation and (c) wet-chemistry.

thermally activated conductivity processes, i.e., electron hopping, oxygen vacancy displacements, etc.

Actually, the complex dielectric response of magneto-electric composites with this 2–2 type connectivity has been described with Maxwell–Wagner relaxation model [38], in which the relaxation time was demonstrated that changed over a wide range of values by varying the volume fractions of the component phases. This effect must be also present in the trilayers, though in this case the perovskite phase dominates the dielectric response in the measuring frequency range. The change to a multilayer configuration with thinner piezoelectric elements and a larger number of interfaces modified the characteristic relaxation times and increases the Maxwell–Wagner extrinsic contribution to the dielectric response.

Nevertheless, the ferroelectric transition can be observed in the low-frequency data at the same $T_C \sim 470$ °C for both multilayers. This is particularly clear for the multilayer prepared with the ferrite by wet-chemistry, as shown in figure 7(b), in which the lack of shift of the temperature of maximum K' with frequency (in the low-frequency region) indicates it is the ferroelectric transition. Lower permittivity values were obtained for the multilayer prepared with the ferrite by mechanosynthesis, yet the transition is still visible as shown in figure 7(c). This sample showed perovskite grain sizes at the submicron range approaching the nanoscale. The most noticeable effect of decreasing the grain size of a high

sensitivity piezoelectric ceramic into the submicron range, and further down to the nanoscale on its dielectric properties is a progressive broadening and depletion of the maximum in permittivity at the temperature of the ferroelectric transition [36].

The ferroelectric hysteresis loops are typically good indicators of the ability for poling a ferroelectric material, and therefore useful to validate the achievement of an effective poling in the multilayer composites. It should be mentioned that the hysteresis loops usually reported for laminate composites with a direct bonding between the component phases resembles that of a lossy dielectric material with leakage currents [29]. Figure 8(a) shows the P–E curves along with their current density curves with increasing the applied fields up to $5\,\mathrm{kV\,mm^{-1}}$, for the trilayer prepared at 900 °C with the ferrite by wet-chemistry. This field amplitude is high enough to achieve a square loop and saturation of polarization, from which P_r of $\sim 35\,\mu\mathrm{C\,cm^{-2}}$ resulted. The field evolution of the curves holds a strong resemblance with the typical ones of BSPT ceramics, for which a maximum P_r of $\sim 40\,\mu\mathrm{C\,cm^{-2}}$ was achieved (see figure 3), indicating that the ferrite sandwiched between two perovskites behaves as a good conductive material, and highlighting the high quality of the interfaces achieved.

Figure 8(b) shows the P–E and current density curves for the analogous trilayer prepared with the ferrite by mechanochemical activation. In this case, square loops and saturation

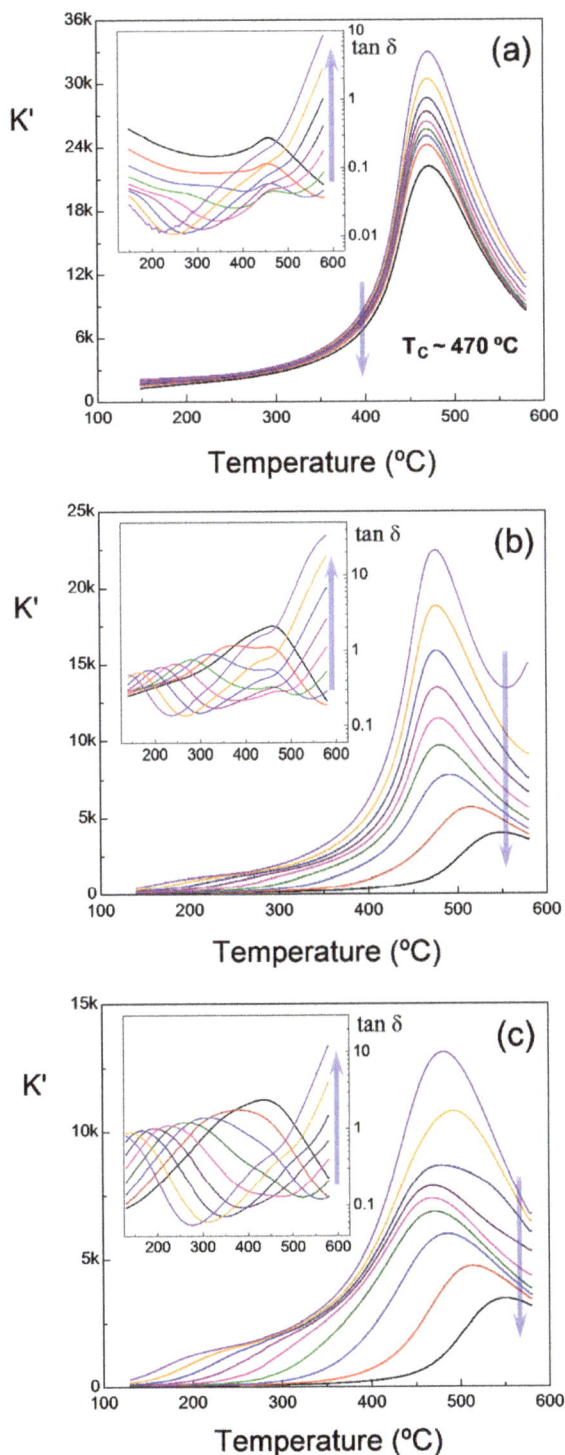

Figure 7. Relative permittivity K' and dielectric losses (tan δ) as a function of temperature at several frequencies (0.1, 0.5, 1, 5, 10, 50, 100, 500, 1000 kHz; arrows indicate increasing frequency) for (a) the trilayer prepared at 900 °C with the ferrite obtained by wet-chemistry; and (b), (c) multilayers prepared at 900 °C with ferrites obtained by (b) wet-chemistry and (c) mechanochemical activation.

of polarization was also achieved, with a P_r of $\sim 32\,\mu C\,cm^{-2}$. However, the coercive field of the material increases from $2.3\,kV\,mm^{-1}$ in the former case to $3.0\,kV\,mm^{-1}$ in this last case, an increase that is attributed to the wide grain size gradient described in figure 5. This can be clearly observed in the current density curves, in which the evolution of the switching with increasing fields can be followed. Two peaks can be well differenciated on the current curves for applied fields of 3 and $4\,kV\,mm^{-1}$, indicating that switching is not homogeneous. These peaks are attributed to the volume fractions of submicron-sized and micron-sized grains with different coercitivity in this material. The increase of the effective coercitivity as the grain size decreases was previously described. Finally, the curves evolve to achieve near saturation of polarization with a single but wider coercive field distribution for the highest applied field, dominated by the coercitivity of grains with the smaller sizes. The increase of the effective coercitivity may be a drawback to achieve an efficient poling of the composites.

Figure 9(a) shows the P–E and current density curves for the multilayer prepared at 900 °C with the ferrite prepared by wet-chemistry. Remarkably, the field evolution of the curves holds a strong resemblance with those of the analogous trilayer, saturation of polarization was practically achieved with the same P_r of $\sim 35\,\mu C\,cm^{-2}$ and E_c of $2.2\,kV\,mm^{-1}$ as trilayers. The comparison of the hysteresis and their current curves between the multilayers prepared at 900 °C with ferrites by wet-chemistry and mechanochemical activation is shown in figure 9(b). The former (labelled **2** in the plot) shows values close to that of the coarse grained ceramic (labelled **1**), although the current density curve indicates a wider distribution of coercive field and thus of grain size, as expected. Note this material also presents submicron-sized grains at the interfaces, yet for a distance of only 5 μm. On the other hand, the hysteresis loops for the latter (labelled **3**) resemble that reported for nanostructured BSPT ceramics [18]. The current curve indicates an extremely wide distribution of coercive fields associated with the small grain sizes, so higher fields are needed to achieve saturation. Nevertheless, a P_r of $\sim 28\,\mu C\,cm^{-2}$ was achieved, in good agreement with results in figure 3 for the similar grain sizes.

Poling was accomplished under the maximum field attained during loop measurements at a very low frequency (0.01 Hz), and after removing the field just before completing the loop. Berlincourt d_{33} coefficients were measured on the poled composites and included in table 1, which summarizes the macroscopic properties of the trilayers and multilayers processed at different synthesis and SPS conditions. In summary, composites prepared with the ferrites by wet-chemistry show higher d_{33} coefficients than those with the ferrite by mechanochemical activation, due to the better grain size distributions in the former. Besides, the higher the SPS temperature, the higher the d_{33} coefficient, as expected due to the increase in grain size. The d_{33} coefficients of the multilayers were three times lowers than those of trilayers under same synthesis and sintering conditions, despite them showing very similar hysteresis loops with full switching of the polarization. Besides, the best d_{33} for the composites were also lower

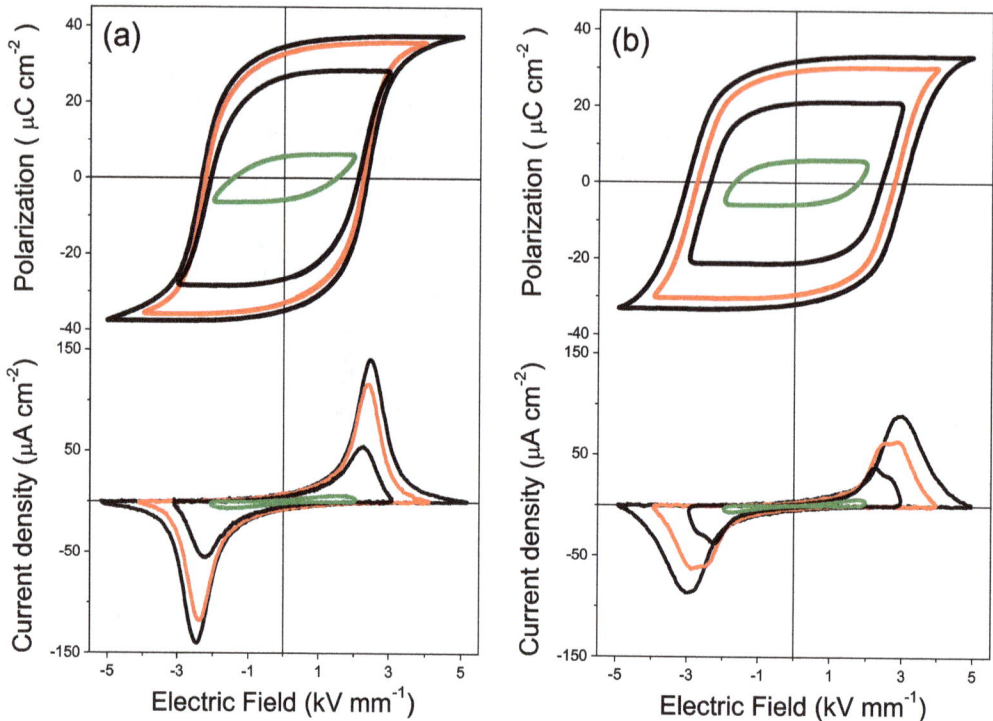

Figure 8. Ferroelectric hysteresis loops and current density curves with increasing applied electric fields for trilayers prepared at 900 °C with ferrites obtained by (a) wet-chemistry and (b) mechanochemical activation.

than those for the BSPT ceramics with similar grain sizes. It should be taken into account that these coefficients should not be considered to be real values of the piezoelectric layers, but just considered as effective coefficients of the composites. It is difficult to know, for instance, which is the effect of the ac mechanical strain of the Berlincourt system on the ferrite layers, but they can be used for comparison between the different composites.

The magnetization of the trilayers was also measured as a function of the magnetic field. Figure 10(a) shows the RT M–H hysteresis loops for the trilayer prepared at different SPS temperature with the ferrite by wet-chemistry. Similar data was obtained with the ferrite by mechanochemical activation. The saturation magnetization reaches the value of $M_{sat} = 51$ emu g^{-1} for the trilayer processed at 1000 °C, very close to that reported for bulk NiFe$_2$O$_4$ [39], and the remnant magnetization M_r and coercive magnetic field H_c were found to be 1.8 emu g^{-1} and 10 Oe, respectively. These are typical values for this ferrimagnetically ordered material, in which magnetic domain switched under very low applied magnetic field. With decreasing the SPS temperature two effects appear as a consequence of the reduction of the spinel grain size: (i) M_{sat} decreases and the material is increasingly far from reaching saturation of the magnetization, analogous to the results of the P–E curves in the perovskite, and (ii) M_r and H_c continuously increases, indicating that the spin canting contribution to the magnetization starts to dominate. M_r of 3 and 4.1 emu g^{-1} and H_c of 27 and 70 Oe were found for materials

prepared at 900 and 800 °C, respectively. These results are in a good agreement with reports on NiFe$_2$O$_4$ nanocrystalline particles [39].

For magnetoelectric composites, the dependence of the magnetoelectric voltage coefficient (α_{31}) on an applied dc magnetic field H_{dc} is tightly related with the effective piezomagnetic coefficient of the spinel phase obtained from the curve of in-plane magnetostriction λ as a function of H_{dc}. The peak value in α_{31} is related to the position of maximum piezomagnetic coefficient [40]. However, some information can also be extracted from the magnetization hysteresis curves, for magnetostriction is correlated with the magnetization as $\lambda \propto M^2$ [41]. A relation can be established between the low-field magnetization saturation behaviour and the position of the magnetoelectric α_{31} peak appearing at low magnetic dc field. Figure 10(b) shows the $\partial M^2 / \partial H$ curves as a function of H_{dc} to illustrate the typical response that would be expected in the composites considering only the quality of the ferrite. Larger magnetoelectric response at lower H_{dc} would be expected with increasing the SPS temperature. Of course, the quality of the interfaces and the multilayers will influence the final response.

Figure 11(a) shows the magnetoelectric voltage coefficients as a function of dc magnetic field in the longitudinal-transverse field mode ($H_{ac} = 10$ Oe at 10 kHz) of trilayers prepared at different temperatures with the ferrite by wet-chemistry. The curves were found to be closely related to the trend of those in figure 10(b), although the maximum α_{31}

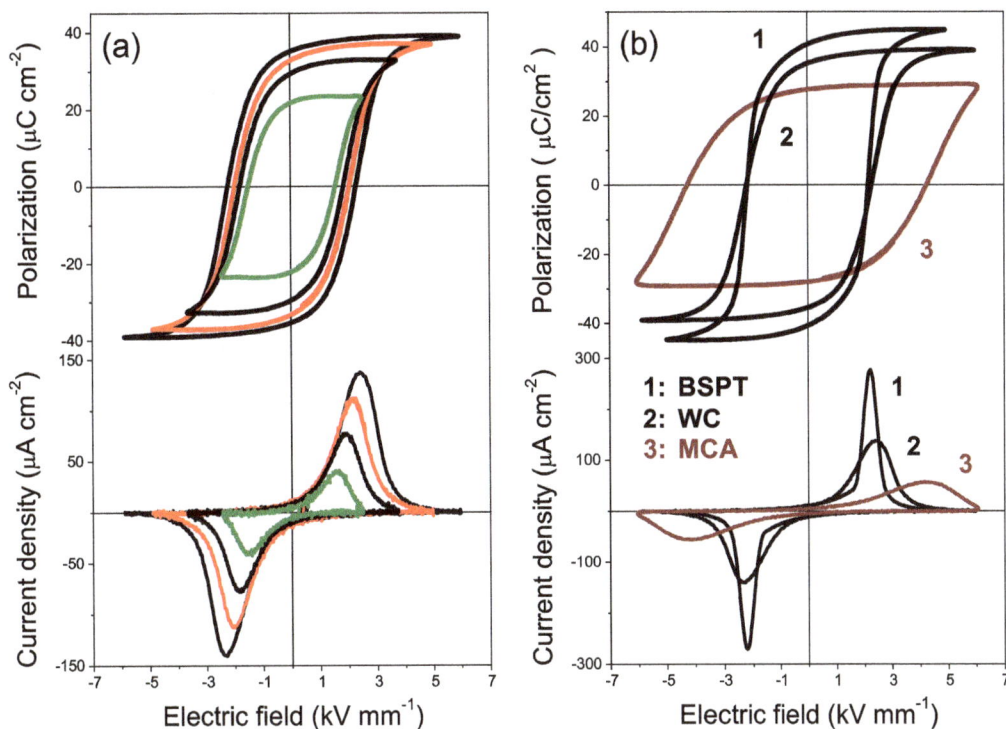

Figure 9. Ferroelectric hysteresis loops and current density curves for (a) a multilayer with the ferrite obtained by wet-chemistry; and (b) comparison between the coarse grained BSPT ceramic and multilayers with ferrites obtained by wet-chemistry (WC) and mechanochemical activation (MCA). All prepared by SPS at 900 °C.

Table 1. Correlation between synthesis/processing conditions and macroscopic properties of trilayer (PSP) and multilayer (ML) ceramic composites. Berlincourt piezoelectric coefficient (d_{33}) and maximum magnetoelectric voltage coefficient (α_{31}) at the dc magnetic field (H_{dc}).

Perovskite BiScO$_3$–PbTiO$_3$	Spinel NiFe$_2$O$_4$	SPS temp.	Comp.	d_{33} (pC N^{-1})	α_{31} (mV cm^{-1} Oe^{-1})	H_{dc} (Oe)
Mechanosynthesis	Mechanosynthesis	800 °C	PSP-trilayer	28 ± 3	5.0 ± 0.5	350 ± 50
		900 °C		230 ± 7	29 ± 3	255 ± 30
		1000 °C		240 ± 5	25 ± 1	200 ± 20
		900 °C	ML	65 ± 2	57 ± 1	245 ± 30
	Wet-chemistry	800 °C	PSP-trilayer	50 ± 3	12 ± 1	345 ± 45
		900 °C		300 ± 7	36 ± 3	235 ± 25
		1000 °C		315 ± 5	32 ± 2	190 ± 25
		900 °C	ML	105 ± 5	108 ± 1	205 ± 25

coefficient of about 35 mV cm^{-1} Oe^{-1} was obtained for the trilayer prepared at 900 °C. The peak voltage coefficient was obtained at relatively low dc fields (about 200–230 Oe), and increases with decreasing SPS temperature following the anticipated trend. Despite the fact that the inherent properties of the constituent phases for the trilayer at 1000 °C were better, the α_{31} coefficient was lower due to the poorer elastic coupling between two phases, emphasizing the role of the strain continuity across the interface. The α_{31} coefficient and the H_{dc} for the peak position in trilayers and multilayers are also included in table 1.

The α_{31} coefficients as a function of H_{dc} for the multilayers prepared at 900 °C with the ferrites by both wet-chemistry and mechanochemical activation, are shown in figure 11(b). In the first case α_{31} reaches up to 108 mV cm^{-1}

Oe^{-1}, which is almost double the value achieved for the latter. This improvement of the magnetoelectric performance can be attributed to the enhanced inherent properties of the perovskite, with a more homogeneous microstructure and tailored interfaces. The output voltage measured is lower than the direct magnetoelectric signal induced from the piezoelectric layers, because of the presence of the poorly insulating ferrite layers. Indeed, internal electrodes have been introduced between piezoelectric and magnetic layers, which can directly collect the output charges produced from the piezoelectric layers, and thus improve the magnetoelectric response [29, 42]. The results here presented demonstrate that direct bonding by SPS of the component phases without any internal electrode layer is suitable to obtain magnetoelectric composites.

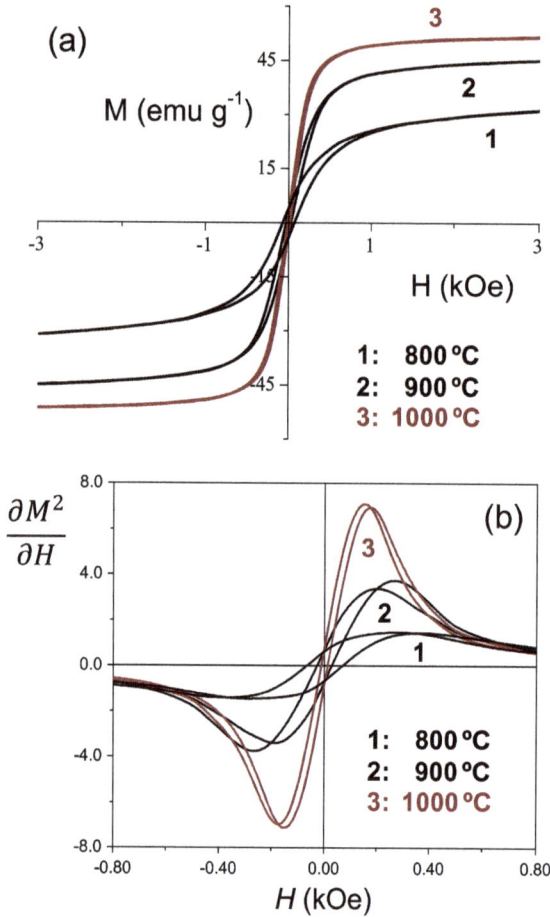

Figure 10. (a) Isothermal magnetization vs magnetic field loops and (b) derivative of square magnetization curves as a function of magnetic field, for trilayers prepared by SPS at different temperatures with the ferrite obtained by wet-chemistry.

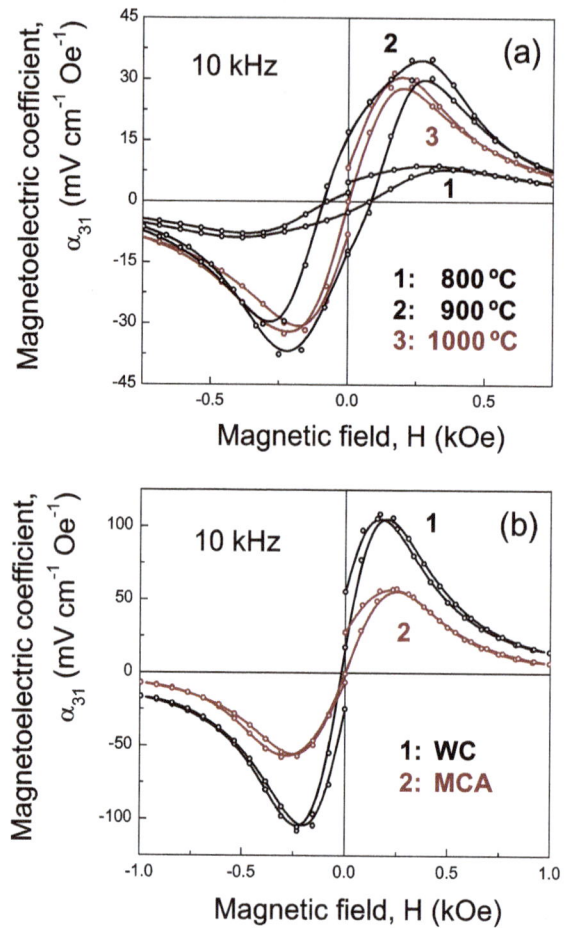

Figure 11. Magnetoelectric voltage coefficients (α_{31}) as a function of dc magnetic field in the L–T mode for (a) trilayers prepared at different temperatures with the ferrite obtained by wet-chemistry, and (b) multilayers with ferrites obtained by wet-chemistry (WC) and mechanochemical activation (MCA).

4. Conclusions

This work demonstrates the suitability of the high-sensitivity MPB piezoelectric perovskite $BiScO_3$–$PbTiO_3$ in combination with the ferrimagnetic spinel $NiFe_2O_4$ for the preparation of laminate magnetoelectric composites with tailored interfaces. The use of the SPS technique and highly reactive nanopowders obtained by mechanochemical activation and wet-chemistry routes were the key to minimize chemical reactions at and interdiffusion across the interface. This low-temperature processing approach makes possible to obtain fully dense ceramic composites with high-quality direct bonding between the component phases. However, a significant grain size reduction also results, and thus, grain size effects on properties across the submicron range and approaching the nanoscale become an issue for functionality. Therefore, a compromise should be found between the benefit of a fine microstructure to decrease thermal expansion mismatch between the different components and the advantage of micrometer grain sizes to achieve efficient poling and good properties, which both determine the magnetoelectric voltage response. The most noticeable effect of decreasing the grain size of high sensitivity piezoelectrics into the submicron range is the increase of the effective coercivity, a drawback to achieve an efficient poling of the composites. Once this compromise has been reached, a magnetoelectric coefficient α_{31} of $108\ mV\ cm^{-1}\ Oe^{-1}$ has been obtained in multilayer composites, comparable to best values reported. The piezoelectric perovskite $BiScO_3$–$PbTiO_3$ is demonstrated to be especially suitable for cofired magnetoelectric composites owing to its better down-scaling behavior than $Pb(Zr,Ti)O_3$.

Acknowledgments

This work has been funded by the Spanish MINECO through projects   MAT2011-23709   and   AIB2010PT-00332.

Collaboration between ICMM and CEMES is framed within the COST Action MP0904. Serviciencia S L (Spain) participation in the design and built-up of a novel magnetoelectric measurement system is acknowledged. HA thanks the Ramón y Cajal Programme for financial support. Technical support from Ms I Martínez and Ms M M Antón, both at ICMM-CSIC, is also acknowledged. The authors are grateful to Professor R Moreno and Ms T Molina (ICV-CSIC) for support in the processing of the materials by tape-casting.

References

[1] Jaffe B, Cook W R and Jaffe H 1971 *Piezoelectric Ceramics* (London: Academic)

[2] Uchino K 1998 *Acta Mater.* **46** 3745

[3] Haertling G H 1999 *J. Am. Ceram. Soc.* **82** 797

[4] Setter N 2005 *Piezoelectric Materials and Devices* (Lausanne: Ceramics Laboratory, EPFL Swiss Federal Institute of Technology)

[5] Anton S R and Sodano H A 2007 *Smart Mater. Struct.* **16** R1

[6] Jia Y, Luo H, Zhao X and Wang F 2008 *Adv. Mater.* **20** 4776

[7] Noheda B, Gonzalo J A, Cross L E, Guo R, Park S E, Cox D E and Shirane G 2000 *Phys. Rev. B* **61** 8687

[8] Damjanovic D 2009 *IEEE Trans. Ultrason. Ferroelectr. Freq. Control* **56** 1574

[9] Damjanovic D 1998 *Rep. Prog. Phys.* **61** 1267

[10] Leontsev S O and Eitel R E 2010 *Sci. Technol. Adv. Mater.* **11** 044302

[11] Grinberg I, Suchomel M R, Davies P K and Rappe A M 2005 *J. Appl. Phys.* **98** 094111

[12] Inaguma Y, Miyaguchi A, Yoshida M, Katsumata T, Shimojo Y, Wang R and Sekiya T 2004 *J. Appl. Phys.* **95** 231

[13] Algueró M, Ricote J, Hungría T and Castro A 2007 *Chem. Mater.* **19** 4982

[14] Castro A, Correas C, Peña O, Landa-Cánovas A R, Algueró M, Amorín H, Dollé M, Vila E and Hungría T 2012 *J. Mater. Chem.* **22** 9928

[15] Algueró M, Ramos P, Jiménez R, Amorín H, Vila E and Castro A 2012 *Acta Mater.* **60** 1174

[16] Eitel R E, Randall C A, Shrout T R and Park S E 2002 *Jpn. J. Appl. Phys.* **41** 2099

[17] Zou T, Wang X, Wang H, Zhong C, Liu L and Chen I W 2008 *Appl. Phys. Lett.* **93** 192913

[18] Algueró M, Amorín H, Hungría T, Galy J and Castro A 2009 *Appl. Phys. Lett.* **94** 012902

[19] Amorín H, Jiménez R, Ricote J, Hungría T, Castro A and Algueró M 2010 *J. Phys. D: Appl. Phys.* **43** 285401

[20] Chaigneau J, Kiat J M, Malibert C and Bogicevic C 2007 *Phys. Rev. B* **76** 094111

[21] Fiebig M 2005 *J. Phys. D: Appl. Phys.* **38** R123

[22] Nan C W, Bichurin M I, Dong S, Viehland D and Srinivasan G 2008 *J. Appl. Phys.* **103** 031101

[23] Srinivasan G 2010 *Annu. Rev. Mater. Res.* **40** 153

[24] Ma J, Hu J, Li Z and Nan C W 2011 *Adv. Mater.* **23** 1062

[25] Scott J F 2012 *J. Mater. Chem.* **22** 4567

[26] Wang Y, Li J and Viehland D 2014 *Mater. Today* **17** 269

[27] Zhai J, Xing Z, Dong S, Li J and Viehland D 2008 *J. Am. Ceram. Soc.* **91** 351

[28] Zhai J, Cai N, Shi S, Lin Y and Nan C W 2004 *J. Appl. Phys.* **95** 5685

[29] Park C S and Priya S 2011 *J. Am. Ceram. Soc.* **94** 1087

[30] Zhou Y, Yan Y and Priya S 2014 *Appl. Phys. Lett.* **104** 232906

[31] Amorín H, Hungría T, Landa-Cánovas A R, Torres M, Dollé M, Algueró M and Castro A 2011 *J. Nanopart. Res.* **13** 4189

[32] Martin D V J L, Lopez-Delgado A, Vila E and Lopez F A 1999 *J. Alloys Compound.* **287** 276

[33] Amorín H, Santacruz I, Holc J, Thi M P, Kosec M, Moreno R and Algueró M 2009 *J. Am. Ceram. Soc.* **92** 996

[34] Hungría T, Galy J and Castro A 2009 *Adv. Eng. Mater.* **11** 615

[35] Prior D J *et al* 1999 *Am. Mineral.* **84** 1741

[36] Amorín H, Jiménez R, Deluca M, Ricote J, Hungría T, Castro A and Algueró M 2014 *J. Am. Ceram. Soc.* **97** 2802

[37] Amorín H, Jiménez R, Ricote J, Castro A and Algueró M 2015 *Nanoscale Ferroelectrics and Multiferroics: Key Processing and Characterization Issues, and Nanoscale Effects* ed M Algueró, J M Gregg and L Mitoseriu (New York: John Wiley & Sons Ltd.) (forthcoming)

[38] Bichurin M and Petrov V 2014 *Modeling of Magnetoelectric Effects in Composites (Springer Series in Materials Science* vol 201) (Dordrecht: Springer)

[39] Sepelak V, Bergmann I, Feldhoff A, Heitjans P, Krumeich F, Menzel D, Litterst F J, Campbell S J and Becker K D 2007 *J. Phys. Chem. C* **111** 5026

[40] Ferisov Y K, Petrov V M and Srinivasan G 2007 *J. Mater. Res.* **22** 2074

[41] Ito S, Aso K, Makino Y and Uedaira S 1980 *Appl. Phys. Lett.* **37** 665

[42] Islam R A, Ni Y, Khachaturyan A G and Priya S 2008 *J. Appl. Phys.* **104** 044103

18

Improved compaction of ZnO nano-powder triggered by the presence of acetate and its effect on sintering

Benjamin Dargatz[1], Jesus Gonzalez-Julian[1] and Olivier Guillon[1]

Friedrich Schiller University of Jena, Otto Schott Institute of Materials Research, Löbdergraben 32, D-07743 Jena, Germany

E-mail: B.Dargatz@fz-juelich.de

Abstract

The retention of nanocrystallinity in dense ceramic materials is still a challenge, even with the application of external pressure during sintering. The compaction behavior of high purity and acetate enriched zinc oxide (ZnO) nano-powders was investigated. It was found that acetate in combination with water plays a key role during the compaction into green bodies at moderate temperatures. Application of constant pressure resulted in a homogeneous green body with superior packing density (86% of theoretical value) at moderate temperature (85 °C) in the presence of water. In contrast, no improvement in density could be achieved if pure ZnO powder was used. This compaction behavior offers superior packing of the particles, resulting in a high relative density of the consolidated compact with negligible coarsening. Dissolution accompanying creep diffusion based matter transport is suggested to strongly support reorientation of ZnO particles towards densities beyond the theoretical limit for packing of ideal monosized spheres. Finally, the sintering trajectory reveals that grain growth is retarded compared to conventional processing up to 90% of theoretical density. Moreover, nearly no radial shrinkage was observed after sinter-forging for bodies performed with this advanced processing method.

Keywords: sintering, green body processing, nano-powder, zinc oxide, water, zinc acetate, zero radial shrinkage

1. Introduction

Zinc oxide (ZnO) is an n-type semiconductor that is extensively used in electronic and optoelectronic systems due to its combination of piezoelectric properties, gas reactivity and wide band gap of 3.3 eV [1]. ZnO is one of the most promising ceramics for challenging applications in the fields of ultrafast nanolasers, cost efficient solar cells, nanogenerators for self-powered devices and thin film gas sensors [2–6]. Nevertheless, prior to the expected outstanding final properties, understanding of the mechanisms involved in the coarsening of nanocrystalline ZnO is required as excessive grain growth leads to deterioration of properties and functionality. As most physical properties are affected by the residual porosity, it is essential to obtain nearly full densification. Unfortunately, the sintering of nanocrystalline, dense bulk ZnO with grain sizes <100 nm is still a challenging task. Roy et al [7] sintered nanocrystalline ZnO powder pellets, but the final grain size was 1.5 μm. Mazaheri et al [8] fully densified ZnO and retarded grain growth by means of two-step sintering, but ended up with a mean grain size of 1.4 μm. Hynes et al [9] attained grains with size approximately 500 nm but achieved only 95% of theoretical density (% TD) by hot pressing. In general, the retention of nanocrystallinity depends on the competition between densification and grain

[1] Present address: Institute of Energy and Climate Research: Materials Synthesis and Processing (IEK-1), Forschungszentrum Jülich, Wilhelm-Johnen-Straße, D-52425 Jülich, Germany

growth, the driving forces for which both depend on reciprocal grain size [10, 11]. The most promising approaches to suppress grain growth are pinning of grain boundaries (by the addition of a second phase), application of external pressure [8], very fast heating [12–14] or two-step sintering [8, 15]. However, particle dispersion and green body processing are key steps to achieve this goal. Thus, the whole processing chain needs to be considered beginning with a homogeneous and high green density sample prior to sintering at elevated temperature [10, 11]. The shaping of green bodies from nanocrystalline powders by dry pressing typically results in compacts containing agglomerates and 50–61% TD [8, 16, 17], although values of 73% TD [18] for cold isostatic pressing are reported. Alternative shaping technologies overcome the problems of dry pressing utilizing suspensions and benefit from lowered friction between the particles. Wet processing, e.g. gel casting, slip casting or pressure filtration, results in a homogeneous packing with typical green densities between 50 and 65% TD [10, 15, 19]. However, well-dispersed stable suspensions are required, which can be difficult to achieve with nanoparticles. This study presents a new approach for the processing of ZnO nanoparticle compacts resulting into homogeneous green bodies with an extremely high density.

2. Experimental procedures

2.1. Materials and methods

The standard ZnO powder utilized for the experiments is referred to as ZinCox10 (IBU-tec advanced materials AG, Weimar, Germany) with a purity of >99.00 wt% and mean particle size of 10 nm from the data sheet of the producer. An additional ZnO powder (NG20, Nanogate AG, Quierschied-Göttelborn, Germany) with a purity of >99.99 wt% and a primary particle size between 20 and 50 nm was used as the reference powder. Furthermore, a third powder was used for the compaction, which was produced by mixing the pure ZnO powder (NG20) with a total amount of 4.2 wt% zinc acetate dehydrate $Zn(H_3C–COO)_2$ (Sigma Aldrich, St. Louis, USA). A total amount of 0.4 g powder was stored in a glass beaker inside an environmental chamber (KBF 240, Binder GmbH, Tuttlingen, Germany). The storage conditions were defined by temperatures between 20 and 85 °C and absolute moistures between 1.0 and 140 g m^{-3}. In particular, 85 °C with 140 g m^{-3} moisture (40% relative humidity), 85 °C with 1 g m^{-3} moisture (0.3% relative humidity) and 20 °C with 14 g m^{-3} moisture (85% relative humidity) will be further referred to 'humid warm', 'dry warm' and 'humid cold' conditions, respectively. The NG20, ZinCox10, and NG20 + acetate powders were each poured into a hard metal pressing die (5 mm diameter and 8.5 mm initial height), after storage of one hour under a particular environmental condition. Then, the ZnO powders were pressed either for 20 h within the environmental chamber or by conventional cold dry pressing for 1 min. In the former procedure the pressing tool was installed inside the environmental chamber under the same

respective storage condition, and a uniaxial pressure of 50 MPa was applied for 20 h. The axial displacement of the metal punches was measured every second during the pressing step by a laser triangulation sensor (LK-G10, KEYENCE Corp., Osaka, Japan) with ±0.1 mm spatial resolution. The axial displacement was correlated to the change in density (as the powder is compressed in a stiff mold and as compaction takes place only vertically). In comparison, green bodies were produced with dried ZinCox10 powder by conventional dry cold pressing for 1 min at 50 MPa. Therefore, ZinCox10 powder had been dried for 24 h prior to the pressing step by storing in a drying cabinet at 105 °C with ≪0.5 g m^{-3} moisture (≪0.1% relative humidity). The pressed ZnO bodies were ejected after the pressing step and installed in the sinter-forging device for further compaction (see next section).

2.2. Sintering study

Green bodies of humid warm and conventionally processed ZinCox10 powder were sintered in air. First, samples were heated up to 400 °C for 20 min isothermal time with a rate of 5 K min^{-1} in order to burn out residual organic content. The samples were further heated to 700 °C with a rate of 20 K min^{-1} and an isothermal time of one hour. Afterwards, the sintered sample was cooled down with 10 K min^{-1} to room temperature. The furnace for sintering was a modified sinter-forging setup, which is composed of a vertical split furnace fixed on a mechanical testing machine (Model 5565, Instron Corp., Norwood, USA). The samples were sintered freely or under an external pressure of 3, 15 and 40 MPa. The actual height and diameter of the samples were measured in situ by a laser system (162-100; BETA Laser Mike Inc., Dayton, USA) every 2.5 s (by averaging 128 values) with a resolution of 0.5 μm. Thus, the true strain during sintering could be calculated. The densification curves were calibrated by the density of sintered specimens measured at room temperatures by the Archimedes method.

2.3. Characterization

Microstructure of compacted and sintered ZnO bodies was investigated by transmission electron microscopy (TEM) and by high resolution scanning electron microscopy (HRSEM). TEM was performed with a JEM-3010 (JEOL, Akishima, Japan) operated at 300 kV. HRSEM was performed with a field emission microscope Auriga60 (Carl Zeiss AG, Oberkochen, Germany). The grain size measurement was carried out on polished samples using the line interception method with the software 'Lince' (v. 2.31, Ceramics Group, TU Darmstadt) using a factor of 1.56, evaluating at least 300 grains [20]. Phases for the powder and compacted specimens were characterized by means of x-ray diffraction (XRD). XRD measurement was carried out using a D8-Discover (Bruker AXS, Billerica, USA) with Cu-K$_\alpha$ radiation at $\lambda = 1.540\,56$ Å, operated at 40 kV and 40 mA, a step size of 0.02° and a counting time of 1.6 s. Crystallite sizes were determined using the Scherrer equation after fitting the $(10\bar{1}0)$, $(11\bar{2}0)$ and (0002) Bragg peaks with the pseudo-

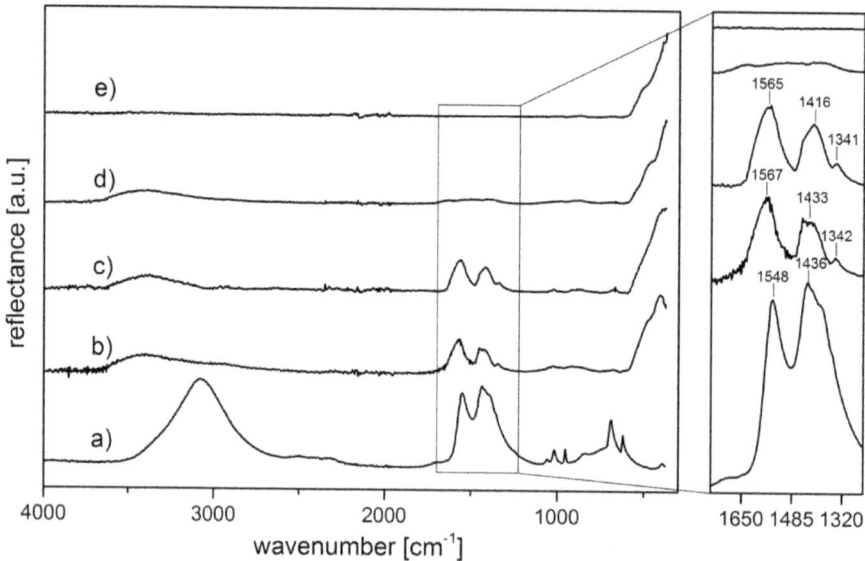

Figure 1. FTIR spectra of (a) zinc acetate dihydrate as-received non-compacted powder, (b) as-received non-compacted ZinCox10, (c) ZinCox10 powder compacted 20 h under humid warm condition (85 °C, 140 g m^{-3} moisture) (d) as-received non-compacted NG20 powder and (e) ZinCox10 powder fired at 800 °C. The spectra are offset for clarity.

Voigt function. A shape factor $K = 0.94$ was applied for the Scherrer equation [21] and full width at half maximum was corrected for instrumental broadening. Fourier-transform infrared spectroscopy (FTIR) measurements were carried out with an 'Alpha' spectrometer (Bruker Optics, Billerica, USA) using the absolute total reflection method. The pressed samples and the raw material zinc acetate were ground up before FTIR measurements. Each measurement was performed twice with approximately 20 mg powder and a resolution of 4 cm^{-1}. Thermogravimetric analysis was performed with a 'DTA/TG-92' (SETARAM Instrumentation, Lyon, France) on 5 mg dried ZnO powder which was heated up to 750 °C in a platinum container at a rate of 10 K min^{-1}.

3. Results and discussion

3.1. Analysis of untreated ZnO powder

The particle size of both commercial powders was investigated prior to compaction study. The high purity powder (NG20) showed a mean particle size of 26 ± 9 and 33 nm for TEM and XRD observation, respectively. The standard powder (ZinCox10) exhibited a mean particle size of 16 ± 9 and 17 nm for TEM and XRD investigation, respectively. Thus, the mean crystallite size given by XRD corresponds to the particle size observed by TEM and is about twice as large for NG20 powder compared to ZinCox10 powder. A previous study [22] proved by TEM and XRD that the ZnO particles are isometric. Further, TEM confirms the presence of isometric particles with polyhedral shape. XRD analysis on both powders confirmed that all Bragg reflexes are attributed to the hexagonal wurtzite phase of ZnO. ZinCox10 powder was

studied for its compaction and sintering behavior. A previous investigation [22] confirmed the presence of organic content (zinc acetate) inside the ZinCox10 powder by comparative infrared spectroscopy investigation with acetate dihydrate powder, whereas the coarser NG20 powder showed negligible impurity. The presence of zinc acetate is proven by FTIR analysis, which is of major interest for the explanation of the compaction behavior of the ZnO powder. FTIR spectra in reflectance mode are illustrated in figure 1 for zinc acetate dihydrate, NG20 and ZinCox10 powders. The broad band reaching from 2700 to 3600 cm^{-1} is present for all ZnO samples except for the sintered ZnO and corresponds to adsorbed water and OH groups [23]. The band at 470 cm^{-1} is found for all the ZnO specimens and can be attributed to the vibrational stretching mode of Zn–O bounding [24] (figures 1(b)–d)). The two bands around 1420 and 1560 cm^{-1} correspond to the symmetrical and asymmetrical stretching vibrations of carboxylate group, respectively [23, 25]. These bands are found for zinc acetate dihydrate powder as well as for the as received and the humid warm processed ZinCox10 powder. Furthermore, no acetate is present in the NG20 powder or after sintering of the ZinCox10 powder. Thus, the zinc acetate is only present in the ZinCox10 prior to sintering. It seems conclusive that residual zinc acetate is still present in the ZinCox10 powder, as zinc acetate is typically used as precursor for the synthesis of ZnO [26–30]. Thermogravimetric analysis of the dried ZinCox10 powder reveals a mass loss of 3.3 ± 0.2 wt% resulting from water and organic removal. If the entire mass is correlated to the molar amount of carbon present in acetate salt, approximately 4.16 wt% of zinc acetate should be contained in the ZinCox10 powder. Spitz et al [31] showed that acetate is covering the surface of ZnO particles. The theoretical thickness of a hypothetical

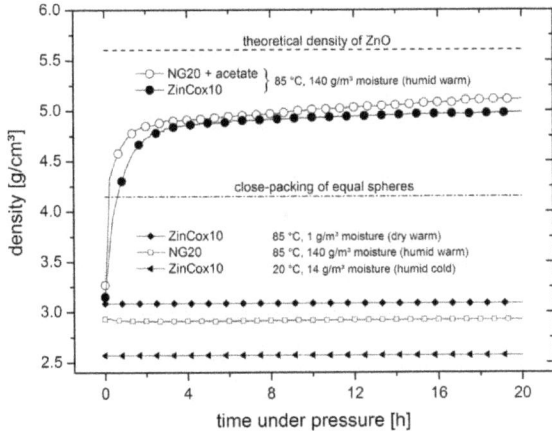

Figure 2. Density as a function of time for bodies under constant uniaxial pressure of 50 MPa under defined temperature and humidity. The dashed and dashed dotted lines give the theoretical density of pure ZnO and the density for close-packing of equal spheres, respectively. The compaction curves are represented by one measured data point each second.

comparable initial nanoparticle size. Ewsuk *et al* [33] achieved similar green density with 62% TD by granulation, uniaxial and isostatic pressing at 140 MPa for both, macroscopic and nanoscale powder. Alternative methods, e.g. pressure filtration, slip casting of ceramic slurries, result into a density of 50–70% TD [10, 15, 19]. The slightly elevated temperature of 85 °C may raise the reorientation process and the presence of water may lower the friction between the ZnO particles. However, these findings are in contradiction with the model for the highest packing of ideal monosized spheres, whereas the packing density is theoretically restricted to 74% TD [34]. On top of that, particles were found to show polyhedral instead of spherical shape, which should result in an even lower maximum theoretical value. Thus, not solely reorientation, but another process discussed in the next section seems responsible for the huge gain in green density. In contrast, no change in density was measured during the compaction step neither under dry warm condition (85 °C, 1 g m^{-3} moisture) nor under humid cold condition (20 °C, 14 g m^{-3} moisture). Thus, a combination of temperature and humidity is required to increase the green body density, if acetate is present in the compacted ZnO powder.

continuous zinc acetate layer covering ZnO nanocrystals has been calculated by assuming a TD of 1.74 g cm^{-3} as well as ideally spherical ZnO nanoparticles with a mean diameter of 17.2 nm and a TD of 5.606 g cm^{-3}. Such a homogeneous coating of zinc acetate is estimated to be 0.37 ± 0.02 nm thick, which is difficult to highlight by TEM for example.

3.2. Processing of ZnO green body

3.2.1. Compaction behavior.
The two nanocrystalline ZnO powders were investigated with respect to their compaction behavior. The absolute density is given as a function of time during compaction in figure 2. Interestingly, the initial density was around 3 g cm^{-3} (55% TD) at the beginning of compaction for both ZnO powders at 85 °C. In contrast, the compact under humid cold condition shows a constant density of 2.57 g cm^{-3} (45.8% TD). The pure NG20 powder was pressed to a density of 2.93 g cm^{-3} (52.3% TD) under humid warm condition and shows no further increase in density with time. Densities around 45–57% TD for green compacts are typical even for dry, cold pressing at 50 MPa [10, 16, 32]. In contrast, the density of the ZinCox10 powder strongly increases within the first hour up to a value of approximately 4 g cm^{-3} if compacted under humid warm condition. The densification rate further decreases successively until a constant density of approximately 4.99 g cm^{-3} is achieved after 20 h of compaction. The measured absolute density of ZinCox10 powder compacts after compaction needs to be corrected in order to take into account the presence of acetate. The corrected density is 85.9% TD, which is extremely high. The addition of zinc acetate to pure NG20 powder results into the same compaction behavior under humid warm condition. To our knowledge, this study presents the highest reported value by uniaxial pressing of ceramic powders. Mazaheri *et al* [8] achieved 61% TD by cold uniaxial pressing at 200 MPa for pure ZnO powder with

3.2.2. Microstructural analysis.
Figure 3 shows SEM micrographs of fracture surfaces of samples, which were compacted in an environmental chamber under humid warm (85 °C, 140 g m^{-3} moisture) and dry warm (85 °C, 1 g m^{-3} moisture) conditions. The sample kept under the humid warm condition shows a regular distribution of particles and only small pores are found. This finding indicates a homogeneous and high packing of the ZnO crystallites (figure 3(a)). The median particle size was estimated from the fracture surface by SEM and it increased to 25 ± 7 nm (growth factor of 1.51 compared to the initial state) under the humid warm condition. In comparison, the SEM micrograph of the dry compacted sample shows a more heterogeneous particle packing and reveals the presence of pores in the scale of 10–100 nm (figure 3(b)). However, under the dry warm condition the mean particle size stays constant, while the absolute density remains constantly low. A former investigation [22] showed that massive coarsening of ZinCox10 particles takes place under the same environmental conditions, if the powder is stored for long time. Thus, preferential crystal growth along *c*-axis higher than 400% was observed within 24 h of storage. In contrast, the present study shows that particle coarsening is strongly retarded and no preferred crystal growth occurs if the powder is simultaneously compacted under 50 MPa, although the contacts between particles are fostered by pressure. This surprising compaction behavior only occurs for the ZinCox10 powder, but not for the NG20 powder. Therefore, the presence of zinc acetate in the ZnO powder is the key trigger for better compaction. Figure 4 shows TEM images from a sample compacted for 20 h under the humid warm condition (85 °C, 140 g m^{-3} moisture). The TEM micrograph in figure 4(a) gives an impression of grown ZnO particles. No indication was found for neck formation or grain boundaries between the ZnO nanocrystals by TEM and

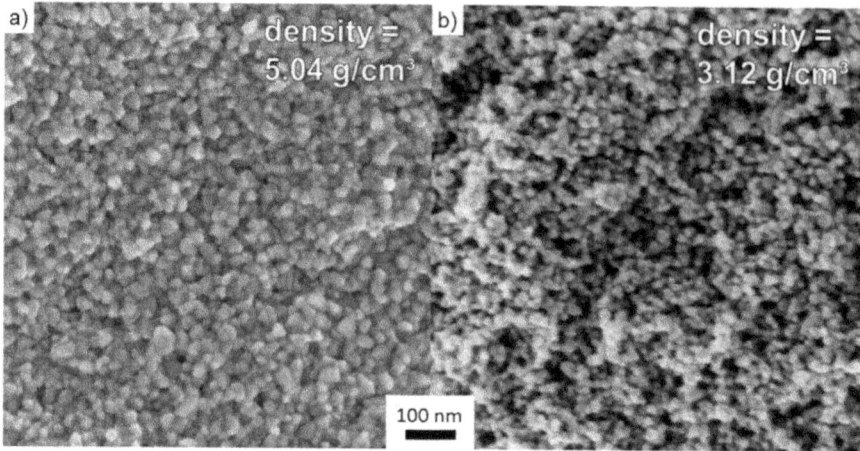

Figure 3. Fracture surfaces of ZinCox10 powder compacted 20 h under (a) humid warm condition (85 °C, 140 g m^{-3} moisture) and (b) dry warm condition (85 °C, 1 g m^{-3} moisture).

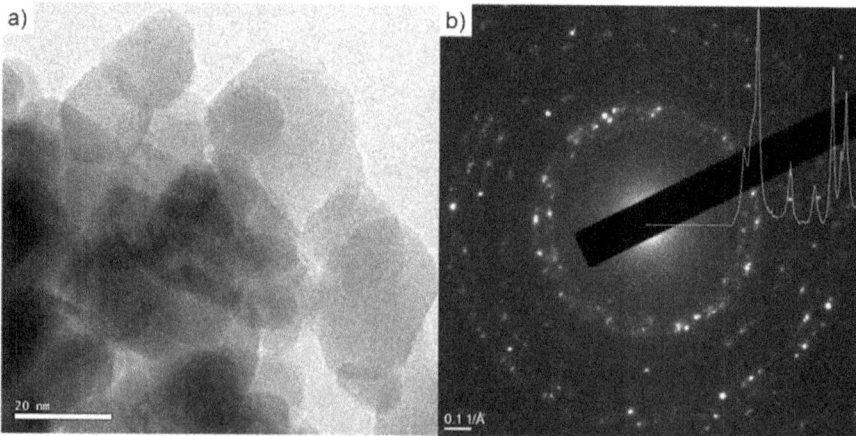

Figure 4. TEM investigation of ZinCox10 powder compacted 20 h under humid warm condition (85 °C, 140 g m^{-3} moisture) with (a) microstructure and (b) diffraction pattern.

HRTEM investigation. The median primary crystallite size of 26 ± 8 nm confirms the results from SEM measurements (this represents an increase in crystallite size of 1.61 times). Figure 4(b) shows the TEM diffraction pattern with the intensity as a function of diffraction angle. The position of reflexes confirms the presence of the wurtzite phase and gives no hint for other phases. XRD analysis also gives no evidence for any additional phase except for the zincite phase of ZnO (figure 5). Neither TEM diffraction pattern nor XRD revealed the presence of phases other than wurtzite ZnO modification (figure 4(b)). The absence of possible additional phases by TEM analysis might be explained by a local temperature increase of the sample under the electron beam and zinc acetate might already be removed, as temperatures about 100 °C initiate strong degradation [35].

On the other hand, a small nanosized layer of acetate would hardly be detected by HRTEM [36]. In general, there are two possibilities for how mass transport is taking part

towards consequent reduction of pore volume under compaction without the necessary activation temperature for sintering. The lowest sintering temperature found for sintering of ZnO is reported to be $0.2 \times T_m$ at an external pressure of 50 MPa [37], i.e. 400 °C, which is well above the temperature found in the environmental chamber. However, the threshold temperature for diffusion of zinc interstitials is below an absolute temperature of 130 K [38]. Moreover, this diffusion process was suggested for coarsening of ZnO particles at room temperature [39]. On top of that, acetate ion interaction with the ZnO particle surface should be taken into account. Thus, dissolved acetate ions can result into the formation of surface defects or enhance dissolution of the particle surface, as pH value should be decreased. As a consequence, the dissolution is resulting in an enrichment of dissolved zinc species, surrounding the ZnO particles. Thus, the aqueous phase can be assumed to be enriched with zinc species and hydroxide ions. Such an aqueous layer is enabled to flow

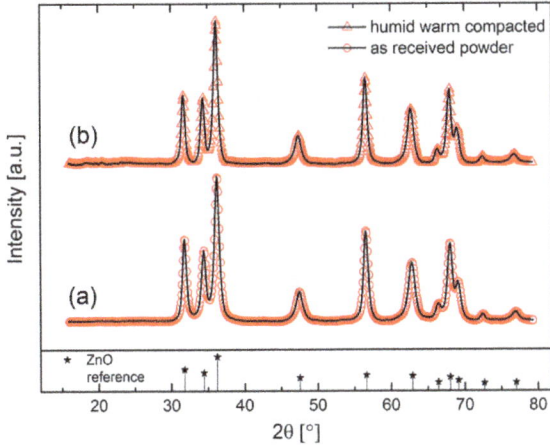

Figure 5. XRD pattern of ZinCox10 for (a) as-received non-compacted powder and (b) compacted powder after 20 h under humid warm condition (85 °C, 140 g m^{-3} moisture). The Bragg reflections for ZnO wurtzite structure [51] are represented below the measured data by stars. The measured data are offset for clarity and only one out of four data points is given.

between ZnO particle interspaces and precipitate again. Recently it was reported [25, 35] that formation of ZnO occurred by thermal degradation of zinc acetate at temperatures of 85 °C. Figure 6 illustrates the compaction process for polyhedral shaped ZnO particles beginning with dry pressed state (figure 6(a)). The application of higher external pressure breaks agglomerates and leads to reorientation process (figure 6(b)). Aqueous layer with dissolved zinc species is pulled between particle contacts by capillary forces at the same time. Effective contact area can be decreased by the coarsening of particles (figure 6(c)).

The resulting effect is shown in figure 6(d) via combination of reorientation and coarsening. In addition, activated solid state diffusion based matter transport could be activated by pressure resulting into further densification. Coble [40] described the grain boundary diffusion controlled creep process for polycrystalline materials. Under stress, grain boundaries act as sources and sinks for vacancies under tensile and compressive stress, respectively. Here, dissolved zinc ions can additionally interact with sinks for vacancies under compressive stress and further participate in the overall compaction process. However, neither grain boundaries nor particle necks were found to indicate a diffusive sintering process. The absence of grain boundaries seems to be conclusive, as the sintering temperatures of ZnO for free sintering are reported between 700 and 1200 °C [41–43].

3.3. Solid state sintering

3.3.1. Pressureless densification.
The sintering of humid warm compacted ZnO samples seems promising in comparison to conventionally processed samples, as packing density is already very high prior to sintering step. Thus, a modified sintering trajectory is expected which should be shifted towards smaller grain sizes at a given density

compared to dry processed green bodies. The densification behavior for freely sintered ZnO is illustrated in figure 7. The relative density is plotted as a function of temperature (during the heating stage) and isothermal time at a maximum temperature of 700 °C. The specimens start with 50 and 85% TD during initial stage of sintering for dry and humid warm conditions, respectively. Besides, the onset of densification with a temperature of 600 °C is 120 °C above the specimen processed under dry condition. Thus, sinterability of humid warm processed ZnO is decreased, as surface energy for humid warm specimen is reduced due to coarsening of particle size by 50% during green body processing. In comparison, particle size remains constant under conventional processing giving a slightly lower initial particle size prior to sintering. Nevertheless, maximum density of 95% was found to be equal for both samples after one hour at 700 °C, although this value was achieved 10 min later for conventionally processed specimen (dry condition) during isothermal dwell time. Hynes *et al* [9] observed similar final density for pressureless sintering of nanocrystalline ZnO powder at 700 °C. Additionally, Hynes *et al* reported about abnormal grain growth with grain sizes ranging from 67 to >530 nm, which is in contrast to the present study. The microstructural development of sintered ZinCox10 specimens is illustrated in figure 8. In general, no abnormal grain growth was observed. Figures 8(a) and (b) show microstructures that were obtained by an interrupted heating experiment for specimen processed under humid warm and dry conditions, respectively. The mean grain size with $d_{50} = 101$ nm is much smaller for humid warm processed specimen compared to conventionally processed ZnO ($d_{50} = 206$ nm), although density is already 6% higher. On the other hand grain size ($d_{50} \sim 300$ nm) and density are similar for both specimens after sintering one hour at 700 °C (figures 8(c) and (d)). This grain size is surprisingly low taking into account that no external pressure was applied during sintering. In comparison, Mazaheri *et al* [8] hot pressed ZnO with equal initial powder particle size under 50 MPa external pressure and achieved a significant higher grain size (>600 nm) for the same density. The sintering trajectory (grain size versus relative density) is given in figure 9 with values that were obtained by interrupted heating experiments by sintering schedule from figure 7. The grain sizes of sintered specimen with conventionally processed ZnO result into more than two times higher values compared to humid warm processing for relative densities below 90%. The grain size increases rapidly for >90% TD, until similar grain sizes are achieved for both processing routes around 95% TD.

3.3.2. Pressure-assisted densification.
In the following, a uniaxial pressure of 15 or 40 MPa was applied during sintering. Table 1 lists the obtained respective grain sizes and relative densities for humid warm and dry processed specimens. The final density achieves 95% TD, when an external pressure of 15 MPa is applied, but grain size increases as well to 600 nm. Dry processed samples show a

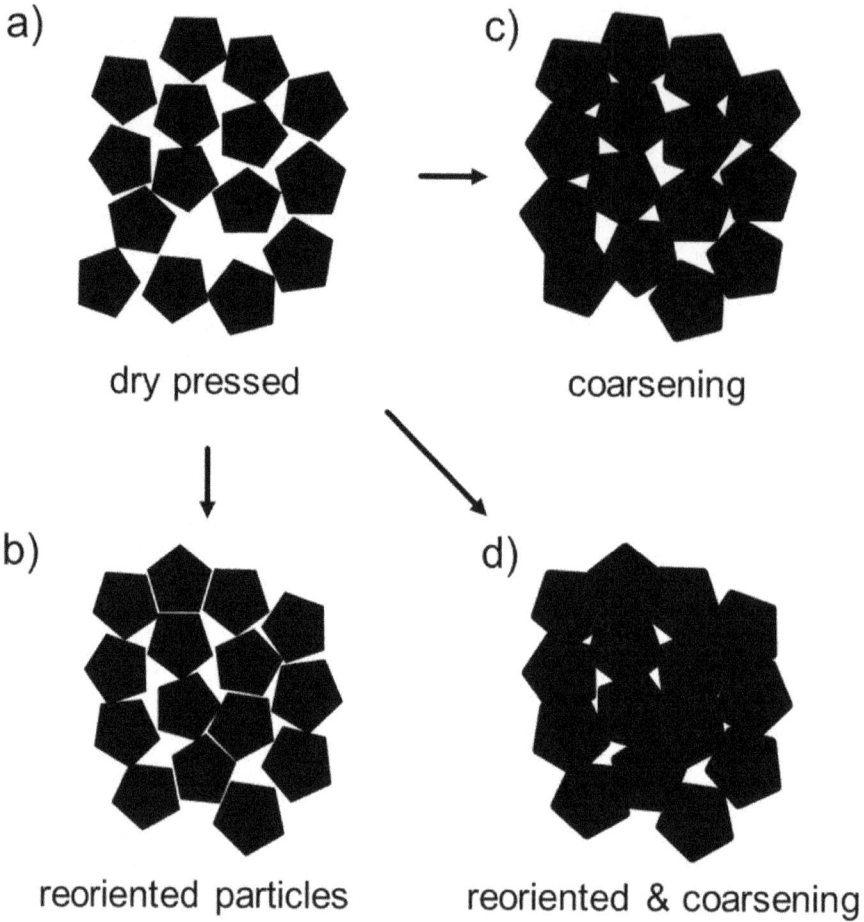

Figure 6. Schematic of the compaction process for polyhedral particles. Dry pressed particles for (a) as-received non-compacted powder, (b) after coarsening, (c) after reorientation and (d) combination of reorientation and coarsening.

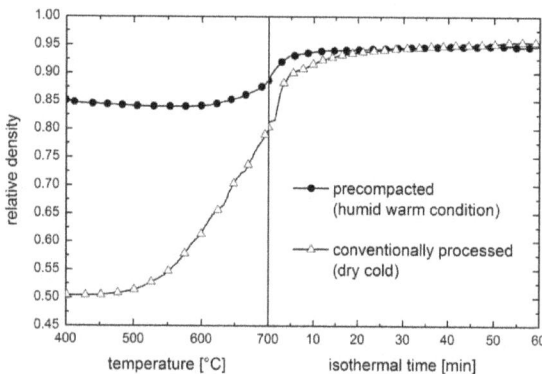

Figure 7. Densification curve for pressureless sintered ZnO bodies processed under humid warm and conventional conditions. Relative density is given as a function of temperature and isothermal time at 700 °C.

comparable behavior. Here, density is stagnating around 95.5% TD and grain size is increasing up to 1.4 μm. The evolution of final relative density is within 0.2%, which lies within the error range. Thus, density is not affected by the

application of an external pressure, whereas the grain size is increasing with increasing external pressure. Interestingly, the grain size is smaller for the dry condition than for humid warm processed bodies at a uniaxial pressure of 15 MPa. Typically, the application of an external pressure retards grain growth in comparison to free sintering of ceramics, as the mechanism of densification is enhanced while the one of coarsening is not affected. Nevertheless, similar results are found in the literature on sintering of alumina [44, 45] and BaLa$_4$Ti$_4$O$_{15}$ [46, 47]. Besson and Abouaf [44, 45] sintered pure alumina by hot isostatic pressing and found a final grain size proportional to the applied stress (up to 200 MPa). They suggested that the applied stress induced point defects and dislocations during densification, which are eliminated by grain growth. Amaral *et al* [46, 47] observed exaggerated grain growth during densification of sintered constrained films of BaLa$_4$Ti$_4$O$_{15}$, which resulted in a much larger grain size for a given density than for freely sintered bulk. Interestingly, the grain size of dry processed ZnO samples remains lower compared to the humid warm condition at an external pressure of 15 MPa (table 1). High strain rates are suggested at the beginning of the sintering in order to

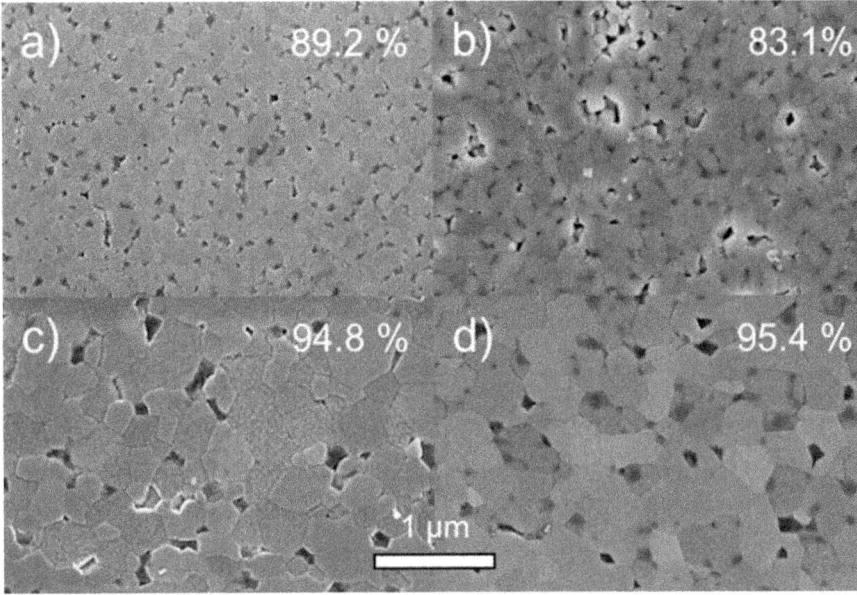

Figure 8. SEM images of polished surfaces for pressureless sintered ZnO samples for (a) and (c) humid warm compacted and (b), (d) conventionally processed green bodies. The samples were sintered at 700 °C (a), (b) without dwell and (c), (d) with dwell of 1 h. The determined relative density is labeled for each image in the right upper corner.

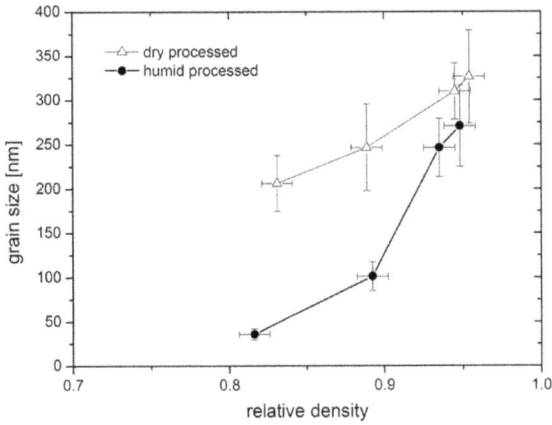

Figure 9. Sintering trajectory for ZnO bodies processed under humid warm and dry conditions and freely sintered.

Table 1. ZnO specimen sintered one hour at 700 °C for external applied pressures between 0 and 40 MPa. The table lists achieved relative density and grain size for specimens processed under humid warm and conventional conditions.

Green body processing	Applied pressure (MPa)	Relative density (%)	Grain size (nm)
	0	94.8	271
Humid warm	15	95.0	600
	0	95.4	326
Dry	15	95.4	342
	40	95.5	1419

eliminate interagglomerate pores prior to the onset of the final stage of sintering, when substantial grain growth occurs [48]. It is already known that neck growth and densification are promoted in the plane perpendicular to its direction under uniaxial loading which may also lead to anisotropy of a sintering body [49]. Figure 10 schematically illustrates the shape evolution of sintered samples from pressureless sintering towards increasing applied external pressure, which corresponds to axial (ε_z) and radial (ε_r) strain. Only 2.5% radial strain was detected for the pressureless sintered specimen under humid warm processing (figure 10(a)). Under dry processing, radial and axial strains are nearly equal (24%) for pressureless sintering, which was already reported before for free sintering (figure 10(b)) [18]. Thus, strains in both directions are equal (axial versus radial strain ratio $\varepsilon_z/\varepsilon_r \sim 1$), whereas a strain ratio $\varepsilon_z/\varepsilon_r$ of 2.5 was found for the humid warm specimen, but with absolute strain nearly one order of magnitude lower than under the dry condition. In general, the evolution of the strain ratio strongly depends on the applied external pressure. Rahaman et al [18] have already reported for sintered ZnO in 1991, that the strain ratio $\varepsilon_z/\varepsilon_r$ increases from ~ 1 to ~ 1.56 by the application of a uniaxial pressure. Moreover, they found that the strain ratio $\varepsilon_z/\varepsilon_r$ regains with a value of ~ 1.2 by a strong increase in initial green density (73% TD). Thus, an increase of initial green density will decrease the radial stain and accordingly a lower pressure is required for zero radial strain. This finding is convenient with the results of the present study, whereas the strain ratio is lower for humid warm processed ZnO (~ 5 g cm^{-3} initial density) in comparison to the dry condition (~ 3 g cm^{-3} initial density). In addition the surface of the dry sintered samples

a) humid processing

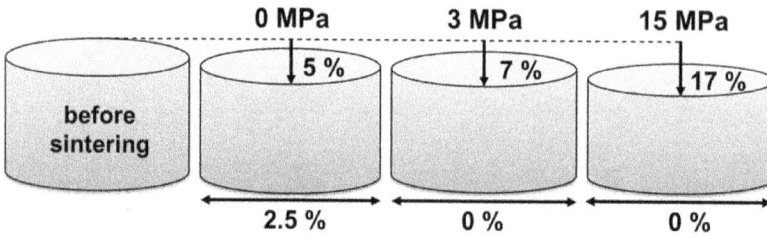

b) dry processing

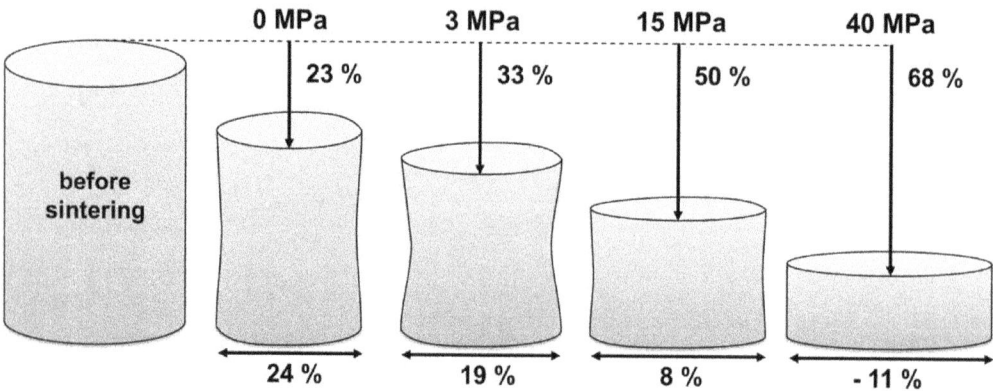

Figure 10. Shape evolution of compacted ZnO bodies before and after firing at 700 °C for 1 h for (a) humid and (b) dry processed green bodies. Axial and radial shrinkage are labeled for each applied external pressure.

shows a concave curvature. On the one hand, the degree of curvature is decreasing with increasing pressure, which is conclusive. On top of that, the radial strain is negative (−11%) under a uniaxial pressure of 40 MPa for dry processed specimens, resulting in a radial expansion, whereas similarly high pressures of 40–100 MPa would lead to a radial expansion for the humid warm specimen as well. Finally, zero radial shrinkage was already achieved by the application of a uniaxial pressure of 3 MPa. In the case of the dry processed specimen, linear interpolation of radial strain data from figure 10 gave a uniaxial pressure of 27 MPa for zero radial strain. It seems conclusive that the higher initial density of the humid warm processed specimen governs a lower sintering stress in axial and radial directions. Bordia *et al* [50] showed that a compressive uniaxial pressure between 20 MPa and 30 MPa during the intermediate stage of sintering is necessary to achieve a zero radial strain rate for sintering submicron alumina. This value, which depends on the viscous parameters of the sintering material, is in good agreement with the presented results of dry processed ZnO in this study.

4. Conclusions

The presented shaping methodology results in green compacts with densities superior to those obtainable by dry or standard wet shaping processes. In general, this new compaction behavior of ZnO offers a promising approach for retaining nanocrystallinity due to a strongly increased green density prior to the sintering step. Acetate was identified to activate this advanced shaping methodology for ZnO green bodies, if compacted under uniaxial pressure in presence of moisture and very low temperature. In contrast, pure ZnO showed no further compaction, but subsequent addition of acetate to pure ZnO powder results in high packing density in the same way. A final density of 95% TD was achieved not only for the humid processing, but also for the dry processing. In addition, nearly zero radial shrinkage was observed without the application of an external pressure. In contrast, an external pressure of 27 MPa is necessary to avoid radial shrinkage by means of conventional processing. Retention of grain growth is superior for densities lower than 92% TD for the presented shaping methodology. The burnout of the organic component hinders densification at high densities. Thus, a retarded grain growth is expected if the amount of acetate is properly decreased in order to decrease the influence of trapped gases. Further, it is likely that compaction and growth behavior of other metal oxide nanocrystals, synthesized on the basis of acetate or even other organic precursors, is similarly enhanced. However, further experiments with other metal oxides are required to validate the above mentioned role of acetate combined with water and temperature.

Acknowledgments

This work was partially supported by the 'Deutsche Forschungsgemeinschaft' (Emmy Noether Program GU993-1). The authors thank Martin Seyring for nano-beam diffraction analysis and Susanne Sandkuhl for sample polishing. We gratefully acknowledge the partial financial support of the Deutsche Forschungsgemeinschaft (DFG), grant reference INST 275/241-1 FUGG, and the Thüringer Ministerium für Bildung, Wissenschaft und Kultur (TMBWK), grant reference 62-4264 925/1/10/1/01.

References

[1] Özgür Ü, Alivov Y I, Liu C, Teke A, Reshchikov M A, Doğan S, Avrutin V, Cho S J and Morkoç H 2005 *J. Appl. Phys.* **98** 041301
[2] Viswanath R N, Ramasamy S, Ramamoorthy R, Jayavel P and Nagarajan T 1995 *Nanostruct. Mater.* **6** 993
[3] Nanto H, Minami T and Takata S 1986 *J. Appl. Phys.* **60** 482
[4] Sidiropoulos T P H, Roder R, Geburt S, Hess O, Maier S A, Ronning C and Oulton R F 2014 *Nat. Phys.* **10** 870
[5] Keis K, Bauer C, Boschloo G, Hagfeldt A, Westermark K, Rensmo H and Siegbahn H 2002 *J. Photochem. Photobiol. A* **148** 57
[6] Xu S, Hansen B J and Wang Z L 2010 *Nat. Commun.* **1** 93
[7] Roy T K, Bhowmick D, Sanyal D and Chakrabarti A 2008 *Ceram. Int.* **34** 81
[8] Mazaheri M, Hassanzadeh-Tabrizi S A and Sadrnezhaad S K 2009 *Ceram. Int.* **35** 991
[9] Hynes A P, Doremus R H and Siegel R W 2002 *J. Am. Ceram. Soc.* **85** 1979
[10] Binner J and Vaidhyanathan B 2008 *J. Eur. Ceram. Soc.* **28** 1329
[11] Mayo M J 1996 *Int. Mater. Rev.* **41** 85
[12] Anselmi-Tamburini U, Garay J E and Munir Z A 2006 *Scr. Mater.* **54** 823
[13] Guo-Dong Z, Kuntz J, Julin W, Garay J and Mukherjee A K 2003 *J. Am. Ceram. Soc.* **86** 200
[14] Chaim R 2006 *J. Mater. Sci.* **41** 7862
[15] Schwarz S and Guillon O 2013 *J. Eur. Ceram. Soc.* **33** 637
[16] Chinelatto A S A and Tomasi R 2009 *Ceram. Int.* **35** 2915
[17] Chen P L and Chen I W 1997 *J. Am. Ceram. Soc.* **80** 637
[18] Rahaman M N, De Jonghe L C and Chu M-Y 1991 *J. Am. Ceram. Soc.* **74** 514
[19] Szepesi C J, Cantonnet J, Kimel R A and Adair J H 2011 *J. Am. Ceram. Soc.* **94** 4200
[20] Mendelson M I 1969 *J. Am. Ceram. Soc.* **52** 443
[21] Zak K A, Majid A W H, Abrishami M E and Yousefi R 2011 *Solid State Sci.* **13** 251
[22] Dargatz B, Gonzalez-Julian J and Guillon O 2015 *J. Cryst. Growth* **419** 69
[23] Zeleňák V, Vargová Z and Györyová K 2007 *Spectrochim. Acta A* **66** 262
[24] Nyquist R A and Kagel R O 1997 *Handbook of Infrared and Raman Spectra of Inorganic Compounds and Organic Salts* (San Diego, CA: Academic) p 221
[25] Vithal Ghule A, Lo B, Tzing S-H, Ghule K, Chang H and Chien Ling Y 2003 *Chem. Phys. Lett.* **381** 262
[26] Wang Y, Li Y, Zhou Z, Zu X and Deng Y 2011 *J. Nanopart. Res.* **13** 5193
[27] Wong E M, Bonevich J E and Searson P C 1998 *J. Phys. Chem. B* **102** 7770
[28] Hsieh C H 2007 *J. Chin. Chem. Soc.* **54** 31
[29] Hu Z, Oskam G and Searson P C 2003 *J. Colloid Interface Sci.* **263** 454
[30] Baruah S and Dutta J 2009 *Sci. Technol. Adv. Mater.* **10** 013001
[31] Spitz R N, Barton J E, Barteau M A, Staley R H and Sleight A W 1986 *J. Phys. Chem.* **90** 4067
[32] Langer J, Quach D V, Groza J R and Guillon O 2011 *Int. J. Appl. Ceram. Technol.* **8** 1459
[33] Ewsuk K G, Ellerby D T and DiAntonio C B 2006 *J. Am. Ceram. Soc.* **89** 2003
[34] Bjork R, Tikare V, Frandsen H L and Pryds N 2012 *Scr. Mater.* **67** 81
[35] Biswick T, Jones W, Pacuła A, Serwicka E and Podobinski J 2009 *Solid State Sci.* **11** 330
[36] Ueno N, Maruo T, Nishiyama N, Egashira Y and Ueyama K 2010 *Mater. Lett.* **64** 513
[37] Schwarz S, Thron A M, Rufner J, Benthem K and Guillon O 2012 *J. Am. Ceram. Soc.* **95** 2451
[38] Erhart P and Albe K 2006 *Appl. Phys. Lett.* **88** 201918
[39] Ali M and Winterer M 2010 *Chem. Mater.* **22** 85
[40] Coble R L 1963 *J. Appl. Phys.* **34** 1679
[41] König M, Höhn S, Hoffmann R, Suffner J, Lauterbach S, Weiler L, Guillon O and Rödel J 2011 *J. Mater. Res.* **25** 2125
[42] Olevsky E A, Tikare V and Garino T 2006 *J. Am. Ceram. Soc.* **89** 1914
[43] Takata M, Tsubone D and Yanagida H 1976 *J. Am. Ceram. Soc.* **59** 4
[44] Besson J and Abouaf M 1989 *Mater. Sci. Eng. A* **109** 37
[45] Besson J and Abouaf M 1991 *Acta Metall. Mater.* **39** 2225
[46] Amaral L, Jamin C, Senos A M R, Vilarinho P M and Guillon O 2013 *J. Eur. Ceram. Soc.* **33** 1801
[47] Amaral L, Jamin C, Senos A M R, Vilarinho P M and Guillon O 2012 *J. Am. Ceram. Soc.* **95** 3781
[48] Mayo M J 1996 *Grain Growth in Polycrystalline Materials II* vol 204 (Zurich: Transtec) pp 389–97
[49] Zuo R, Aulbach E, Bordia R K and Rödel J 2003 *J. Am. Ceram. Soc.* **86** 1099
[50] Bordia R K, Zuo R, Guillon O, Salamone S M and Rödel J 2006 *Acta Mater.* **54** 111
[51] Kihara K and Donnay G 1985 *Can. Mineral.* **23** 647

19

Long-term aging of Ag/a-C:H:O nanocomposite coatings in air and in aqueous environment

Martin Drábik[1], Josef Pešička[2], Hynek Biederman[3] and Dirk Hegemann[1]

[1] Empa, Swiss Federal Laboratories for Materials Science & Technology, Laboratory for Advanced Fibers, Lerchenfeldstrasse 5, 9014 St.Gallen, Switzerland
[2] Charles University in Prague, Faculty of Mathematics and Physics, Department of Physics of Materials, Ke Karlovu 5, 121 16 Prague 2, Czech Republic
[3] Charles University in Prague, Faculty of Mathematics and Physics, Department of Macromolecular Physics, V Holešovičkách 2, 180 00 Prague 8, Czech Republic

E-mail: martin.drabik@gmail.com

Abstract

Nanocomposite coatings of silver particles embedded in a plasma polymer matrix possess interesting properties depending on their microstructure. The film microstructure is affected among others also by the RF power supplied during the deposition, as shown by transmission electron microscopy. The optical properties are characterized by UV–vis–NIR spectroscopy. An anomalous optical absorption peak from the Ag nanoparticles is observed and related to the microstructure of the nanocomposite films. Furthermore, a long-term aging of the coatings is studied in-depth in ambient air and in aqueous environments. It is shown that the studied films are not entirely stable. The deposition conditions and the microstructure of the films affect the processes taking place during their aging in both environments.

Keywords: aging, nanocomposites, plasma polymerization, silver, sputtering

1. Introduction

Metal/plasma polymer composite coatings attract a long-term interest in research due to a wide range of possible applications as a result of a great potential in tuning their mechanical, electrical and optical properties in various different directions according to the special requirements of any particular application [1–5]. Such films are composed of metal particles embedded in a matrix of a plasma polymer and often denoted as M/C:X, where M stands for a metal, C for carbon and X for any element present in the plasma polymer matrix (typically hydrogen, oxygen or fluorine). The term plasma polymer denotes a material that is formed as a result of an electric discharge in an organic gas, plasma-enhanced chemical vapor

deposition (PE-CVD). Different to conventional polymers, a plasma polymer means a new class of material that consists of short chains that are randomly branched and terminated with a high cross-linking density. These nanocomposite films have already been identified as a suitable type of material for sensor or optical filter applications [6, 7]. Lately, composites containing nanoparticles of silver have been particularly widely studied due to the well-known antibacterial activity of silver ions [8–10]. The intention has been to use them in biomedicine as a wound dressing or in bone cements [11]. The use of silver nanoparticles is nowadays widespread in medical use and also in our everyday lives. However, not all the questions regarding the toxicity of so called 'nanosilver' (and nanoparticles generally) in the human body and in nature have been so far properly answered [12]. Especially, there are concerns about the Ag nanoparticles (Ag^0) themselves, as well as the long-term exposure to Ag^+ ions which leach from the particles in the presence of water [13, 14]. It has been shown that the Ag^+ ion release occurs not only from free Ag particles dispersed in water but also from particles bound within a polymer matrix. In this case, the Ag^+ release is influenced by the Ag^0 content, as well as by the nanoparticle morphology and distribution [15]. Generally, the amount of metal determines the properties of such a composite material. When the amount of metal in the film is low, metal nanoparticles are separated from each other in the plasma polymer matrix and the coatings behave mostly as electrical insulators. On the other hand, the separation vanishes with a high amount of metal inclusions and the character of the films resembles conductors. The gradient intermediate phase between these two limiting structural regimes is called the percolation threshold. The structure of a metal/plasma polymer nanocomposite coating is typically described by a so-called 'filling factor' f (volume fraction ratio). It is defined as the volume of metal inclusions embedded in a plasma polymer matrix in the total volume of the composite material.

A commercial application of metal/plasma polymer films might require either long-term stability of the coating or its controlled aging over the period of application, e.g. the mentioned metal ion release [15], or matrix degradation [16]. The concerns about potential instability of the structure of metal/plasma polymer coatings can be considered as one of the main reasons why these materials have failed to be widespread in industrial applications requiring stable coatings [17]. Both the plasma polymer matrix and the metallic component of the composite are prone to aging. The plasma polymer matrix typically undergoes several structural changes when exposed to the ambient atmosphere after deposition due to the presence of free radicals, structural defects or residual stress [18]. Oxidation of a hydrocarbon matrix is among the most common processes. Reviews by Hlidek *et al* give a summary of various oxidation processes taking place in Ag-containing nanocomposite films which are also important for the initiation of Ag^+ release [19, 20]. Particularly, oxygen and water vapor were shown to be the most active oxidative agents. The embedded metal nanoparticles are often affected by changes in their size and shape distributions. There were several mechanisms of their restructuring described: atomic diffusion along the particle surface (recrystallization and coalescence) and through the matrix (Ostwald ripening) and particle migration [1, 21]. All of these might be further enhanced by thermal activation during high temperature treatment or by laser or electron irradiation. On the other hand, controlled aging of a composite coating requires proper selection of metal and matrix materials and precise tuning of their structure. This is of a particular importance in the potential biomedical applications of antibacterial silver-containing composites with controlled Ag^+ ion release [15, 22]. The various processes taking place in water and aqueous media are rather complex. The initiation of the Ag^+ ion release requires a Ag^0 particle to be oxidized [23, 24]. The oxidation process in turn requires the presence of both dissolved oxygen and protons. It produces reactive oxygen intermediates and might even proceed to a complete reactive dissolution under some conditions and disappearance of the solid nanoparticle phase [25]. The essential mechanism

pathway can be written as follows:

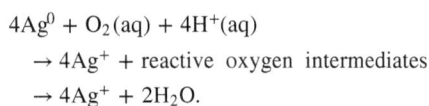

$$4Ag^0 + O_2(aq) + 4H^+(aq)$$
$$\rightarrow 4Ag^+ + \text{reactive oxygen intermediates}$$
$$\rightarrow 4Ag^+ + 2H_2O.$$

The reaction of a silver nanoparticle with H_2O_2 is even faster than with O_2 so the initial nanoparticle oxidation via reaction with oxygen is the rate limiting factor. Rates of Ag^+ ion release increase with temperature and decrease with increasing pH. The value of pH can be further influenced by the potential degradation of the plasma polymer and its resulting products. The released Ag^+ ions can readsorb on silver nanoparticle surfaces (and be reduced by them, i.e. Ostwald ripening in the case of plasma polymer nanocomposites), so even simple colloids contain at least three forms of silver: solids (Ag^0), free Ag^+ ions or their soluble complexes and surface-adsorbed Ag^+ with the possible occurrence of Ag oxides.

Overall, the tendency to nanostructural changes (their rate and overall measure) is to a great extent determined by the matrix material, its chemical structure, stability, glass transition and crosslinking. Especially, the wettability and ability of the matrix to uptake water is crucial for both the aging of the polymer (possible hydrolysis of the backbone chain) as well as for the structural modifications of the metal nanoparticles. The silver ion release was found to increase exponentially with the maximum water absorption by the polymer matrix and the diffusion coefficient of water within the matrix [22, 26]. Further, the Ag^+ release rate decreases with increasing crosslinking of the matrix [27], and also its efficiency is higher in hydrophilic polymers compared to hydrophobic [28]. An increasing crystallinity of the matrix was found to decrease the Ag^+ release [29]. All of these parameters affect the diffusion coefficients of water molecules as well as the Ag^+ ions through the polymer matrix. In the case of the plasma polymer nanocomposites, the changes in the structure and properties of the silver-containing coatings during aging in air or in water and have been previously studied both on amorphous hydrophobic matrices (hydrocarbon, Ag/a-C:H, or fluorocarbon, Ag/a-C:F) [30–34], as well as on hydrophilic matrices (oxygenated hydrocarbon, Ag/a-C:H:O, or organosilicon, Ag/a-C:Si:O) [15, 35]. Apart of the matrix material, also the overall film architecture plays an important role. For example, plasma polymer multilayer films with metal nanoparticles buried in between typically show a good long-term stability [1], particularly for Au/C:F systems [36].

In this contribution, we present a detailed study of structure and optical properties of Ag/a-C:H:O nanocomposite coatings with a broad range of microstructures around the percolation threshold. The aging (in air or in water) of different a-C:H:O plasma polymer films has recently been investigated [16, 37, 38]. A rather stable plasma polymer matrix deposited from CO_2/C_2H_4 discharges has thus been selected. Therefore, this work particularly focuses on the long-term aging of Ag nanoparticles embedded in such plasma polymer films in ambient air and on the influence of an aqueous environment on their structure and optical properties. Finally, a particular suitable application is suggested for each type of nanocomposite structure.

2. Experimental section

2.1. Deposition of Ag/a-C:H:O coatings

The studied nanocomposite coatings of silver particles embedded in a hydrophilic amorphous oxygenated hydrocarbon plasma polymer matrix (Ag/a-C:H:O) were deposited by a combination of PE-CVD and physical vapor deposition processes within a low pressure glow discharge. A plane-parallel capacitively coupled plasma reactor ($\varnothing = 30$ cm $\times$ $L = 5$ cm) was used. Ag electrode ($\varnothing = 13$ cm) was mounted on the top wall inside the plasma reactor and insulated by a ceramic plate. The Ag sputtering target (intrinsic purity of 99.9 vol%) was connected to an RF power generator (13.56 MHz, Cesar 133, Advanced Energy, USA) through a matchbox. The smaller diameter of the driven electrode, as compared to the grounded base plate, yielded strong asymmetrical conditions and built up a high (negative) bias voltage. The chamber and the silver sputtering target were mechanically cleaned prior to each experiment to remove all by-products of the plasma deposition process and thus maintain the same well-defined starting experimental conditions. The chamber was pumped through the bottom electrode (grid) using a combination of roots and rotary vane pumps. More details on the experimental chamber can be found elsewhere [15]. Working gases were introduced into the chamber through a gas shower system in the top chamber wall which enabled homogeneous plasma conditions. Argon was used for sputtering of Ag, while ethylene (C_2H_4) was the source hydrocarbon monomer gas for the plasma polymer matrix. In addition, a reactive carbon dioxide (CO_2) was added to the working gas mixture resulting in oxidation processes in the gas phase and incorporation of oxygen-functional groups within a crosslinked hydrocarbon plasma polymer matrix (C:H:O) [15, 39]. All the gases were used as obtained from Carbagas, Switzerland, without any further purification (intrinsic purity of 99.99 vol%). The gas flow rates (in standard cubic centimeters per minute (sccm)) were measured by mass flow controllers (MKS, Germany). The working gas flow rate ratio $Ar:CO_2:C_2H_4$ and the total gas pressure were held constant in all of the deposition processes at 50:6:1 sccm and 5 Pa, respectively. The gas mixture was previously found to influence the degree of the functionality of the coatings [40]. Similarly, the silver content in the composite increases with the increasing $CO_2:C_2H_4$ ratio due to the reduced deposition rates of the plasma polymer matrix (at a constant power). Finally, the RF power input was varied between 20 and 100 W to obtain nanocomposite films with different filling factors.

The composition and stability of the plasma discharge was monitored *in situ* during every deposition process by optical emission spectroscopy (OES) using an AvaSpec-ULS2048-USB2 spectrometer (Avantes, Netherlands). The spectrometer was connected to the chamber through an optical fiber. The emission spectra were recorded in the wavelength range of 300–820 nm. The resolution of the spectrometer set-up was 0.4 nm. The obtained data was analyzed using the atomic spectra database provided by the National Institute of Standards and Technology (NIST) [41], and other literature [2, 42].

The Ag/a-C:H:O nanocomposite films were deposited on various substrates as required by different analytical techniques for characterization of the properties of the films. They were all placed on the grounded bottom electrode. After finishing the deposition process, the vacuum chamber was left being pumped for another 30 min before opening it to atmosphere to de-activate the radicals trapped in the plasma polymer matrix. This leads to substantially slower aging process in air [43]. All the studied nanocomposite films were of thicknesses (21 ± 2) nm as measured by a surface profiler Dektak 150 (Veeco, USA). The values of thicknesses were averaged at least from seven different positions on the glass substrates.

2.2. Microstructure of Ag/a-C:H:O coatings

The structure of the Ag/a-C:H:O nanocomposite films deposited on carbon foils supported by Cu grids S160 (Plano, Germany) was studied by transmission electron microscopy (TEM, Jeol 2000FX, 200 kV, Jeol, Japan). The obtained micrographs were used to characterize the size and the shape of the silver nanoparticle inclusions and their distributions. The obtained digital micrographs were corrected for contrast and/or brightness where necessary. Images were then statistically processed by ImageJ v.1.43q software [44]. The size of the embedded Ag nanoparticles was described in terms of 'equivalent nanoparticle diameter', d, defined as the diameter of an equivalent circular area. The measurement error of d can be estimated to be 3 pixels which corresponds to about 1 nm at the magnification of 200 000× used for analysis. The dispersion of the nanoparticle sizes was expressed in terms of the standard deviations of the size distributions. Further, 'shape factor' S of the projections of the Ag nanoparticles was studied. It is often called 'circularity' and defined as $S = \frac{4\pi A}{U^2}$, where A is the area of the particle projection as observed by TEM and U is the particle circumference. It is used to measure how close a projection of a particle is to a perfect circle. The values for the particle shape factor lie in the range $0 \leqslant S \leqslant 1$, where $S = 1$ corresponds to a perfect circle [1]. Nanoparticle size and shape distributions were obtained from statistical processing of sets of several micrographs (magnification 200 000×). The size distributions were fitted by logarithmic normal (log-normal) distribution functions $f_{LN}(d)$ from which modal values of diameters d_m were obtained. Further, the crystal structure of the embedded silver nanoparticles was estimated from electron diffraction patterns obtained by selected-area diffraction (SAD) analysis according to the database provided by the International Centre for Diffraction Data (ICDD) [45].

2.3. Optical properties of Ag/a-C:H:O coatings

The optical properties of the Ag/a-C:H:O nanocomposite films deposited on microscopic glass slides were characterized by ultraviolet–visible–near-infrared (UV–vis–NIR) spectrophotometer Lambda 9 (Perkin-Elmer, USA) in the spectral range 300–2500 nm. The optical properties of the nanocomposites were characterized on samples deposited in the same deposition process as the corresponding films for the characterization of the microstructure by TEM. The transmittance spectra were collected using an integration sphere and certified reflectance standard Spectralon USRS-99-010 (SphereOptics, Germany) as a light spectral reference. The obtained spectra were smoothed for noise removal around the wavelength of 830 nm which is the position of the detector change. The UV–vis–NIR spectra are represented in terms of absorbance, which was calculated according to the relation between transmittance T and absorbance A: $A = \log_{10} \frac{1}{T}$. Absolute measurement error of the characterization of optical properties was estimated to be 2%.

2.4. Aging of Ag/a-C:H:O coatings

Aging of Ag/a-C:H:O coatings was studied in ambient air and in distilled water. In both cases, the films were left to age for a given time in the dark at a room temperature of 20 °C. In the case of measurement by TEM, the films were characterized right after the deposition (as-prepared) and after 2, 10, 18 and 23 months from the deposition for aging in air and after 1 h and 1 day in distilled water. Further, the structure of the films stored in water for 1 day was also characterized by TEM after 2 months. In the case of the measurement of the optical properties, the films were characterized right after the deposition (as-prepared) and after 1, 5, 9 and 22 months of aging in air and after 1 h and 1 day of aging in distilled water.

3. Results and discussion

3.1. Control of the deposition process

The composition and stability of the plasma discharge during the deposition of Ag/a-C:H:O nanocomposite films was controlled *in situ* using OES. An example of a typical spectrum with identification of the emitting species can be found elsewhere [7]. Major emitting species comprise Ag, Ar and H atoms and CO and CH radicals. Especially important are the atomic emission lines of Ag species ($^2P_{3/2}$ and $^2P_{1/2}$) originating from excited states after sputtering from the metal target that can be found at 328.1 nm and at 338.3 nm, respectively. The intensity of the emission line of Ag is proportional to the sputtering rate and can thus be related to the amount of the deposited metal in the resulting nanocomposite [31, 46]. As can be seen in figure 1(a), the ratio of the intensities of Ag and Ar emission lines increases linearly with the increasing RF power of the discharge. The ratio was calculated from values of the intensities of the selected emission lines averaged over the first minute of each of the depositions. The reasons for this will be explained further. Monitoring the level of intensity of one of the emission lines of silver is a convenient means for controlling the deposition rate of silver, which is a factor determining the final filling factor of the deposited nanocomposite film. A fine control of the silver sputtering rate is important for the preparation of films with repeatable quality and composition.

The stability of the deposition process was controlled through monitoring of intensities of emission lines of several important species. An example of a time evolution of emission lines of Ag (328.1 nm), Ar (420.1 nm), H (656.3 nm), CH (430.8 nm) and CO (450.5 nm) during the deposition process at 50 W can be seen in figure 1(b). As can be seen, the emission lines of Ar, CH and CO species are almost constant throughout the whole deposition process which reflects the stable energetic and gas composition conditions. On the other hand, we can observe a strong decrease in the intensities of Ag and H emission lines throughout the deposition process. The decrease in the intensity of the Ag signal reflects the decrease of the Ag deposition rate. This is the result of an effect commonly known as 'target poisoning' that occurs during sputtering in an atmosphere composed of reactive gases [47]. The fragments of the reactive gases react with the surface of the target and continuously passivate it which in result lowers the metal sputter yield. Note, however, that the embedded Ag nanoparticles in the growing film are metallic despite using oxidative plasma [15]. It was shown that silver oxide (Ag_2O) can only be produced in a pure oxygen plasma at very low growth rates ($\sim$0.3 nm min^{-1}) [48], while the growth rate for Ag was well above 1 nm min^{-1} for all conditions as used in this work (with a less oxidative plasma at the same time). Nevertheless, the target poisoning leads to a vertically inhomogeneous structure of the nanocomposite film. A typical vertical gradient in the amount of metal in the nanocomposite films can be seen on cross-sectional TEM micrographs previously published by Körner *et al* [15]. We did not compensate anyhow for this gradual decrease of sputtering rate of Ag in our experiments. As a result, a kind of barrier layer with a lower amount of metal was formed on the top of the anticipated nanocomposite film. The advantage of these gradients lies mainly in the lowered initial burst ion release in aqueous environments as was previously reported by Körner *et al* [15]. Furthermore, these structures avoid loss of nanoparticles supporting the formation of stable nanocomposites and serve as reservoirs of silver ions. In any case, the metallic target was cleaned before each experiment so that the starting conditions were the same in each deposition (i.e. 'metal' sputtering mode).

Due to the vertical inhomogeneity of the studied nanocomposite films which are about 20 nm thick, we did not calculate exact values of their filling factors f although it is the basic parameter characterizing a metal/plasma polymer composite. However, f is calculated as an average value. As such, it is suitable only for a homogeneous structure and in our case might be misleading. Thus, all the properties of the films will be further discussed in the dependence on the particular RF power supplied during their deposition. Nevertheless, we will refer to films with 'higher' or 'lower'

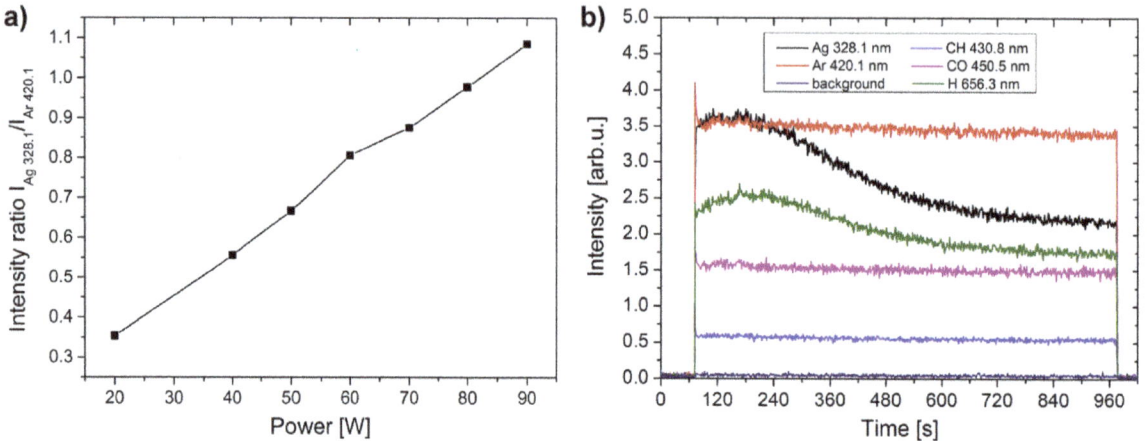

Figure 1. Ratio of intensities of emission lines of Ag (328.1 nm) and Ar (420.1 nm) as recorded during sputtering of Ag in Ar/CO$_2$/C$_2$H$_4$ atmosphere (gas flow rates of 50/6/1 sccm, respectively) at a pressure of 5 Pa and various RF powers (a). Time evolution of selected emission lines during deposition at 50 W (b).

filling factors where appropriate. The atomic ratio of Ag in Ag/a-C:H:O films deposited at various RF powers was reported previously [7].

3.2. Microstructure of Ag/a-C:H:O films

3.2.1. As-prepared films. The microstructure of Ag/a-C:H:O nanocomposite films was studied on films deposited at various RF powers delivered to the silver target. As could be expected, a broad range of powers supplied leads to nanocomposite films with considerably different microstructures. Typical TEM micrographs of selected Ag/a-C:H:O nanocomposite films deposited at various powers as measured after the deposition can be seen in figure 2. All of the films are formed of island-like structures where the silver nanoparticles are embedded in the plasma polymer matrix. In the case of the bright field images (the left-hand side micrograph for each power), the silver nanoparticles are displayed as the dark gray regions and the plasma polymer matrix surrounding them is represented by the light gray space. The nucleation and growth of metal-polymer nanostructures is governed by surface diffusion coefficient of metal atoms [49]. The size of the silver inclusions increases, while the average value of the shape factor S_a decreases with the increasing total amount of metal in the films (corresponds to an increasing power supplied). At the same, both distributions become broader. As can be seen, small ($d_m = 6.2$ nm) and almost circular individual silver nanoparticles homogeneously distributed within the plasma polymer are formed at the low power of 30 W, i.e. below percolation threshold. Their size increases with the increasing power. The particles start to touch their closest neighbors and coalesce into small islands ($d_m = 8.4$ nm) already at the power of 40 W. This can be seen also on the decreasing shape factor, although the islands are still convex. Nevertheless, the particles are still to a great extent separated by the plasma polymer matrix. As the power slightly increases to 45 W, the

formed islands grow further ($d_m = 11.0$ nm) and start to be more and more interconnected, i.e. reach percolation threshold. At this point, structures of almost any shape can be find in the film ($0.05 < S \leqslant 1$). Even though most of the particles are still almost circular, some concave island can be found. The depositions at further increased powers lead to formations of larger metal inclusions in the composite and its structure is above the percolation threshold. The modal value of the equivalent nanoparticle diameter d_m of the nanoparticles in the nanocomposite films deposited at 60 W was estimated to be 60.2 nm. At the same time, the structure of the islands becomes more and more irregular and the shape factor gradually decreases down to about 0.1. However, it is interesting to note that the depositions at 55 and 60 W lead (apart from larger nanoparticles) also to the formation of larger 'interstitial space', or gaps, where the separation of the island-like structures is larger than can be observed at the lower deposition powers. After a closer investigation and magnification of the micrographs, it can be seen that the plasma polymer matrix in this interstitial space contains a considerable number of very small, almost circular Ag nanoparticles with $d_m \sim 3$ nm. They are represented with the blue color in the histograms of equivalent nanoparticle diameters and shape factors in figure 2. These nanoparticles are most probably located close to the film–substrate interface where they were deposited in the early stages of film growth [15]. Their further growth into large particles and/or islands was hindered by the growing plasma polymer matrix. These nanoparticles will be further discussed in the following sections.

The dark field images are displayed at the right-hand side for each power. In this case, the diffracting crystals are represented by the bright spots on the dark background. The images reveal the monocrystalline structure of the individual silver nanoparticles. The islands merged from monocrystalline nanoparticles of various sizes and crystal orientations grown at higher powers are typically multicrystalline.

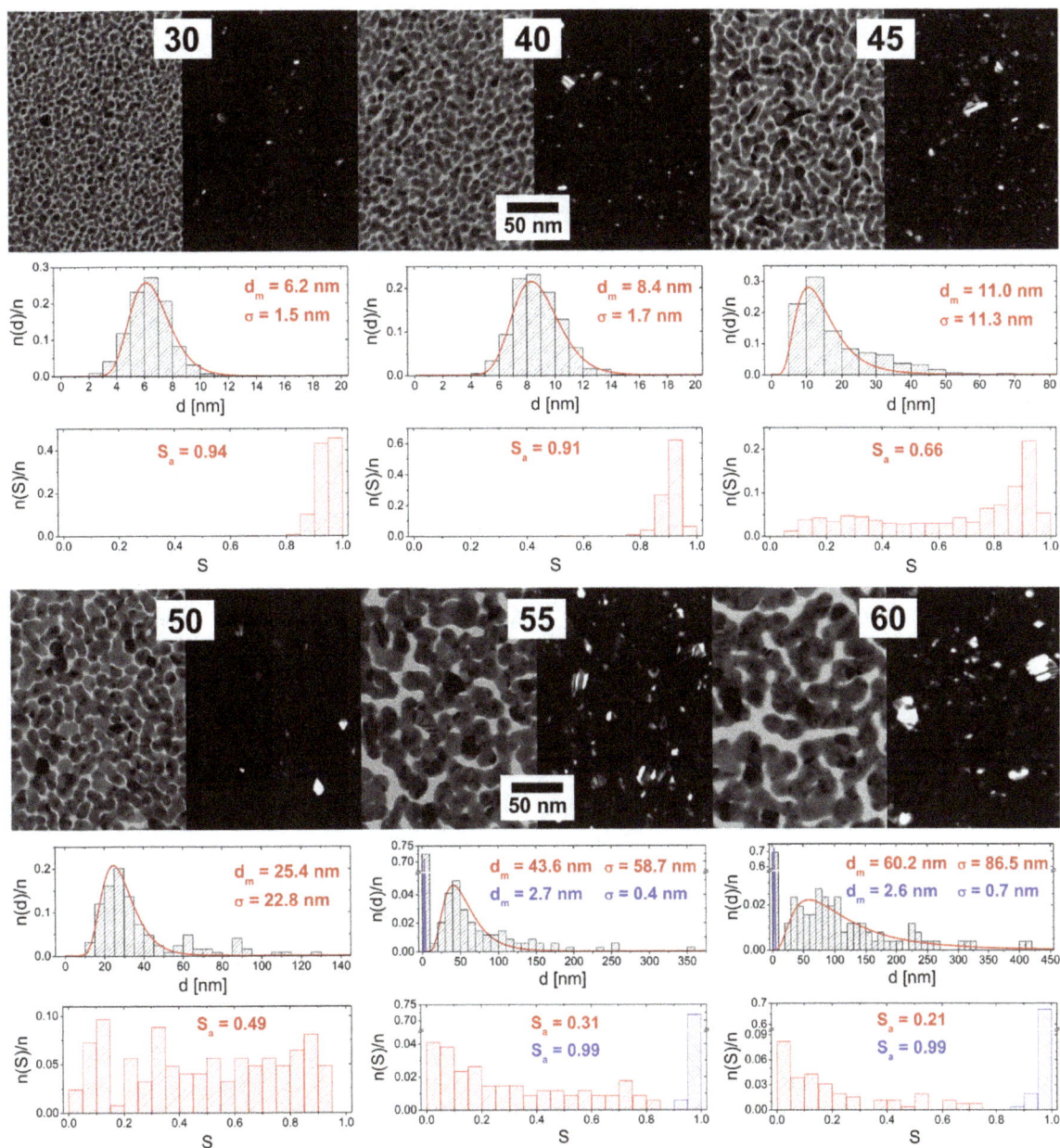

Figure 2. TEM micrographs of Ag/a-C:H:O nanocomposite films deposited at different RF powers (W) as measured right after their deposition. In each case, a bright field image (left) is displayed together with the corresponding dark field image of the same spot (right). For each micrograph, a distribution histogram of equivalent nanoparticle diameters d with its log-normal fit and modal value of nanoparticle diameter (d_m) and its standard deviation (σ) are displayed (top). The corresponding histogram of nanoparticle shape factor S and the average value of shape factor (S_a) are displayed (bottom).

3.2.2. Aging in ambient air. The characterization of microstructure of the Ag/a-C:H:O nanocomposite films was repeated after the samples aged in ambient air for several days and months after the deposition. TEM micrographs of nanocomposite films deposited at powers of 30 and 60 W, as observed right after the deposition, and 2 and 18 months from the deposition can be seen in figure 3. These two films

were selected because they represent the two limiting cases in our study: from the point of view of the discharge power (below and above the percolation threshold), as well as the behavior during aging. All the other characterized films will be discussed further below.

The changes in the structure of the films deposited at the lower power of 30 W that take place during the aging of the

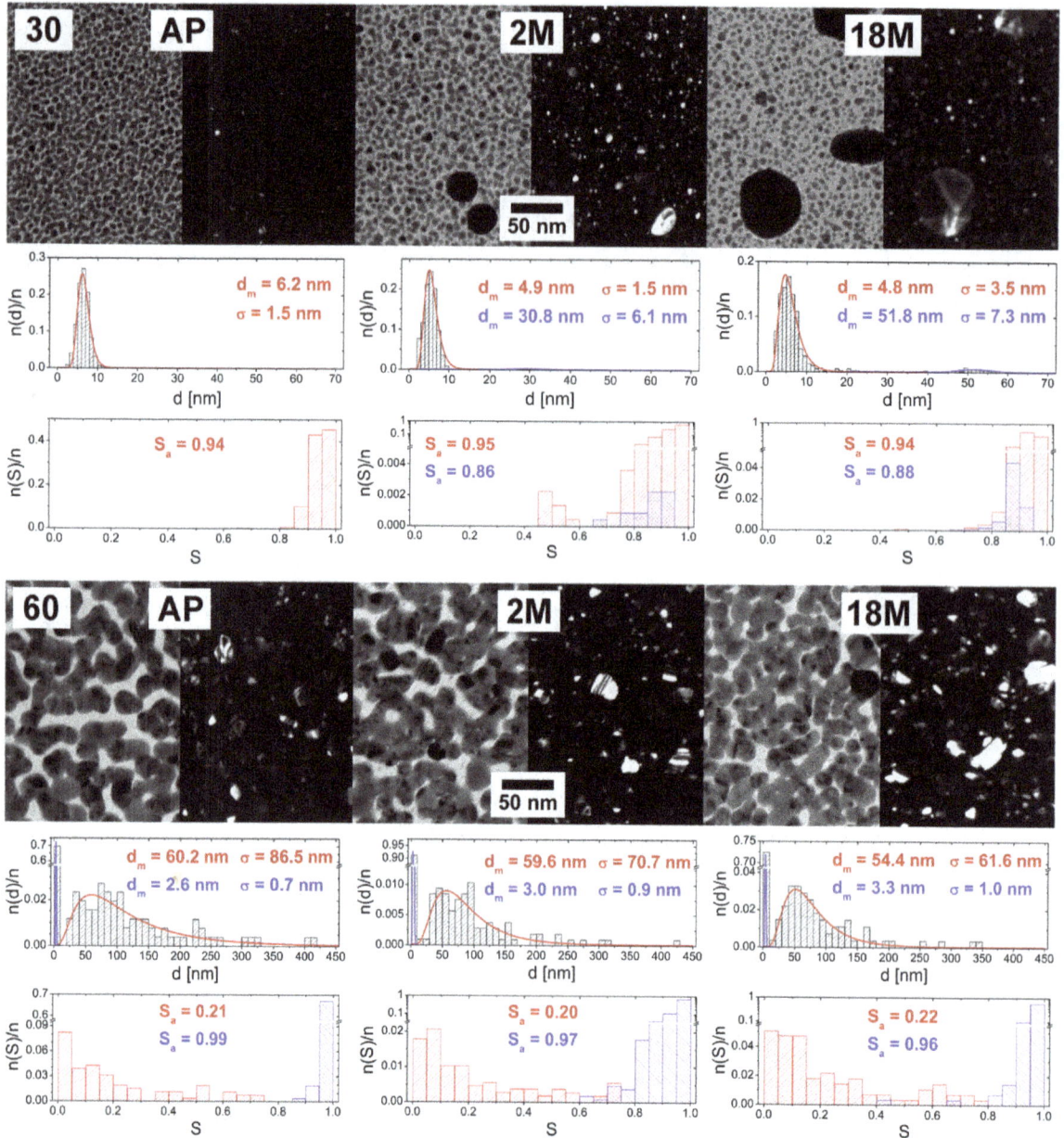

Figure 3. TEM micrographs of Ag/a-C:H:O nanocomposite films deposited at RF power of 30 W (top) and 60 W (bottom) as measured right after the deposition (left), 2 months after the deposition (middle) and 18 months after the deposition (right). In each case, a bright field image (left) is displayed together with the corresponding dark field image of the same spot (right). For each micrograph, a distribution histogram of equivalent nanoparticle diameters d with its log-normal fit and modal value of nanoparticle diameter (d_m) and its standard deviation (σ) are displayed (top). The corresponding histogram of nanoparticle shape factor S and the average value of shape factor (S_a) are displayed (bottom).

films in air are obvious at first sight: a gradual growth of several nanoparticles into larger objects at the expense of the other small nanoparticles which serve as the source material for their growth. The modal value of the equivalent nanoparticle diameter d_m of the intrinsic small nanoparticles slightly decreases from about 6 nm down to about 5 nm, while d_m of the grown large crystals (marked with blue color in the graphs) increases to about 31 and 52 nm after 2 months and 18 months of aging in air, respectively. It is important to note that the projections of all the observed nanoparticles remain mostly circular. However, the images of the dark field and the SAD analysis discussed further below suggest that all the particles are monocrystals. Their growth most probably proceeds through atom/ion diffusion (Ostwald ripening) rather than a particle migration and coalescence. Ostwald ripening and the atom/ion diffusion through the matrix are driven by the concentration gradient around particles of different size which is the kinetics factor for the diffusion. Particularly, the concentration next to larger particles is lower than around those smaller and in result, larger particles grow at the expense of smaller ones. The growth of larger particles is further affected by minimization of the surface energy, where the larger particles also become more energetically stable [1]. The water vapor from the ambient air penetrating the hydrophilic C:H:O plasma polymer matrix most probably supports this structural change by enabling the Ag^+ ion release from the silver nanoparticles and even enhancing the ion mobility. Since the nanocomposite is not in a direct contact with water, the silver ions are not lost and rather contribute to the growth of some of the nanoparticles. It is interesting to note that while the size of the small intrinsic nanoparticles remains almost the same in time, their number obviously decreases. Since their modal diameter is about 6 nm and their size distribution relatively narrow (no particles larger than 10 nm were observed) and the film thickness is about 20 nm, it can be expected that most of the nanoparticles are fully embedded within the plasma polymer matrix. Further, it can be assumed that the changes occur at first with the nanoparticles which are at or close to the film surface and only afterwards proceed deeper in the film [50]. Therefore, the nanoparticles close to the film–substrate interface remain untapped within the first two years of aging in air. It also important to mention that the small particles located at the substrate surface are better stabilized, because they are in fact only half-spheres (i.e. have smaller surface than the structures fully embedded in the plasma polymer matrix) and also the plasma polymer matrix tends to be more crosslinked in the early stages of the film growth [51]. The growth of metal nanoparticles in Ag/C:H composites, together with the increase in the nearest-neighbor distance within the first few days of their aging in air, was reported also by Biederman *et al* [30].

The Ag/a-C:H:O nanocomposite films deposited at the high power of 60 W are relatively stable in air when compared to those prepared at the lower power. All the micrographs of the aging of this film might well seem identical at first sight. Although the composites deposited at higher powers contain more silver and complex structures and therefore the particular changes in the films are difficult to spot and follow, a careful analysis of the micrographs brings more insight. Even though the small nanoparticles situated in between the islands (marked in blue in the graphs) remain almost the same with d_m of about 3 nm, a slight disruption of the percolation structure of the nanocomposite film can be observed after 18 months of aging. The nanoparticles previously fully merged in the island-like structure are in the end more sharply delineated and even individual nanoparticles can be observed at several positions. Even though the wide distribution of the shape factor remains unchanged, the disruption of the percolation structure is supported by the statistical processing of the equivalent nanoparticle diameter, where d_m decreases from 60.2 nm for

the as-deposited films down to 59.6 nm after 2 months and to 54.4 nm after 18 months. This suggests that the island-like structure of the nanocomposite loses material, particularly from the 'neck' interconnections of the nanoparticles. The causes of the aging might be the water vapor that penetrates the hydrophilic matrix and enhances the silver release. However, it is not entirely clear what happens with the lost silver atoms when the films are stored in air. It is possible that the silver ions released from the 'necks' of the island-like structures are reduced on the top of the spherical nanoparticles which thus grow in the vertical direction. This is a process that cannot be detected from the TEM 2D projections and a cross-sectional analysis would be necessary.

From the micrographs in figure 3, it might seem that the aging processes taking place in the film deposited at 60 W are much less pronounced than in the case of the film deposited at the low power of 30 W. Particularly, that the film with the higher filling factor deposited at 60 W loses less silver. However, this observation might be misleading because the TEM micrographs provide only a 2D projection of the structure. No conclusion on the total release of silver can be made. On the contrary, it has already been shown that the intrinsic amount of silver influences the Ag^+ ion release: the higher the initial silver content, the higher the total release of silver was observed [15]. Although the mentioned study was performed in water, we can expect that a similar effect might take place here due to the water vapor, even though on a smaller scale. On the other hand, the properties of the plasma polymer matrix contribute to the stability of the composite. The deposition at higher powers results in a more crosslinked plasma polymer which slows down both the penetration of water molecules and possibly also the resulting atom/ion diffusion through the matrix [22]. Also, a high surface area might greatly enhance the Ag release with respect to the total Ag^0 content.

3.2.3. Aging in distilled water. The microstructure of the Ag/a-C:H:O nanocomposite films was studied after their immersion into distilled water for 1 h and 1 day. Körner *et al* found in the previous study that most of the Ag^+ release from similar samples with structures below the percolation threshold of a comparable thickness of 25 nm occurs within the first day of storage in the water bath [15]. The TEM characterization was repeated again two months after the samples were taken out from the water bath to control if the aqueous environment triggers any further changes in the structure of the films. TEM micrographs of nanocomposite films deposited at powers of 30 and 60 W, as observed right after the deposition, and after 1 h and 1 day of storage in distilled water are displayed in figure 4.

The soaking of the Ag/a-C:H:O film deposited at 30 W into a water bath for a few hours results in two eye-catching effects. On the one hand, it is the fast growth of large crystals and on the other the vanishing of the intrinsic small nanoparticles. While the modal value of the size of the small nanoparticles is about 6 nm, distribution of their shape factor becomes broader in time. Large, to a great extent circular,

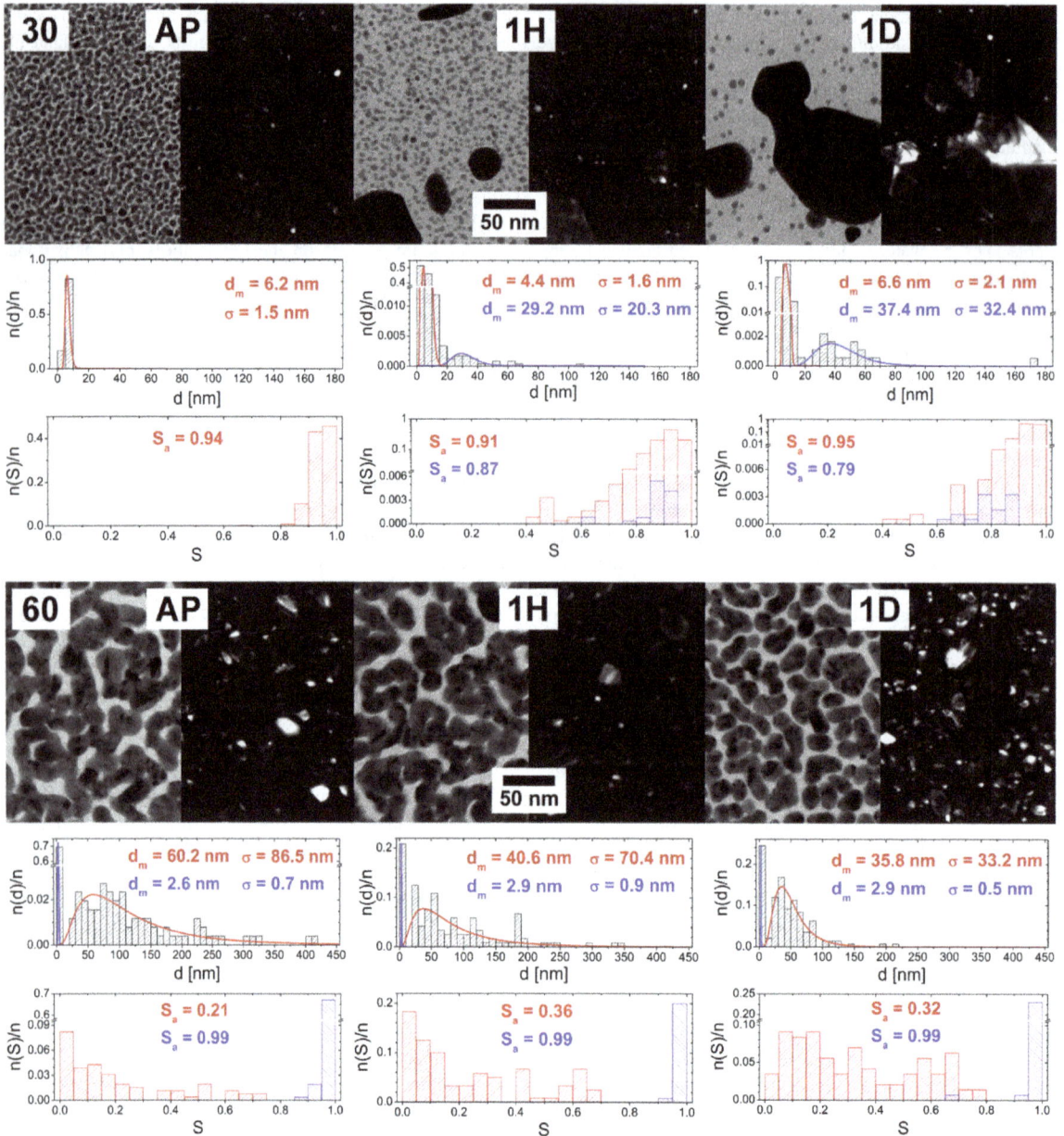

Figure 4. TEM micrographs of Ag/a-C:H:O nanocomposite films deposited at RF power of 30 W (top) and 60 W (bottom) as measured right after the deposition (left), after 1 h in distilled water (middle) and after 1 day in distilled water (right). In each case, a bright field image (left) is displayed together with the corresponding dark field image of the same spot (right). For each micrograph, a distribution histogram of equivalent nanoparticle diameters d with its log-normal fit and modal value of nanoparticle diameter (d_m) and its standard deviation (σ) are displayed (top). The corresponding histogram of nanoparticle shape factor S and the average value of shape factor (S_a) are displayed (bottom).

crystals with diameter of about 29 nm appear in the composite film already after 1 h in the water. Their size grows further and reaches about 37 nm after 1 day in water. Based on the SAD analysis discussed further, we cannot expect that all of the nanoparticles are monocrystals. Also, the growth of the large crystals in the later stages can be accompanied by the coalescence of contacting particles, as can be seen in figure 4

top right. It is important to stress that the films are about 20 nm thick and therefore the large crystals grown are most probably sticking out of the nanocomposite film. The crystals are thus in a direct contact with water solution of the free Ag$^+$ ions. The released Ag$^+$ ions can be readily adsorbed and reduced on their surfaces which enhances the growth of the particles. In the case of the Ag/a-C:H:O film deposited at

60 W, no growth of larger nanoparticles is found in the presence of water. Rather a disruption of the complex percolated structure can be observed in time: d_m decreases from the original 60.2 nm down to 40.6 nm after 1 h and even down to 35.8 nm after a whole day in water. The randomly shaped, mostly concave, metal islands separate into individual particles with more circular shape: the shape factor distribution gradually shifts to higher values. On the other hand, the small almost circular silver nanoparticles in between the island-like structure remain unchanged within the first day of aging in water. This observation supports the previously stated assumption that they are located close to the substrate and therefore are not affected in this early period of aging in water. Despite this, the film deposited at the higher power of 60 W obviously releases silver similar to the film deposited at the lower power of 30 W; it is not clear why the silver material is completely lost in the solution and does not contribute to the growth of larger particles as is observed in the case of the film with the lower filling factor.

When we compare the structural changes induced by water with the aging of the Ag/a-C:H:O composites taking place in air, we can see that both reveal the same characteristics. However, the changes are much more pronounced and occur at much shorter time scales in the aqueous environment. While the changes induced in water within the first hour are roughly equivalent to the aging for 2 months in air in the case of the film deposited at 30 W, they are almost equivalent to the aging for 18 months in the case of the film deposited at 60 W. This will be further discussed in the following section.

3.2.4. Aging of Ag/a-C:H:O films.

Apart from the Ag/a-C:H:O nanocomposite films deposited at 30 and 60 W discussed in detail above, composites with other filling factors deposited at powers between these two limiting cases were also studied. The changes in the basic microstructural parameters, i.e. modal values of the equivalent nanoparticle diameter, d_m, standard deviation of distribution of diameters, σ, and average value of circularity of nanoparticles, S_a, as recorded during the aging of the films in air or in distilled water are displayed in the dependence on the deposition power in figure 5.

As can be seen in figure 5(a), the sizes of the as-prepared metallic structures and the widths of their distributions in the Ag/a-C:H:O nanocomposites increase with the increasing deposition power, while the corresponding shape factor decreases. In the case of high powers of 55 and 60 W, we can observe a bi-modal distribution of the particle sizes: small nanoparticles distributed in between the island-like structure, as reported above. The studied films are not only of different intrinsic microstructures, also their aging in air proceeds in different directions. The most obvious changes occur with the nanocomposite films deposited at the low powers of 30 and 40 W that contain mostly individual circular silver nanoparticles separated by the plasma polymer matrix. While the size of the original as-prepared small nanoparticles does not change within about the first two years, large and almost circular particles start to grow in the films during the first

months of their aging. The changes are fastest in the first months when they reach sizes of 30–50 nm. Despite this it seems that the onset of the growth of the large particles is postponed in the case of the film deposited at 40 W (no larger crystals were observed in the film after 2 months of aging in air); finally the crystals grow larger after almost 2 years (up to about 60 nm) most probably due to more silver material available for their growth. On the other hand, the nanocomposite films deposited at powers of 45 W (i.e. at the percolation threshold) and higher seem to be more stable in air. The size of the as-prepared nanoparticles is almost constant in these cases and no growth of large crystals was observed in them within 23 months of their aging in air. However, a slight decrease in the d_m of the island-like nanostructures in the film deposited at 60 W can be found after 23 months, particularly from the original size of about 60 nm down to about 55 nm. It seems that the percolated structures of the films with higher filling factors deposited at 50 and 60 W start to be disrupted, as it was discussed in detail in the section above. This process is evidenced by the narrowing size distributions and slightly increasing values of their shape factors (circularity). On the other hand, the small nanoparticles at the substrate–film interface seem to be intact during the aging of the films. Overall, it is most probable that the studied films are not in equilibrium after almost two years from the deposition and their aging in air will continue.

Figure 5(b) shows the aging in water of similar Ag/a-C:H:O nanocomposites prepared at the same experimental conditions in the dependence on the deposition power. Similarly to the aging in air, also two distinct structural changes were observed in the case of the aging in water. A fast growth of large crystals occurs in the films deposited at the low powers of 30 and 40 W already within the first hour in water, when the particles reach diameters of about 20–30 nm. The formed crystals grow further during the next one day in water. However, this growth is much slower than in the beginning, most probably due to loss of silver material into the water bath which cannot contribute to the growth. The massive leaching of Ag^+ in water might also be the reason why the sizes of the grown metal crystals most probably do not reach the size of the crystals grown during the aging in air. There is simply not enough material for the growth of the larger nanoparticles, while its loss during the aging in air is limited and the Ag^+ leaching induced by the air humidity promotes Ostwald ripening and recrystallization. On the other hand, we did not observe any growth of particles in water in the case of the films deposited at higher powers. Rather a degradation of the island-like structure during their aging in water is obvious. While in the case of the aging in air, the degradation was difficult to recognize and could be spotted only in the Ag/a-C:H:O nanocomposite deposited at 60 W, it is clearly visible in the TEM micrographs in figure 4 (and also evidenced by their statistical processing) of the films deposited at powers higher than 45 W after their aging in water for 1 h. The equivalent nanoparticle diameter decreases from 60.2 nm down to 40.6 nm for the film deposited at 60 W, from 43.6 to 29 nm for the 55 W film, and finally from 25.4 to 14.9 nm for the 50 W film. In correlation to the decrease in the

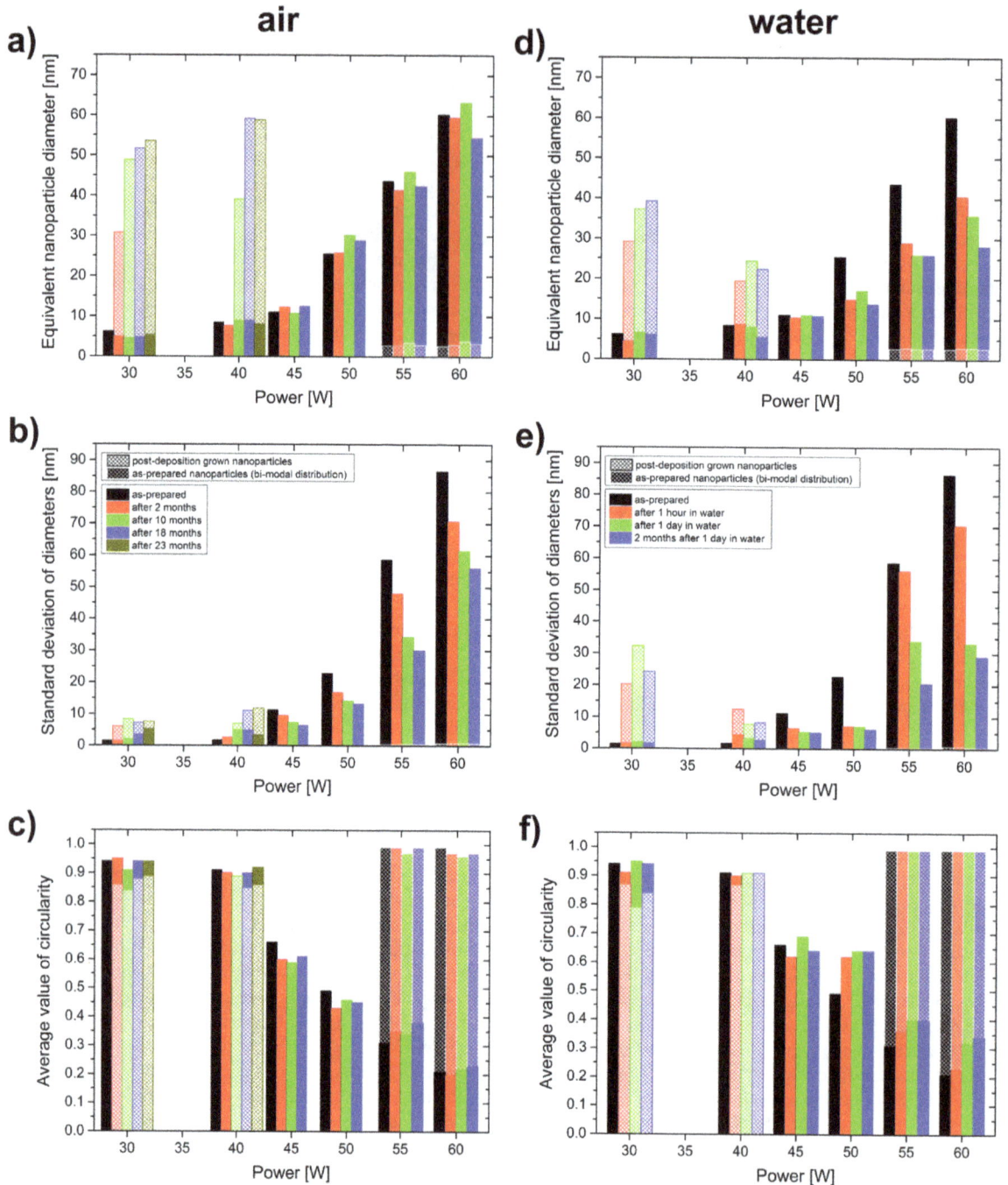

Figure 5. Time evolution of the microstructural parameters of Ag/a-C:H:O nanocomposite films (modal value of equivalent nanoparticle diameter d_m, (a) and (d), standard deviation of diameters distribution σ, (b) and (e), and average value of shape factor of nanoparticles S_a, (c) and (f)) deposited at different RF powers during their aging in ambient air (a)–(c) and in distilled water (d)–(f).

diameter, the nanoparticle size distributions become narrower and the average values of shape factors increase. The small nanoparticles at the substrate interface are again left intact. Further, only slight changes were observed in the structure of the film deposited at 45 W. For these four films, not only the

silver release is higher for the film deposited at the higher power, but also the silver release continues for a longer time. While in the case of the film deposited at 60 W, the degradation continues also during the first day in water and clearly proceeds even during its further aging in air for 2 months, in the case of the film deposited at 50 W, the degradation is very limited after the first hour in water. This observation seems to be correlation to the previous study of Körner et al [15], where a clear dependence of the Ag^+ release on the initial silver content was shown: an increase in the Ag content in the coating results in an increasing Ag^+ release in distilled water. From the TEM characterization, it is difficult to give any similar comparison of total silver released also to the two films deposited at 30 and 40 W, because at least part of the released ions contribute back to the growth of the crystals in the films. When the films were taken out from the water bath after the one day, their microstructure did not change much and it seems that all the processes that could have proceeded, already took place in the water bath. No similar changes, as those observed on the previous series of films within their first 2 months of aging in air, were found. The only difference is the case of the film deposited at the highest power of 60 W, at which we can still track the continuing disruption of its percolated structure.

To summarize this section, we can say that two aging processes were observed during aging of the Ag/a-C:H:O nanocomposites in air or in water: nanoparticle growth and disruption of the percolated structure. Overall, these are much enhanced by the direct presence of water when compared to air (time order of hours versus months). The a-C:H:O plasma polymer matrix contains various oxygen functional groups [52], and thus allows good water penetration. The water contact angle was previously measured to be between 37° and 42° depending on the silver content [40]. It is obvious that d_m of structures in the films with higher filling factor decreases much faster than in those with lower filling factors. This supports the previous measurements that the total silver release from the films with higher filling factors is much higher although visually the structures of these films look less disrupted when compared to the films with lower factors in which mostly only the large grown crystals are left in the end [15]. One of the reasons of the low measured Ag^+ release might be the contribution of the leached silver ions to the growth of the crystals. However, it is not clear why the crystals do not grow also in the case of films with higher filling factors. Out of all the studied Ag/a-C:H:O nanocomposite films, the one deposited at the moderate power of 45 W seems to be the most stable. No changes were observed in the structure of the film during its aging in ambient air or in distilled water. This suggests that there is a good balance of concentration of Ag atoms/ions around the silver nanostructures in the films at the percolation threshold.

3.2.5. Crystal structure. The deposited Ag/a-C:H:O nanocomposite films were studied by the SAD method to reveal the nature of the crystal structure of the metallic inclusions by analysis of the electron diffraction patterns. Typical diffractograms of selected films with corresponding Miller indices are displayed in figure 6. The diffraction patterns are composed of both rings and spots. While the diffraction rings are attributed to small nanoparticles, the bright spots represent larger crystals. The SAD analysis

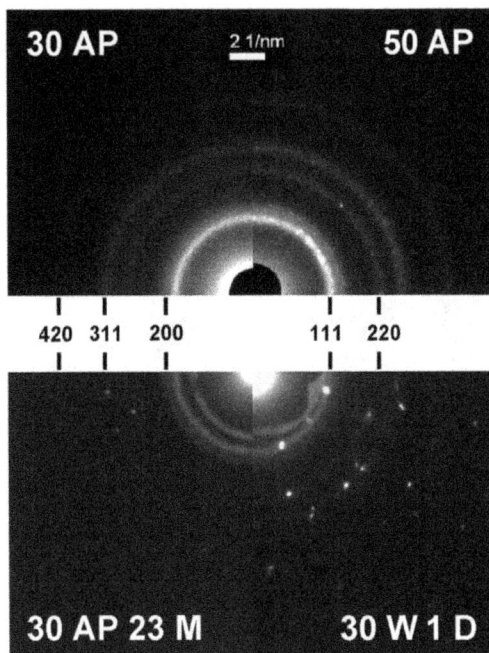

Figure 6. Electron diffraction patterns of Ag/a-C:H:O nanocomposite films deposited at the RF power of 30 W (top left) and 50 W (top right) as measured right after the deposition and Ag/a-C:H:O nanocomposite film deposited at RF power of 30 W, as measured after 23 months of aging in ambient air (bottom left) and after 1 day of aging in distilled water (bottom right). Each pattern is displayed with corresponding Miller indices.

corresponds to the conclusion from the analysis of the bright field: the size of the silver crystalline inclusions in the nanocomposites increases with the deposition power. From the analysis of the diffractograms, it can be further concluded that all the inclusions in all of the studied films are of pure Ag with the face centered cubic structure with the lattice parameter $a = 4.086$ Å, which corresponds well to the data obtained from ICDD [45]. However, the SAD analysis does not principally exclude the possibility of presence of thin oxidized layers on surfaces of the Ag nanoparticles. The growth of the nanocomposite film takes place under non-equilibrium conditions, where the deposited energy enables surface mobility and bond opening/formation. Thus, metallic Ag nanoparticles are formed within a plasma polymer matrix. In equilibrium, however, Ag metal nanoparticles get rapidly oxidized at the surface [53]. Furthermore, Ag was found to form oxygen bonds on polymers without any plasma activation and clearly more of them after an O_2 plasma treatment [54]. From this, we can expect that the metallic Ag nanoparticles have oxidized surfaces which partly form bonds to the hydrocarbon-based plasma polymer matrix (Ag–O–C bonds).

It is interesting to note that the crystal structure of the silver nanoparticles does not change during the aging of the Ag/a-C:H:O nanocomposite films in air or in water. No signs of oxidation of the silver nanoparticles were observed. This

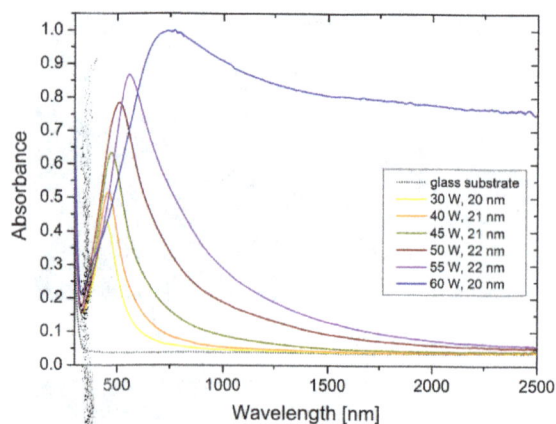

Figure 7. UV–vis–NIR absorbance spectra of Ag/a-C:H:O nano-composite films deposited at different RF powers as measured right after their deposition. The optical spectrum of the glass substrate is displayed for comparison. Each of the films is listed with its corresponding thickness.

degradation process is typically reported for the aging of Ag nanoparticles in water [55]. On the other hand, the size and the amount of the silver particles in the Ag/a-C:H:O nanocomposite film deposited at 30 W increases during its aging, while the small intrinsic nanoparticles almost disappear. Further, a detailed analysis of the aging films revealed that the large grown Ag particles can contain several crystallographic domains.

3.3. Optical properties of Ag/a-C:H:O films

3.3.1. As-prepared films.
The optical properties of the as-deposited Ag/a-C:H:O nanocomposites were characterized within the UV–vis–NIR spectral ranges on samples prepared in the same deposition process as the corresponding films for the characterization of the microstructure by TEM. The absorbance spectra of the nanocomposite films deposited at different RF powers as measured right after their deposition are displayed in figure 7. As can be seen, the spectra of the Ag/a-C:H:O films deposited at powers lower than 60 W reveal a strong anomalous absorption roughly around 500 nm with a long tail of gradually decreasing absorbance towards the NIR spectral range. This absorbance peak is typical for composite nanomaterials containing metallic nanoparticles. It appears due to the well-known phenomenon of surface plasmon resonance (SPR) as a result of collective oscillations of conduction electrons after an interaction between the nanoparticle and the electromagnetic field of the light [56]. As can be seen in the spectra presented in figure 7, three characteristics of the absorbance peaks change with the different powers during the deposition: the intensity of the absorbance, the position of the absorbance maximum and also the full width at half maximum (FWHM) of the absorbance peak. Particularly, the intensity of the absorbance increases from 0.43 for the film deposited at 30 W up to 0.87 for the film deposited at 55 W, the position of the SPR gradually

shifts to the longer wavelengths, from 447 nm for the film deposited at 30 W up to 557 nm for the film deposited at 55 W, and finally the FWHM increases from 133 nm for the narrowest absorption peak of the film deposited at 30 W up to 353 nm for the film deposited at 55 W. These three absorbance peak parameters are closely connected to the microstructure of the coatings as will be discussed in the following section in detail. In short, the increasing intensity of the absorption is caused mainly by the increased size of nanoparticles, while the red-shift of the peak originates in the decreasing shape factor [1, 56]. Also, the absorption peaks are much broader with the increasing deposition power due to the increasing inhomogeneity of the sizes of the silver nanoparticles (broader distributions) which gives rise to different plasmon modes [57]. Further, it can be seen that the spectrum of the nanocomposite film deposited at the power of 60 W differs from the others. The exact position of the SPR peak is difficult to locate due to the broadness of the peak and the film has rather high absorbance of about 0.8–0.9 within the whole NIR spectral range. This phenomenon can be ascribed to the absorption of light by the free electrons (inter- and intraband electron excitations) due to the percolated structure of this nanocomposite. The absorbance spectrum resembles that of a continuous silver thin film rather than the spectra of discontinuous nanocomposites with separated nanoparticles or with island-like structures revealing SPR.

3.3.2. Aging of Ag/a-C:H:O films.
The aging of the Ag/a-C:H:O nanocomposite films in air or in water induces changes in their optical spectra as can be seen in figure 8. The dominant feature in the spectra of the aging films is the decrease in the intensity of the absorbance peak which is especially striking in the case of aging in water. This decrease is caused by the loss of silver material. However, an interesting effect can be observed in the NIR part of the spectrum of the film deposited at 50 W. While the values of absorbance of the films both with low (30–45 W) and high (60 W) filling factors decrease in this region during aging, the absorbance of the film deposited at 50 W increases. The exact cause of this effect is not known, but it might be connected with the special percolated structure of this film and the leaching of Ag^+ ions as was discussed above. It can be expected that not all of the ions released from the silver nanoparticles also leave the matrix. They might remain (temporarily) bound within the plasma polymer and affect the permittivity of the matrix or virtually increase the percolation of the film. Since these are only distributed ions they could not be detected in the TEM characterization. Further, the positions of the SPR slightly red-shift during aging, even though, the shape factors of the films with higher filling factors increase. The maximum in the spectrum of the 60 W film aged in water cannot be identified. Finally, the FWHM measures of the SPR peaks increase during aging. Overall, the trends of the changes are the same during aging in air or in water. However, it can be seen that their progress is much faster and more pronounced during their aging in water. This

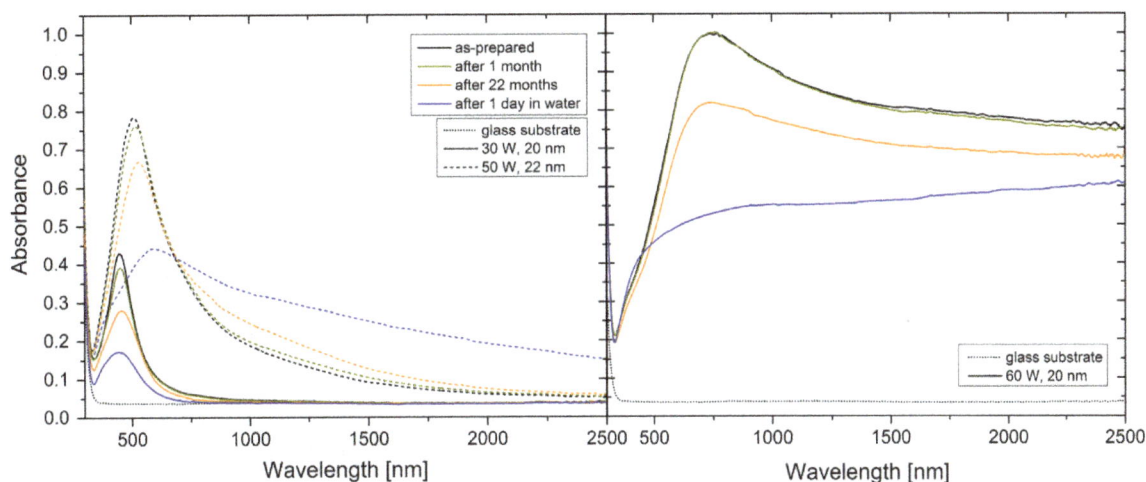

Figure 8. UV–vis–NIR absorbance spectra of the Ag/a-C:H:O nanocomposite films deposited at the RF powers of 30 and 50 W (left) and 60 W (right). The spectra were measured right after the deposition, and after aging in air for 1 month and 22 months and in water for 1 day. The optical spectrum of the glass substrate is displayed for comparison.

observation corresponds to the observed changes in the microstructure of the nanocomposites.

The systematic changes in the three SPR peak parameters can be observed in the following pictures. Particularly, the intensities of the absorbance are summarized in figure 9(a), their positions in figure 9(b) and the FWHM in figure 9(c). As for the absorbance intensity, the values were collected from two points of the spectra: at the position of the SPR maximum in the visible part of the spectrum and at the wavelength of 1000 nm. As was already reported above, the intensity of the SPR of the as-prepared films increases with the deposition power and decreases during aging. The overall decrease of the intensity is faster in water than in air and this rate increases with the higher filling factors of the films: for 30 and 40 W films, the absorbance intensity after aging for 1 h in water is at the level of about 5 months of aging in air, for the 45 W film, the aging for 1 h in water corresponds to about 9 months of aging in air, while the intensity after aging for 1 h in water is lower than after aging in air for about 22 months for the films deposited at 50 and 55 W. The absorbance maximum could not be localized in the case of the film deposited at 60 W because the peak is very broad and lies partly in the region where the noise level is higher due to the change of spectrometer detectors. Nevertheless, the overall decrease of the intensity of the absorbance peak is observed in all of the films. However, the situation is different when the changes in the NIR region are taken into account. The absorbance of the as-prepared films increases with the deposition power also in the NIR spectral range. Particularly, the intensity at the wavelength of 1000 nm increases from almost 0 in the case of films with low filling factors up to about 0.9 in the case of the film deposited at 60 W. The aging of the films however shows different effects. The level of absorption almost does not change (or very slightly decreases) in the case of films deposited up to the power of 45 W, while it increases in the case of the films deposited at 50 and 55 W. The intensity of

absorption of the film with high filling factor decreases during aging in air or in water, while the decrease is much stronger in the latter case.

A gradual shift in the position of the SPR to higher wavelengths can be seen in the spectra of the as-prepared composites upon increasing the deposition power, as discussed above (figure 9(b)). Similarly, a gradual red-shift in the position of the SPR develops also during the aging of the films. However, the changes in the position of the peak induced by aging are not that strong as was previously observed in the case of the intensity of the peak. Nevertheless, the rates of the change are higher for the films deposited at higher powers for aging in air or water which corresponds to the changes observed in the case of the absorbance intensity. There are several microstructural parameters that affect the position of the SPR and will be further discussed in the next section.

The development similar to the position SPR can be observed also in FWHM of the absorbance peak (figure 9(c)). The values of FWHM of the absorbance peak in the spectrum of the as-prepared nanocomposite films increase with the increasing deposition power. The aging of the films in air or in water causes the broadening of the absorbance peak. Especially broad are the absorbance peaks in the cases of the films deposited at 50 and 55 W after their aging for 1 day in water. They resemble the spectrum of the film deposited at 60 W with well-percolated microstructure. It is interesting to notice that on the other hand, the changes in the width of the SPR peak of the 55 W film are only limited during aging in air.

To briefly summarize this section, we can say that the optical properties of the Ag/a-C:H:O nanocomposite films are modified during the aging of the coatings. Both media trigger qualitatively the same aging changes, particularly the decrease in the intensity of SPR, red-shift in the position of SPR and its broadening. However, similar to the case of microstructure, these changes are enhanced by the direct presence of water when compared to aging in air only.

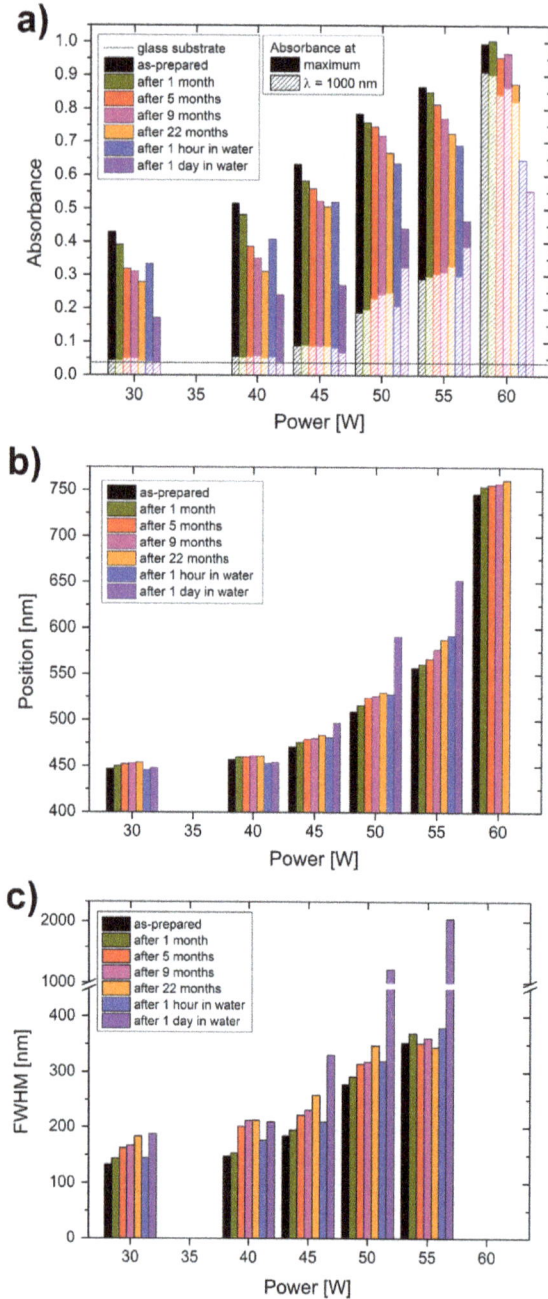

Figure 9. Changes in the parameters of the surface plasmon resonance of Ag/a-C:H:O nanocomposite films deposited at different RF powers during aging of the films in ambient air and in distilled water: (a) intensity of absorbance at the wavelength of the SPR and at the wavelength of 1000 nm, (b) position of the absorbance maximum, (c) width of the SPR peak (FWHM).

3.4. Correlation between microstructure and optical properties of Ag/a-C:H:O films

It is well known that the optical properties of composites containing metal nanoparticles are closely connected to the

characteristics of the respective material component and also microstructure of the composite [1, 56]. All of the microstructural (nanoparticle size, d_m, size dispersion, σ, average shape factor, S_a, permittivity of the matrix, ε_m, and interaction between nanoparticles, NPI) as well as the optical (intensity, position, and FWHM of SPR) parameters are highly dependent on the deposition power, as could be seen in the discussion above or as reported elsewhere. This influence is summarized in table 1. As can be seen, all of them increase with the increasing power as more metal is incorporated into the plasma polymer matrix: more and larger particles (d_m) with broader size distributions (σ) are formed which also increases the interaction between nanoparticles (NPI) due to decrease in the average inter-particle distances and thus enhancement of their dipole moments [58]. Permittivity of plasma polymer films (ε_m) can be expected to increase with the increasing deposition power as was previously shown in other studies [17, 59]. The only exception is the average shape factor S_a which decreases instead due to the formation of island-like structures of highly irregular shapes at higher powers. On the other hand, the observed influence of aging on the film characteristics is more complex. The aging effect on some of the microstructural parameters, particularly d_m, σ, and S_a, and also on the position of SPR, is less pronounced in the case of the films with lower filling factors (can be regarded as constant) while it is more obvious for the films with higher filling factors (see figures 5 and 9). The border between the 'low' and the 'high' filling factor is not exactly drawn and differs from parameter to parameter. It is used here to stress that films with different filling factors might react to the aging with a different rate. Nevertheless, only the actual change in the respective parameter is recorded in table 1 for a better visualization.

The effects of aging on the microstructure were found to be exactly opposite. All parameters decrease with the exception of S_a: the loss of material through leaching of Ag^+ ions results in decreasing of particle size, narrowing of size distribution and also decreasing the inter-particle interaction, while S_a increases as the island-like structure becomes separated into individual nanoparticles with more circular-like projections. The effect of aging on ε_m is not clear and requires an individual research to clarify the consequences. Nevertheless, it has been shown that ε_m of several plasma polymers increases in time due to water vapor penetrating the plasma polymer network [60].

The influence of the respective film microstructural parameters on the SPR is to a great extent interconnected, as can be seen in table 2. The nanoparticle size influences primarily the intensity of the absorbance maximum: the intensity increases with the increasing d_m. On the other hand, increasing d_m causes a decrease in FWHM of the SPR peak which was found opposite when analyzing the spectra of the aging nanocomposites. Nevertheless, the dominant influence on FWHM has the size dispersion, σ [56], which correlates very well with the analysis of the spectra. The increasing shape factor S_a affects only position of SPR by shifting it to lower wavelengths. This was also observed on the spectra of the as-prepared films, however not in the case of aging of the films in air or in water. Nevertheless, the position of SPR is affected also by ε_m and NPI which correlate very well with

Table 1. Observed influence of an increasing deposition power and aging on the microstructural and SPR parameters of the Ag/a-C:H:O nanocomposite films.

	Effect on microstructural parameter					Effect on SPR parameter		
	d_m	σ	S_a	ε_m	NPI	Intensity	Position	FWHM
Increasing power	↑	↑	↓	↑?	↑	↑	↑	↑
Aging	↓	↓	↑	?	↓	↓	↑	↑

Table 2. Theoretical influence of increasing values of the respective microstructural parameters on the characteristics of SPR.

	Effect on SPR parameter		
Increase in microstructural parameter	Intensity	Position	FWHM
d_m	↑		↓
σ			↑
S_a		↓	
ε_m	↑	↑	
NPI		↑	↑

the experiment. The permittivity of the matrix influences also the intensity of the absorption which is in opposite relation in the case of aging of the films (table 1). Nevertheless, the dominant effect on the intensity originates from d_m. As can be seen, the optical properties (particularly the properties of SPR) of the as-prepared Ag/a-C:H:O nanocomposite films correlate very well with their microstructural parameters in the dependence on the deposition power. However, the influence of aging is a rather more complicated and a more precise comparison requires a modeling approach.

4. Conclusion

Nanocomposite coatings of silver nanoparticles embedded in a plasma polymer matrix with different contents of metal can be deposited by simultaneous sputtering from a silver target and plasma polymerization of ethylene and carbon dioxide. The RF power supplied during the deposition has a major effect on the microstructure of the nanocomposite coatings. As shown by TEM, the size of the as-prepared nanoparticles, expressed via their equivalent diameter, increases from about 6 nm to about 60 nm when increasing the power from 30 to 60 W, i.e. small nanoparticles separated by the plasma polymer matrix grow larger and form a percolating island-like structure. In connection, also other microstructural parameters change. Particularly, the nanoparticle size dispersion, expressed in terms of standard deviation of the size distributions, broadens from about 2 nm for the films deposited at 30 W up to about 87 nm in the case of the films deposited at 60 W. On the other hand, the values of the shape factor of the projections of the nanoparticles decrease as particles with various sizes and shapes are deposited at higher powers. The optical properties of the nanocomposite films are closely connected to their microstructure. An anomalous absorption peak of the Ag nanoparticles is observed around 500 nm which is a typical feature of materials containing metal nanoparticles due to the effect of SPR. Similarly to microstructural parameters, also characteristics of the SPR peak depend on the deposition power. Particularly, the intensity of absorbance, the position of SPR, and the FWHM of the peak all increase with the increasing deposition power. It was observed that the as-prepared films are not stable, but rather undergo changes in their structure and optical properties when left for aging in ambient air or in distilled water. The aging processes in both environments are qualitatively the same. However, the processes are much faster and intense in water, where similar changes were induced on a time scale of hours as compared to months when aging in air. The aging of the nanocomposites is driven by water molecules that penetrate the coatings and induce changes in the structure and properties of the plasma polymer matrix as well as cause leaching of Ag⁺ ions from the surface of silver nanoparticles. The deposition conditions and the intrinsic microstructure of the as-prepared films affect the processes taking place during their aging. Three distinct aging regimes were observed. The formation of large silver crystals by Ostwald ripening was observed in the films with lower filling factors (below percolation threshold) deposited at lower powers. Such films with tuned microstructure with respect to Ag⁺ ion release might be suitable for application in biomedicine as antibacterial coatings. However, the growth of the crystallites needs to be taken into account and attention should be paid to the storage of the material prior to its use. Further, the films deposited at moderate power input possess relatively stable microstructure at the percolation threshold with a good balance of the size of the nanoparticles, the size distribution and the interparticle separation which makes them suitable for applications as various sensors. Finally, a degradation of the percolating island-like structure was observed in the films with high filling factors (above percolation threshold) deposited at higher powers. Nevertheless, these films could still be of interest for catalytic applications. In connection, the optical properties of the nanocomposite films are altered during their aging. The dominant feature in the optical spectra of the aging films is the decrease in their absorbance which is especially striking in the case of aging in water. This decrease is a direct consequence of the loss of silver material due to leaching of Ag⁺ ions.

The above discussed extremely thin films can be deposited within a short time and show some other advantages, e.g. they do not suffer from internal stresses and do not influence the mechanical properties of the substrate material. Nevertheless, it might be important to consider the requirements on the thickness of the films for any of their particular industrial applications.

Acknowledgments

Financial support from the Swiss National Science Foundation (SNSF) within the Tec-in-Tex project of the RTD Nanotera.ch programme and the 10.118—ExTraSens project of the Sciex-NMS[ch] programme is acknowledged.

References

[1] Heilmann A 2003 *Polymer Films with Embedded Metal Nanoparticles* (Berlin: Springer)
[2] d'Agostino R 1990 *Plasma Deposition, Treatment and Etching of Polymers* (New York: Academic)
[3] Biederman H, Kudrna P and Slavinska D 2004 *Plasma Polymer Films* ed H Biederman (London: Imperial College Press) chapter 9
[4] Faupel F, Zaporojtchenko V, Strunskus T and Elbahri M 2010 *Adv. Eng. Mater.* **12** 1177
[5] Biederman H, Kylian O, Drabik M, Choukourov A, Polonskyi O and Solar P 2012 *Surf. Coat. Technol.* **211** 127
[6] Hanisch C, Ni N, Kulkarni A, Zaporojtchenko V, Strunkus T and Faupel F 2011 *J. Mater. Sci.* **46** 438
[7] Drabik M, Vogel-Schäuble N, Heuberger M, Hegemann D and Biederman H 2013 *Nanomater. Nanotechnol.* **3** 13
[8] Morones-Ramirez J R, Winkler J A, Spina C S and Collins J J 2013 *Sci. Transl. Med.* **5** 190ra81
[9] Lischer S, Körner E, Balazs D J, Shen D, Wick P, Grieder K, Haas D, Heuberger M and Hegemann D 2011 *J. R. Soc. Interface* **8** 1019
[10] Zanna S, Saulou C, Mercier-Bonin M, Despax B, Raynaud P, Seyeux A and Marcus P 2010 *Appl. Surf. Sci.* **256** 6499
[11] Chaloupka K, Malam Y and Seifalian A M 2010 *Trends Biotechnol.* **28** 580
[12] Lubick N 2008 *Environ. Sci. Technol.* **42** 8617
[13] Nowack B, Krug H F and Height M 2011 *Environ. Sci. Technol.* **45** 1177
[14] Nowack B 2010 *Science* **330** 1054
[15] Körner E, Aguirre M H, Fortunato G, Ritter A, Rühe J and Hegemann D 2010 *Plasma Process. Polym.* **7** 619
[16] Drabik M, Kousal J, Celma C, Rupper P, Biederman H and Hegemann D 2014 *Plasma Process. Polym.* **11** 496
[17] Inagaki N 1996 *Plasma Surface Modification and Plasma Polymerization* (Lancaster: Technomic Publishing Company)
[18] Hollaender A and Thome J 2004 *Plasma Polymer Films* ed H Biederman (London: Imperial College Press) chapter 7
[19] Hlidek P, Biederman H, Choukourov A and Slavinska D 2008 *Plasma Process. Polym.* **5** 807
[20] Hlidek P, Biederman H, Choukourov A and Slavinska D 2009 *Plasma Process. Polym.* **6** 34
[21] Faupel F, Willecke R and Thran A 1998 *Mater. Sci. Eng.* **R22** 1
[22] Damm C and Muenstedt H 2008 *Appl. Phys.* A **91** 479
[23] Schmidt M, Masson A and Brechignac C 2003 *Phys. Rev. Lett.* **91** 243401
[24] Hoskins J S, Karanfil T and Serkiz S M 2002 *Environ. Sci. Technol.* **36** 784
[25] Liu J and Hurt R H 2010 *Environ. Sci. Technol.* **44** 2169
[26] Blanchard N E, Hanselmann B, Drosten J, Heuberger M and Hegemann D 2015 *Plasma Process. Polym.* **12** 32
[27] Lee J-E, Park J-C, Lee K H, Oh S H, Kim J-G and Suh H 2002 *Artif. Organs* **26** 636
[28] Damm C, Münstedt H and Rösch A 2007 *J. Mater. Sci.* **42** 6067
[29] Kumar R and Münstedt H 2005 *Polym. Int.* **54** 1180
[30] Biederman H, Hlidek P, Pesicka J, Slavinska D and Stundzia V 1996 *Vacuum* **47** 1385
[31] Biederman H, Chmel Z, Fejfar A, Misina M and Pesicka J 1990 *Vacuum* **40** 377
[32] Takele H, Kulkarni A, Jebril S, Chakravadhanula V S K, Hanisch C, Strunskus T, Zaporojtchenko V and Faupel F 2008 *J. Phys. D: Appl. Phys.* **41** 125409
[33] Alissawi N *et al* 2012 *J. Nanopart. Res.* **14** 928
[34] Wang X, Somsen C and Grundmeier G 2008 *Acta Mater.* **56** 762
[35] Alissawi N, Peter T, Strunskus T, Ebbert C, Grundmeier G and Faupel F 2013 *J. Nanopart. Res.* **15** 2080
[36] Kay E 1986 *Z. Phys.* D **3** 251
[37] Guimond S, Hanselmann B, Hossain M, Salimova V and Hegemann D 2014 *Plasma Process. Polym.* at press (doi:10.1002/ppap.201400164)
[38] Drabik M, Celma C, Kousal J, Biederman H and Hegemann D 2014 *Thin Solid Films* **573** 27
[39] Körner E, Rupper P, Lübben J F, Ritter A, Rühe J and Hegemann D 2011 *Surf. Coat. Technol.* **205** 2978
[40] Körner E, Fortunato G and Hegemann D 2009 *Plasma Process. Polym.* **6** 119
[41] National Institute of Standards and Technology (NIST) *Atomic Spectra Database* (http://physics.nist.gov/PhysRefData/ASD/lines_form.html) (accessed September 2012)
[42] Pearse R W B and Gaydon A G 1965 *The Identification of Molecular Spectra* (London: Chapman and Hall)
[43] Stundzia V 1997 *Doctoral Thesis* Charles University in Prague, Czech Republic
[44] Rasband W 2009 *National Institute of Health, USA, ImageJ Project, Image Processing Program* (http://rsb.info.nih.gov/ij/)
[45] The International Centre for Diffraction Data *Powder Diffraction File* (www.icdd.com/) (accessed December 2012)
[46] Hanus J, Drabik M, Hlidek P, Biederman H, Radnoczi G and Slavinska D 2009 *Vacuum* **83** 454
[47] Depla D and De Gryse R 2004 *Surf. Coat. Technol.* **183** 190
[48] Bock F X, Christensen T M, Rivers S B, Doucette L D and Lad R J 2004 *Thin Solid Films* **468** 57
[49] Zaporojtchenko V, Strunskus T, Behnke K, Von Bechtolsheim C, Kiene M and Faupel F 2000 *J. Adhes. Sci. Technol.* **14** 467
[50] Chakravadhanula V S K, Kuebel C, Hrkac T, Zaporojtchenko V, Strunskus T, Faupel F and Kienle L 2012 *Nanotechnology* **23** 495701
[51] Hegemann D, Hanselmann B, Blanchard N and Amberg M 2014 *Contrib. Plasma Phys.* **54** 162
[52] Hegemann D, Körner E, Albrecht K, Schutz U and Guimond S 2010 *Plasma Process. Polym.* **7** 889
[53] Henglein A 1998 *Chem. Mater.* **10** 444
[54] Grace J M and Gerenser L J 2003 *J. Dispersion Sci. Technol.* **24** 305
[55] Glover R D, Miller J M and Hutchinson J E 2011 *ACS Nano* **5** 8950
[56] Garcia M A 2011 *J. Phys. D: Appl. Phys.* **44** 283001
[57] Genov D A, Sarychev A K and Shalaev V M 2003 *J. Nonlinear Opt. Phys. Mater.* **12** 419
[58] Stalmashonak A, Seifert G and Abdolvand A 2013 *Ultra-Short Pulsed Laser Engineered Metal-Glass Nanocomposites* (Heidelberg: Springer) chapter 2
[59] Lee S, Woo J, Nam E, Jung D, Yang J, Chae H and Kim H 2009 *Jpn. J. Appl. Phys.* **48** 106001
[60] Sawa G, Ito O, Morita S and Ieda M 1974 *J. Polym. Sci. Polym. Phys. Ed.* **12** 1231

20

Low temperature and cost-effective growth of vertically aligned carbon nanofibers using spin-coated polymer-stabilized palladium nanocatalysts

Amin M Saleem[1,2], **Sareh Shafiee**[1], **Theodora Krasia-Christoforou**[3,4], **Ioanna Savva**[3], **Gert Göransson**[5], **Vincent Desmaris**[1] **and Peter Enoksson**[2]

[1] Smoltek AB, Regnbågsgatan 3, Gothenburg, SE-41755, Sweden
[2] Micro and Nanosystems group, BNSL, Department of Microtechnology and Nanoscience, Chalmers University of Technology, Gothenburg, SE-41296, Sweden
[3] Department of Mechanical and Manufacturing Engineering, University of Cyprus, Nicosia, Cyprus
[4] Nanotechnology Research Center (NRC), University of Cyprus, Nicosia, Cyprus
[5] Department of Chemistry and Molecular Biology, University of Gothenburg, Gothenburg, SE-41296, Sweden

E-mail: amin@smoltek.com

Abstract

We describe a fast and cost-effective process for the growth of carbon nanofibers (CNFs) at a temperature compatible with complementary metal oxide semiconductor technology, using highly stable polymer–Pd nanohybrid colloidal solutions of palladium catalyst nanoparticles (NPs). Two polymer–Pd nanohybrids, namely poly(lauryl methacrylate)-block-poly((2-acetoacetoxy)ethyl methacrylate)/Pd ($LauMA_x$-b-$AEMA_y$/Pd) and polyvinylpyrrolidone/Pd were prepared in organic solvents and spin-coated onto silicon substrates. Subsequently, vertically aligned CNFs were grown on these NPs by plasma enhanced chemical vapor deposition at different temperatures. The electrical properties of the grown CNFs were evaluated using an electrochemical method, commonly used for the characterization of supercapacitors. The results show that the polymer–Pd nanohybrid solutions offer the optimum size range of palladium catalyst NPs enabling the growth of CNFs at temperatures as low as 350 °C. Furthermore, the CNFs grown at such a low temperature are vertically aligned similar to the CNFs grown at 550 °C. Finally the capacitive behavior of these CNFs was similar to that of the CNFs grown at high temperature assuring the same electrical properties thus enabling their usage in different applications such as on-chip capacitors, interconnects, thermal heat sink and energy storage solutions.

Keywords: polymer-stabilized nanoparticles, carbon nanofibers, low temperature growth, cost effective

1. Introduction

Carbon nanotubes (CNTs) were discovered by Sumio Iijima in 1991 [1] and since then, they have received high attention due to their extraordinary mechanical, electrical and thermal properties. Besides that, CNTs have very low density [2] and very high aspect ratio. CNTs consist of graphene sheets coaxially rolled into hollow cylinders, and can be single-walled or multi-walled depending on the number of graphene sheets. Moreover, CNTs can be metallic or semiconducting depending on the twist of the tubes.

In contrast to CNTs, carbon nanofibers (CNFs) comprise cone-shaped graphene layers stacked on top of each other creating completely filled cylinders, which also have excellent mechanical, thermal and electrical properties. At 400 and 7 GPa, the Young modulus and tensile strength of CNFs are higher than those of steel [3]. The thermal conductivity of CNFs (about $1900 \, W \, mK^{-1}$) is five times higher than copper [4] and can carry current density of about $12 \, MA \, cm^{-2}$ which is three times higher than copper [5, 6]. CNFs exhibit many significant advantages over CNTs, since they can be grown at lower temperatures and they are 100% metallic, whereas 2/3 of CNTs are semiconducting when grown in bulk [7]. Because of these excellent properties CNFs have potential applications in electrical interconnects [8], as thermal interface materials, as reinforcement additives in polymer composites and as electrodes for supercapacitors.

The carbon nanostructures, including CNFs, can be synthesized in several ways including electrospinning, chemical vapor deposition (CVD), laser ablation and arc discharge. There are different types of CVD and these are classified on the basis of energy type used during deposition: microwave plasma enhanced CVD, thermal CVD and direct current plasma enhanced CVD (DC-PECVD) [9–11]. Depending on the catalyst and the CVD growth method used, different types of carbon nanostructures are obtained. The DC-PECVD method has an advantage over the other growth methods in that it provides high purity and vertical alignment of the produced CNFs [12].

To grow CNFs using CVD, the deposition of catalytic particles on a substrate is required. Transition metals such as iron, cobalt, palladium and nickel are often used as catalysts [11, 13–15]. The catalysts can be deposited in different ways onto the substrate such as deposition of a pure metal film using sputtering or e-beam evaporation techniques and deposition of a nanoparticle (NP)-containing solution via spin coating, dip-coating or spraying methods [16–18]. The physical vapor deposition (PVD) deposition methods, such as sputtering or evaporation, require expensive equipment, a metal target, clean room laboratory facilities and costly equipment maintenance.

In the present study, catalytic palladium NPs are deposited on silicon substrate via spin-coating using highly stable and inexpensive polymer–Pd colloidal solutions. CNFs are then grown on these catalyst particles at complementary metal oxide semiconductor (CMOS) process-compatible temperature thus allowing CNFs to be used in on-chip thermal, electrical and energy storage solutions.

Two types of polymer–Pd NPs solutions were used to deposit the palladium catalyst particles: (1) poly(vinylpyrrolidone)-Pd (PVP–Pd) NPs stabilized in methanol solution and (2) poly(lauryl methacrylate)-block-poly(2-(acetoacetoxy)ethyl methacrylate) (LauMA$_x$-b-AEMA$_y$/Pd) micellar nanohybrids stabilized in n-hexane. Different types of polymers are used due to their different thermal stabilization temperatures. The polymers are stable up to certain temperature and burn off when heated to higher temperature thus leaving Pd NPs on the surface of the substrate. The stabilization temperature of LauMA$_x$-b-AEMA$_y$ and PVP polymers is around 200 °C and 400 °C respectively [19]. The synthesis and characterization of the aforementioned polymer-stabilized Pd NPs has been reported in other studies [20–23]; however, in this work these polymer-metal nanohybrids have been employed for the growth of CNFs. The use of such catalyst-containing solutions is advantageous since those are nontoxic, cheap and can be easily prepared using low-cost wet-chemistry procedures. In addition, their deposition onto selected substrates does not require the use of expensive equipment. Moreover, the polymers can be easily washed away and the substrate can be reused in case of process failure during NPs deposition. This is a rapid and industry compatible deposition method that can be used to grow a film of CNFs at the backside of a chip for heat sink applications.

Three different PVP–Pd nanohybrid solutions were prepared in methanol, in which the polymer content was kept constant and only the palladium content varied. In the case of the LauMA$_x$-b-AEMA$_y$/Pd micellar systems stabilized in n-hexane, two different block copolymers (polymer A: $x = 120$, $y = 67$; polymer B: $x = 50$, $y = 9$) are employed as the palladium NPs stabilizing agents. The CNFs were grown using the DC-PECVD technique at different temperatures i.e. at a high temperature (550 °C), at a CMOS compatible temperature (390 °C) and even at 40 °C lower (350 °C). The specific capacitance and cyclability of the CNFs were measured using cyclic voltammetry to demonstrate the electrical properties and applicability of these nanostructures as supercapacitor electrodes directly built on CMOS chips.

2. Fabrication, characterization and measurements

The synthetic methodologies followed for the preparation of the PVP–Pd and the LauMA$_x$-b-AEMA$_y$-Pd hybrid solutions are described below.

(1) A typical synthetic methodology followed for the synthesis of the PVP–Pd nanohybrid colloidal systems (moles vinyl pyridine units/moles palladium salt = 9:1) is described as follows [19, 23]: in a round bottom flask equipped with a magnetic stirrer, PVP (1 g, 9×10^{-3} moles of vinylpyrridine units) (Sigma-Aldrich, MW = 1300 000) was dissolved in methanol (10 mL). Subsequently, palladium acetate (0.22 g, 1×10^{-3} moles) (Pd(OAc)$_2$, Sigma-Aldrich, 98%) was added to the polymer solution and the reaction flask was placed under reflux for 2 h. During this period, the color of the solution changed from yellow to dark brown indicating the formation of palladium NPs. After the completion of the reaction, the brown-colored solution was left to cool down to room temperature and it was then stored in sealed glass vials. The solutions were highly stable and

no precipitation was observed even after several months.

The above-mentioned methodology was employed for synthesizing two more PVP–Pd nanohybrid colloidal systems stabilized in methanol, in which the polymer content was kept the same and only the metallic (Pd) content varied i.e. the molar ratio of the vinyl pyridine units to the palladium salt was 18:1 and 38:1.

(2) The Pd-containing LauMA$_x$-b-AEMA$_y$ micellar hybrids have been prepared in n-hexane by following a previously reported methodology [20, 22]. Exemplarily, the procedure followed for the synthesis of the LauMA$_{120}$-b-AEMA$_{67}$/Pd micellar nanohybrids is described as follows: LauMA$_{120}$-b-AEMA$_{67}$ (20 mg, 0.03 mmoles of AEMA units) was dissolved in n-hexane (5 mL). After complete dissolution of the polymer, triethylamine (0.021 mL, 0.15 mmoles) was added. Subsequently, the polymer solution was transferred into a glass vial containing Pd(OAc)$_2$ (3.40 mg, 0.015 mmoles) and the mixture was left to stir at room temperature until complete solubilization of the salt. Upon complexation and solubilization, the color of the solution changed from white to yellow transparent. Finally, the reducing agent namely hydrazine monohydrate (1.46 μL, 0.03 mmoles) was added to the solution upon stirring. This was accompanied by a color change of the solution from yellow to dark brown indicating the reduction of palladium (II) ions into metallic palladium (0) NPs. The resulting solution was left to stir for 24 h at room temperature. The size of the palladium NPs generated within the micellar cores has been already reported [22]. The average diameter of the palladium NPs is 2.3 $\pm$ 0.3 nm for the LauMA$_{50}$-b-AEMA$_9$/Pd(solution A) and 8.8 $\pm$ 1.7 nm for LauMA$_{120}$-b-AEMA$_{67}$/Pd(solution B).

A 2 inch n-type silicon wafer was used as a substrate for the uniform deposition of the polymer-palladium solutions. Titanium/titanium nitride (Ti/TiN) of thicknesses 50 and 100 nm were sputtered on each side of the chip using FHR MS 150 sputter machine in order to have better electrical contact between back and front side of the sample, because of backside probing of the chip for measurements. The polymer-palladium NPs solutions were later spun using standard resist spinners. Vertically aligned CNFs were grown at different temperatures by the DC-PECVD method as described in [8, 24]. A mixture of acetylene and ammonia gases was used where the acetylene is the source gas and the ammonia is a carrier gas. The growth was executed for 2 h on all samples. The temperature of the heater is measured and regulated during the whole growth process to an accuracy of $\pm 2°$ using a built-in thermocouple inside the grounded heating plate.

The chips were weighed before the deposition of the palladium NPs and after the growth of CNFs for determining the weight of the CNFs including catalyst, a parameter used in evaluating the capacitance per gram of such electrodes. The weight measurements were performed using a high precision

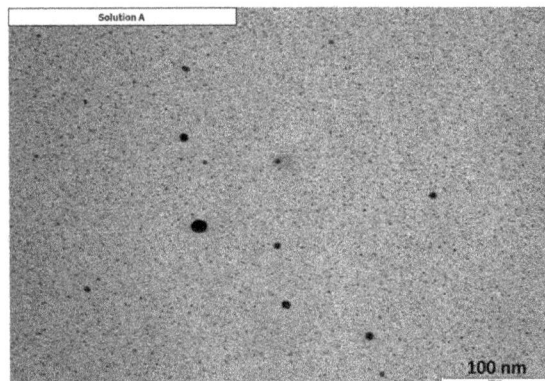

Figure 1. TEM image of the LauMA$_{50}$-b-AEMA$_9$/Pd system (solution A).

balance 'Sartorius Analytical Balance BP211D' with a resolution of 10 μg. Transmission electron microscopy (TEM) analysis performed on the LauMA$_x$-b-AEMA$_y$/Pd (solution A) system was carried out on a 1010 JEOL microscope (200 kV). The suspension of MNPs was dried on a carbon coated copper grid to allow the TEM investigation. The scanning electron microscopy (SEM) analysis of the CNFs was conducted using JEOL JSM-6301F. In order to see the morphology of the CNFs, images were taken at 40° tilt angle as well as top view images at zero tilt.

Cyclic voltammetry (CV) was performed to evaluate the capacitance. A three electrode setup was used with a silver/silver chloride (Ag/AgCl) electrode as the reference electrode and platinum mesh as the counter electrode. The electrolyte used was 1M KOH (99.99%, Aldrich). The limit of measuring the electrode cell was one centimeter diameter disk. The samples were mounted and dipped in the electrolyte solution for a few minutes while nitrogen gas was blowing to wet the CNFs. The CV curve was obtained using Gamry Instruments Reference 600 (Framework 6.1). The voltage sweep ranging from −0.2 to 0.3 V was applied and different voltage scan rates were used. The voltammetry using 10 mV s^{-1} scan rate was carried out for 18 cycles to exclude the doubt of initial surface reactions whereas 5 cycles were run for other scan rates. The same measurement was also performed for 1000 cycles to measure the cyclability at 20 mV s^{-1} scan rate while keeping the other parameters the same. The ideal double layer capacitance behavior is the rectangular shape of the voltammetry characteristics drawn between current and voltage [25].

3. Results and discussion

Information on the palladium NPs content (provided as the molar ratio of the metal-binding units of the polymer to the palladium salt precursor), growth temperature, weight of CNFs and capacitance is provided in table 1, whereas the specific capacitance measured at different voltage scan rates is given in table 2.

Table 1. Properties of the CNFs grown using different polymer-palladium NPs solutions at different temperatures, their capacitance and specific capacitance mF cm^{-2} (per footprint area).

[vinyl pyridine]/[Pd(OAc)$_2$]	Growth tempera- ture (°C)	Weight (μg)	Length (μm)	Capacitance (mF)	Specific capacitance (mF cm^{-2})
Bare chip				0.1	0.12
38:1	550	53	14	0.8	1
18:1	550	245	9	3.1	4
9:1	550	916	5	7.1	9
9:1	390	612	1	2.6	3.3
[AEMA]/[Pd(OAc)$_2$]					
[LauMA$_{120}$-b-AEMA$_{67}$] 2:1	390	654	1	2.1	2.7
[LauMA$_{50}$-b-AEMA$_9$] 2:1	390	115	2	1.4	1.8
[LauMA$_{120}$-b-AEMA$_{67}$] 2:1	350	250	1.4	3.1	4
[LauMA$_{50}$-b-AEMA$_9$] 2:1	350	115	1.6	0.2	0.2

Table 2. Specific capacitance mF cm^{-2} (per footprint area) measured at different voltage scan rates.

[vinyl pyridine]/[Pd(OAc)$_2$]	T (°C)	10 mV s^{-1}	20 mV s^{-1}	50 mV s^{-1}	100 mV s^{-1}
PVP/Pd 38:1	550	1	0.8	0.7	0.6
PVP:Pd 18:1	550	4	2.4	1.6	1.2
PVP/Pd 9:1	550	9.	6	4.6	4
PVP/Pd 9:1	390	3.3	2.6	2	2
[AEMA]/[Pd(OAc)$_2$]					
Solution A	350	0.2	0.2	0.15	0.14
Solution B	390	2.7	2	1.5	1.3
Solution B	350	4	3	2.3	2
Solution A	390	1.8	1.41	1.1	1

A representative TEM image of the solution A system is provided in figure 1. As seen in the image, tiny, spherical palladium NPs can be visualized with average diameters in the range 2.3 ± 0.3 nm. The presence of a few larger aggregates may be due to drying-induced aggregation during TEM sample preparation. In the case of the PVP/Pd systems, HRTEM analysis revealed that the palladium NPs are nanocrystals with average diameters below 10 nm. The crystalline planes (111) and (200) of palladium NP could be visualized with characteristic interplanar distances of 2.27 and 1.97 Å, respectively [26]. Tilted view and top view (inset on each picture) SEM images of CNFs grown at 550 °C using three PVP/Pd solutions having different metallic (palladium) content are shown in figures 2(a)–(c) whereas SEM images of the CNFs grown at 390 °C using the PVP/Pd (1:9) solution is provided in figure 2(d).

Similarly, the tilted view and top view (inset on each picture) SEM images of the CNFs grown at 390 and 350 °C using solution A and solution B are shown in figures 3(b), (d) and (a), (c).

It is clear from the SEM images that the CNFs grown at lower temperature 350 °C are vertically aligned similar to those grown at the highest temperature (550 °C) as shown in figures 2 and 3. In addition, inset of figures 2 and 3 also show the uniform growth of CNFs confirming the uniform

deposition of palladium catalyst NPs using spin-coating, which is in line with [26].

Moreover, in the case of the PVP–Pd systems, the CNF density increases significantly by increasing the palladium content within the PVP–Pd colloidal hybrid solutions i.e. from 38:1 to 18:1 and to 9:1 (figures 2(a)–(c) respectively) thus resulting in an increase in CNF weight, as shown in table 1. In addition, SEM analysis also reveals that under identical growth conditions, the CNFs grown using the highest palladium-containing system (9:1) are shorter in length compared to those grown using the lowest palladium-containing system (38:1). By using the same palladium-containing system (9:1) the CNFs grown at higher (550 °C) temperatures are much longer than those grown at 390 °C, in line with the weight measurements, although less dense than the CNFs grown at 390 °C.

In the case of the LauMA$_x$-b-AEMA$_y$/Pd solutions, the average diameter of the palladium nanocatalyst is 2.3 ± 0.3 nm for solution A and 8.8 ± 1.7 nm for solution B. Despite their size differences both systems are found within the optimum size range enabling CNF growth at a low temperature (350 °C) as seen in the SEM images provided in figures 3(c) and (d). The top view SEM images (inset in figure 3) clearly show that under identical growth temperature, the CNFs grown using solution B are denser and heavier than those grown using solution A, but shorter in length

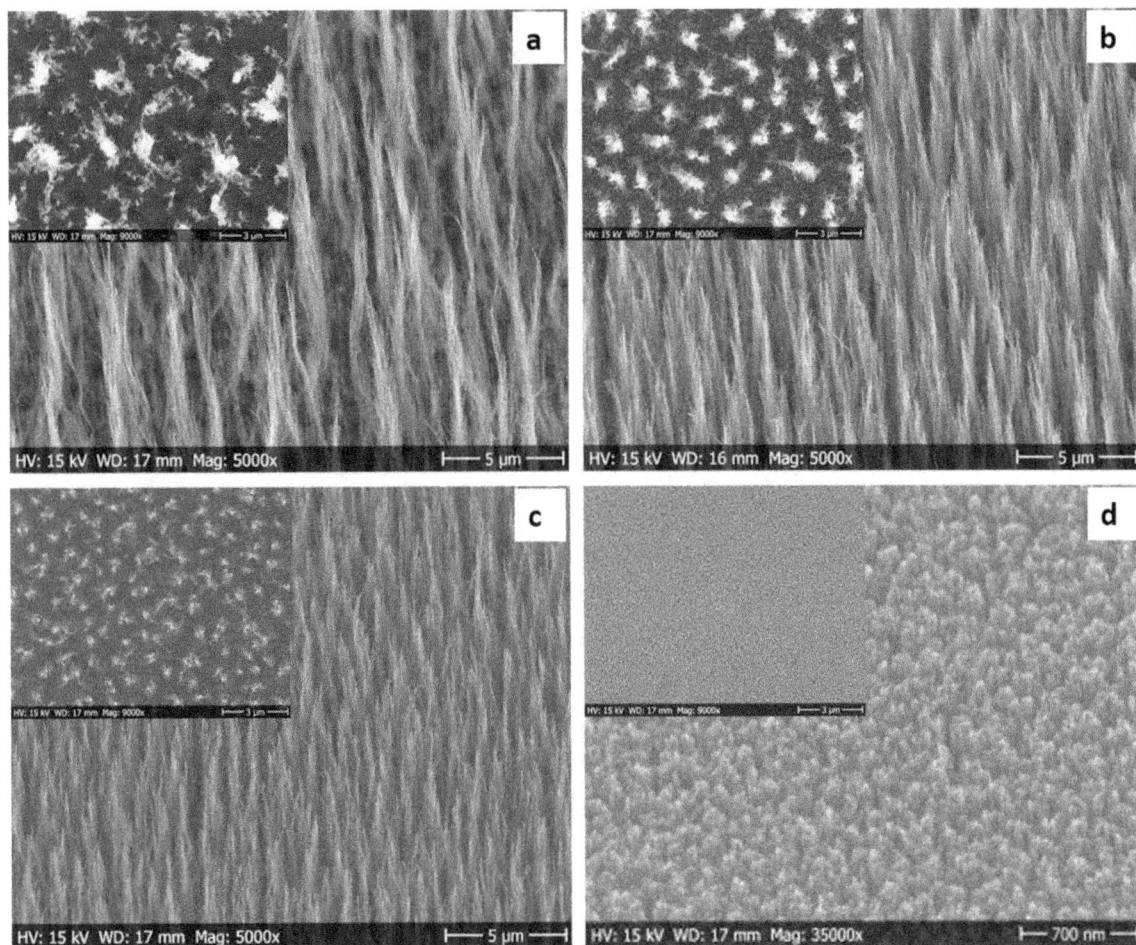

Figure 2. SEM images of CNFs grown by DC-PECVD for 2 h at different temperatures on palladium NPs using various PVP–Pd colloidal solutions given as (a) PVP:Pd 38:1, growth at 550 °C. (b) PVP:Pd 18:1, growth at 550 °C. (c) PVP:Pd 9:1, growth at 550 °C. (d) PVP:Pd 9:1, growth at 390 °C.

compared to those obtained from solution A, as shown in table 1. Furthermore, when the growth is performed at different temperatures, solution B gives longer but less dense CNFs at low temperature 350 °C. On the other hand, solution A gives denser and shorter CNFs at the same growth temperature.

The CV curves for the electrical characterization of the CNFs grown using the PVP–Pd and the LauMA$_x$-b-AEMA$_y$-Pd solutions are provided in figures 4(a), (b) and 5(a). The specific capacitance (mF cm^{-2} footprint area) curves measured at difference voltage scan rate are shown in figures 4(c) and 5(b). The cyclability curves for 1000 cycles are shown in figures 4(d) and 5(c) whereas magnified images of the last 20 cycles from 980 to 999 are shown in figures 4(e) and 5(d).

The voltammograms given in figures 4(a), (b) and 5(a) are fairly rectangular showing the double layer capacitive behavior of the CNFs obtained using both types of polymer–Pd catalyst solutions, however, the peaks at extreme sweep voltages show some pseudocapacitive charge storage which

may arise from the reaction of underlayer with aqueous electrolyte.

At 550 °C growth temperature and 10 mV s^{-1} scan rate, the specific capacitance (per cm^{-2} foot print area) is larger (9 mF cm^{-2}) for higher concentration (9:1) of palladium NPs in the PVP–Pd solution and lower (1 mF cm^{-2}) for lower concentration (38:1) confirming the active capacitive role of CNFs. In fact, shorter but denser CNFs for the (9:1) solution give a higher surface area accessible to electrolyte than longer but less dense CNFs for the (38:1) solution, and the higher surface area results in higher current on voltammogram and hence higher specific capacitance as shown in figure 4(a) and table 1. In addition the specific capacitance is 3.3 mF cm^{-2} for the CNFs grown at 390 °C using the PVP–Pd (9:1) solution.

In the case of the LauMA$_x$-b-AEMA$_y$/Pd solutions the specific capacitance is higher (2.7 and 4 mF cm^{-2}) for CNFs synthesized using solution B than solution A when grown at corresponding temperatures 350 °C and 390 °C which is in line with the weight of the CNFs. Moreover, for the same

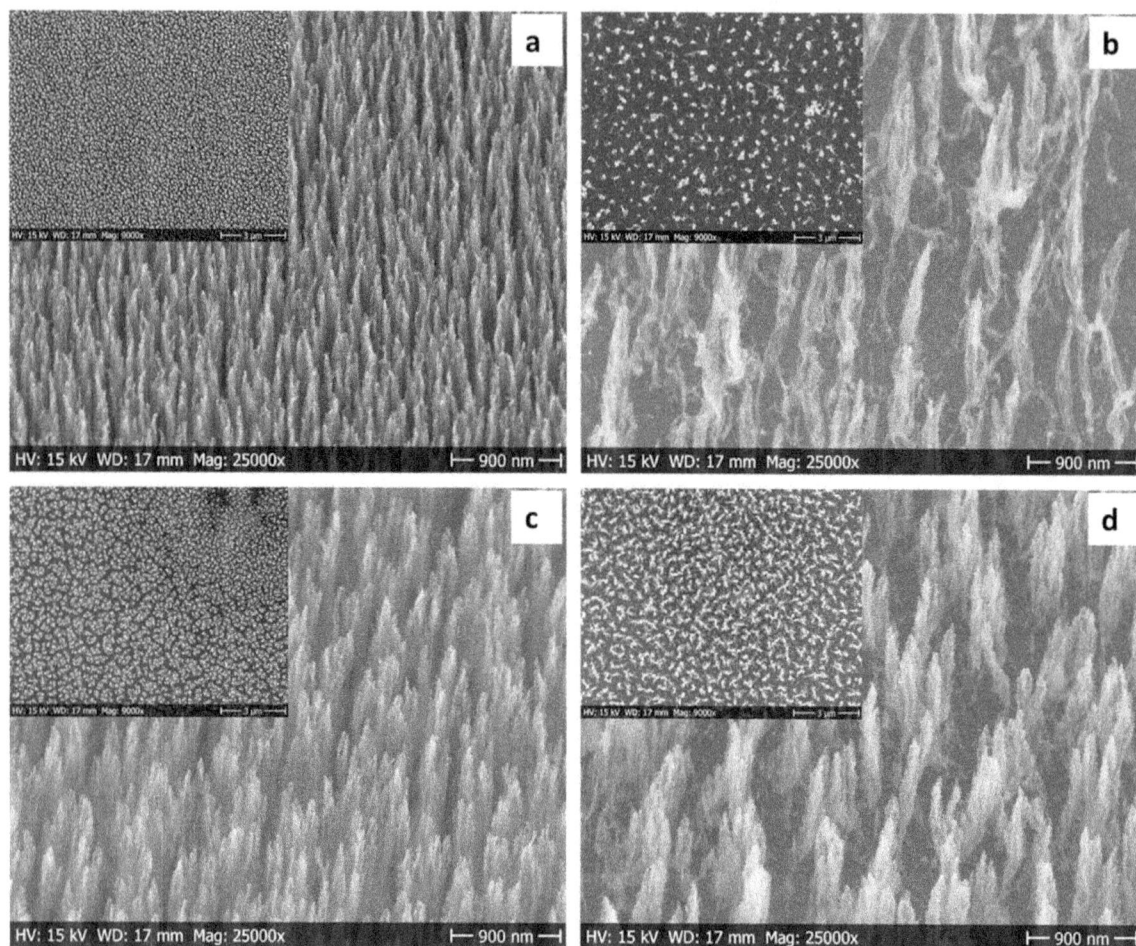

Figure 3. SEM images of CNFs grown by DC-PECVD for 2 h on two different LauMA$_x$-b-AEMA$_y$/Pd micellar solutions at different temperatures given as (a) LauMA$_{120}$-b-AEMA$_{67}$/Pd (solution B) growth at 390 °C. (b) LauMA$_{50}$-b-AEMA$_9$/Pd (solution A) growth at 390 °C. (c) LauMA$_{120}$-b-AEMA$_{67}$/Pd (solution B) growth at 350 °C. (d) LauMA$_{50}$-b-AEMA$_9$/Pd (solution A) growth at 350 °C.

solution, the longer CNFs give a higher surface area which results in higher current on the voltammogram and hence higher specific capacitance, as shown in figure 5(a) and table 1. The capacitive behavior indicates the similar electrical behavior of CNFs grown at both low and high temperatures.

The specific capacitance at 100 mV s^{-1} ranges from 3.9 to 0.140 mF cm^{-2}. This specific capacitance is many folds higher than the recently reported capacitance obtained from the CVD grown vertically aligned CNFs [27]. Furthermore, the maximum specific capacitance at 100 mV s^{-1} scan rate is 3.9 mF cm^{-2} obtained from the CNFs grown at 550 °C, table 2, which is almost equal to the maximum reported capacitance obtained from graphene-CNT based micro-capacitors (3.93 mF cm^{-2}) synthesized on a similar silicon substrate but using a longer voltage scan range [28]. More-over, the specific capacitance obtained from CNFs grown at low temperature 350 °C and 390 °C, table 2, is 1.9–2 mF cm^{-2}, half of the reported specific capacitance mentioned above, nevertheless the growth temperature of

CNFs here is below the CMOS process-compatible tem-perature. In addition, the cyclic voltammetry at different voltage scan rates shows that specific capacitance decreases by increasing the scan rate from 10 to 100 mV s^{-1} for both polymer–Pd types used, figures 4(c) and 5(b), due to a kinetically slow faradaic reaction on the electrode surface which does not take place at high scan rate[5].

Moreover, the cycle life curves of both polymer–Pd solution types show the decrease in specific capacitance with increase in cycles, figures 4(d) and 5(c), which eventually becomes uniform and remains unchanged as shown in the plots of the last 20 cycles, given in figures 4(e) and 5(d). The more than 90% capacitance retention for CNFs grown from the PVP–Pd solution and 80% for the CNFs grown from solution B after 1000 cycles of cyclic voltammetry proves the long cycle life of these materials as supercapacitor electrodes.

[5] www.gamry.com/application-notes/testing-electrochemical-capacitors-part-1-cyclic-voltammetry-and-leakage-current/, accessed successfully on 06-10-2014.

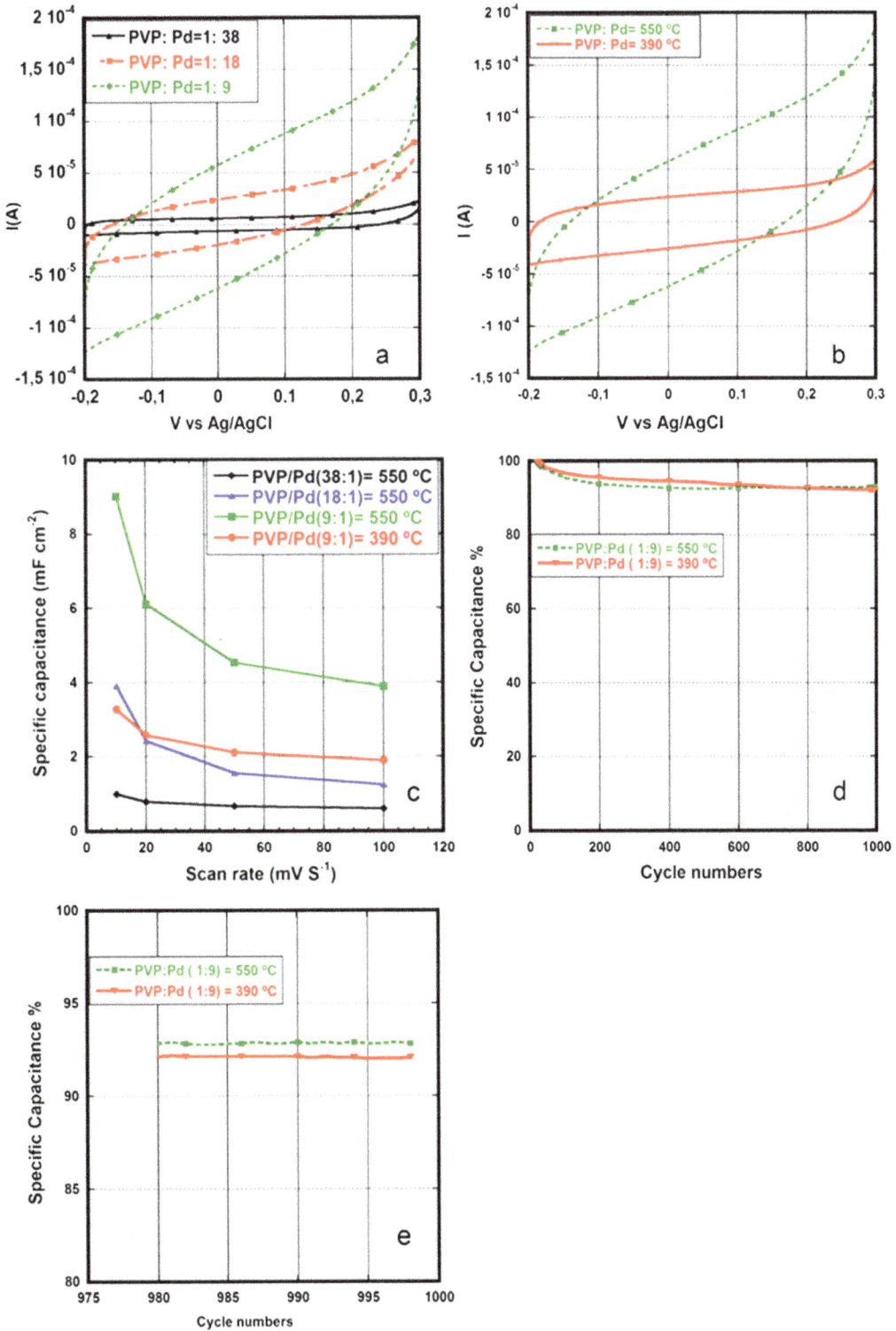

Figure 4. Cyclic voltammetry curves of CNFs. (a) At 550 °C using different PVP–Pd solutions with different palladium content. (b) At 550 °C and 390 °C using PVP:Pd 9:1 solution. (c) Specific capacitance (mF cm^{-2} foot print area) versus voltage scan rate. (d) Cyclability of CNFs grown at 550 °C and 390 °C using PVP:Pd 9:1 solution (normalized by specific capacitance of first cycle). (e) Cycles from 975 to 999.

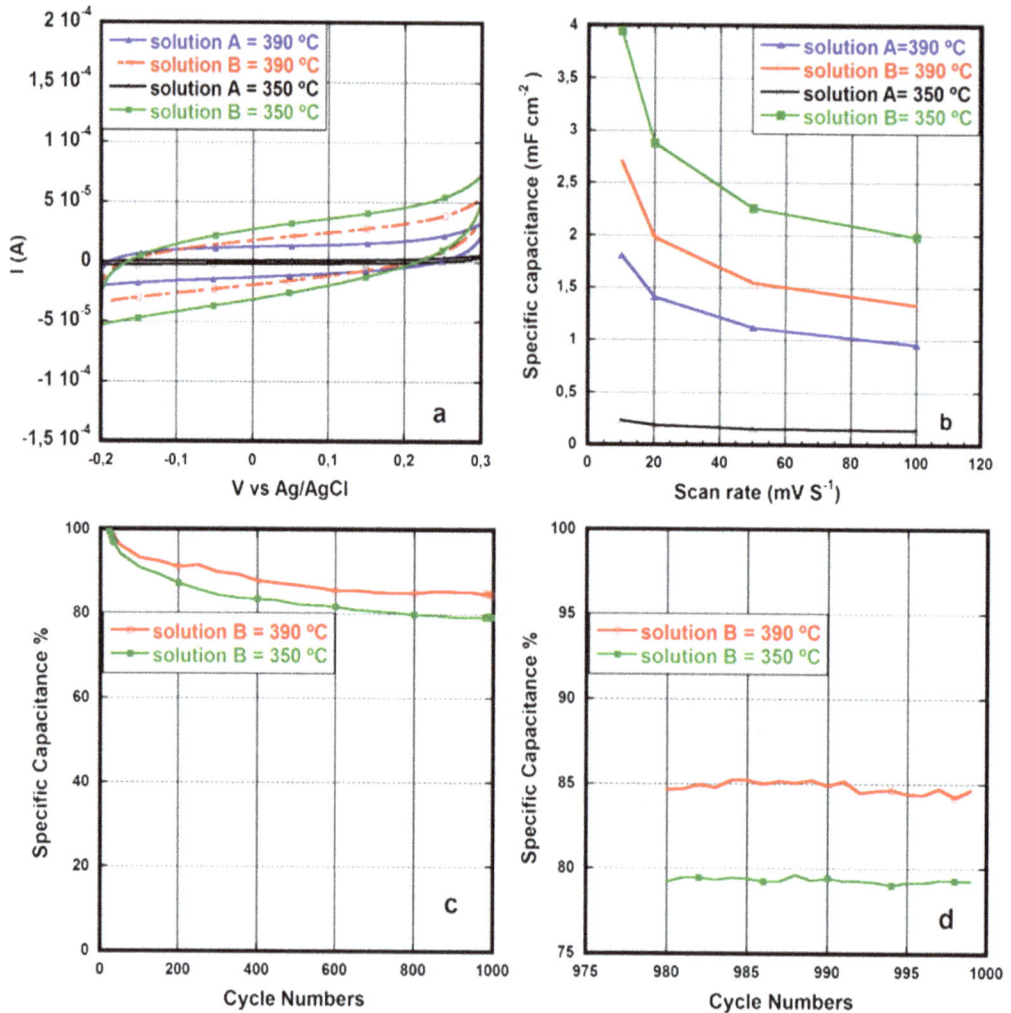

Figure 5. (a) Cyclic voltammetry curves of CNFs grown using solution A and solution B at 350 °C and 390 °C. (b) Specific capacitance (mF cm^{-2} foot print area) versus voltage scan rate. (c) Cycle life of solution B (normalized by specific capacitance of first cycle). (d) Cycles from 980 to 999.

Due to the lower density of CNFs for solution B the electrolyte damages the metal underlayer in between the CNFs thus resulting in lower capacitance after 1000 cycles.

The above results indicate that the CNFs grown from all types of polymer-stabilized Pd nanocatalysts are vertically aligned similar to CNFs grown from a PVD deposited catalyst film. Moreover the density and weight of the CNFs grown from a PVP–Pd solution can be controlled by controlling the palladium contents in the PVP–Pd solution in the tested concentration range. Furthermore CNFs of different diameters can be grown from LauMA$_x$-b-AEMA$_y$/Pd solutions. Finally, the electrical characterization proves that electrical properties of all the CNFs are the same; however, the low capacitance in solution A is due to the lower density of CNFs in the footprint area.

The CNF growth performed at this temperature has created the opportunity to grow the CNFs directly on CMOS

chips. The possibility of CNF growth at temperatures lower (350 °C) than CMOS temperature ensures their usage even if the CMOS temperature drops further.

4. Conclusions

Catalytic palladium NPs of various sizes were deposited on a silicon substrate via spin-coating using two types of highly stable and cost-effective polymer–Pd colloidal solutions. More precisely, PVP or methacrylate-based diblock copolymers of the type LauMA$_x$-b-AEMA$_y$ were employed as steric stabilizers for palladium NPs in organic solvents. Vertically aligned CNFs were synthesized on these palladium nanocatalysts using the direct plasma enhanced CVD technique at different temperatures ranging from 550 °C down to 350 °C, which can be considered as a CMOS-compatible temperature.

The SEM characterization had shown that the CNFs grown at low temperature were vertically aligned similar to those generated at high temperature. The capacitive behavior and long cyclability had proven not only the suitability of the as-grown CNFs as electrodes for supercapacitors but also that the CNFs grown at low and high temperatures had similar electrical behavior. We have demonstrated a CMOS compatible process to build efficient heat sink, interconnects or supercapacitor electrodes directly on the chip based on vertically aligned CNFs as the active material.

Acknowledgments

This work has been performed within the MNT-ERA.NET Carpolcap project supported by Vinnova, the Swedish Governmental Agency for Innovation Systems and the Cyprus Research Promotion Foundation (Grant No. KOINA/MNT-ERA.NET/0311/01). The authors thank Dr Rodica Paula Turcu (National Institute for Isotopic and Molecular Technologies, Cluj-Napoca, Romania) for performing the TEM characterization on the LauMA$_{50}$-b-AEMA$_9$/Pd system and Dr Maria Demetriou for the synthesis of the LauMA-b-AEMA block copolymers used in the present study.

References

[1] Iijima S 1999 Helical microtubules of graphitic carbon *Nature* **354** 56–8

[2] Collins P G and Avouris P 2000 Nanotubes for electronics *Sci. Am.* 62–9

[3] Zhou Z, Lai C, Zhang L, Qian Y, Hou H, Reneker D H and Fong H 2009 Development of carbon nanofibers from aligned electrospun polyacrylonitrile nanofiber bundles and characterization of their microstructural, electrical, and mechanical properties *Polymer* **50/13** 2999–06

[4] Bal S 2010 Experimental study of mechanical and electrical properties of carbon nanofiber/epoxy composites *Mater. Des.* **31** 2406–13

[5] Suzuki M, Ominami Y, Ngo Q, Yang C Y, Cassell A M and Li J 2007 Current-induced breakdown of carbon nanofibers *J. Appl. Phys.* **101** 114307

[6] Desmaris V *et al* 2011 A test vehicle for rf/dc evaluation and destructive testing of vertically grown nanostructures (VGCNS) *NT11: Int. Conf. on the Science and Application of Nanotubes (Cambridge, UK, 10–16 July)*

[7] Nessim G D 2010 Properties, synthesis, and growth mechanisms of carbon nanotubes with special focus on thermal chemical vapor deposition *Nanoscale* **2** 1306

[8] Desmaris V, Saleem A M, Shafiee S, Berg J, Kabir M S and Johansson A 2014 Carbon nanofibers (CNF) for enhanced solder-based nano-scale integration and on-chip interconnect solutions *2014 IEEE 64th Electronics Components and Technology Conf. 2014 (27–30 May)* pp 1071–6

[9] Teng I J, Chuang P K and Kuo C T 2006 One-step growth process of carbon nanofiber bundle-ended nanocone structure by microwave plasma chemical vapor deposition *Diam. Relat. Mater.* **15** 1849–54

[10] Saleem A M, Rahiminejad S, Desmaris V and Enoksson P 2014 Carbon nanotubes as base material for fabrication of gap waveguide components *EUROSENSORS 2014 (Brescia, Italy, 7–10 September)*

[11] Saleem A M, Berg J, Desmaris V and Kabir M S 2009 Nanoimprint lithography using vertically aligned carbon nanostructures as stamps *Nanotechnology* **20** 375302

[12] Sharma S P and Lakkad S C 2011 Effect of CNTs growth on carbon fibers on the tensile strength of CNTs grown carbon fiber-reinforced polymer matrix composites *Composites* A **42** 8–15

[13] Mori S and Suzuki M 2008 Effect of oxygen and hydrogen addition on the low-temperature synthesis of carbon nanofibers using a low-temperature CO/Ar dc plasma *Diam. Relat. Mater.* **17** 999–02

[14] Lin C-C, Pan F-M, Chang K-C, Kuo C-W and Kuo C-T 2009 Mechanitis study of cobalt catalysed growth of carbon nanofibers in a confined space by plasma-assisted chemical vapor deposition *Diam. Relat. Mater.* **18** 1301–5

[15] Vollebregt S, Derakhsh J, Ishihara R, Wu M Y and Beenakker C I M 2010 Growth of high-density self-aligned carbon nanotubes and nanofibers using palladium catalyst *J. Electron. Mater.* **39** 371–5

[16] Seah C-M, Chai S-P, Ichikawa S and Mohamad A R 2013 Growth of uniform thin-walled carbon nanotubes with spin-coated Fe catalyst and the correlation between the pre-growth catalyst size and nanotube diameter *J. Nanopart. Res.* **15** 1371

[17] Murcia A B, Rebrov A B, Cabaj M, Wheatley A E H, Johnson B F G, Robertson J and Schouten I C 2009 Confined palladium colloids in mesoporous frameworks for carbon nanotubes growth *J. Mater. Sci.* **44** 6563–70

[18] Gohier A, Kim K H, Norman E D, Gorintin L, Bondavalli P and Cojocaru C S 2012 Spray-gun deposition of catalyst for large area and versatile synthesis of carbon nanotubes *Appl. Surf. Sci.* **258** 6024–8

[19] Demetriou M and Christoforou T K 2008 Synthesis and characterization of well-defined block and statistical copolymers based on lauryl methacrylate and 2-(acetoacetoxy)ethyl methacrylate using RAFT-controlled radical polymerization *J. Polym. Sci. A Polym. Chem.* **46** 5442–51

[20] Iliopoulos K, Chatzikyriakos G, Demetriou M, Christoforou T K and Couris S 2011 Preparation and nonlinear optical response of novel palladium-containing micellar nanohybrids *Opt. Mater.* **33** 1342–9

[21] Bradley J S, Hill E W, Behal S, Klein C, Duteil A and Chaudret B 1992 Preparation and characterization of organosols of monodispersed nanoscale palladium. Particle size effects in the binding geometry of adsorbed carbon monoxide *Chem. Mater.* **4** 1234–9

[22] Zervos M, Demetriou M, Christoforou T K, Othonos A and Tursu R P 2012 Synthesis of hybrid polymethacrylate–noble metal (M = Au, Pd) nanoparticles for the growth of metal-oxide semiconductor nanowires *RSC Adv.* **2** 4370–6

[23] Savva I, Kalogirou A S, Chatzinicolaou A, Papaphilippou P, Pantelidou A, Vasile E, Vasile E, Koutentis P A and Christoforou T K 2014 PVP-crosslinked electrospun membranes with embedded Pd and Cu$_2$O nanoparticles as effective heterogeneous catalytic supports *RSC Adv.* **4** 44911–21

[24] Kabir M S, Morjan R E, Nerushev O A, Lundgren P, Bengtsson S, Enoksson P and Campbell E E B 2006 Fabrication of individual vertically aligned carbon nanofibers on metal substrates from prefabricated catalyst dots *Nanotechnology* **17** 790–4

[25] Frackowiaka E and Béguinb F 2001 Carbon materials for the electrochemical storage of energy in capacitors *Carbon* **39** 937–50

[26] Cardoso M B T, Lewis E, Castro P S, Dantas L M F, De Oliveira C C S, Bertotti M, Haigh S J and Camargo P H C 2014 A facile strategy to support palladium

nanoparticles on carbon nanotubes, employing
polyvinylpyrrolidone as a surface modifier *Eur. J. Inorg. Chem.* 1439–45

[27] Hsia B, Marschewski J, Wang S, In J B, Carraro C, Poulikakos D, Grigoropoulos C P and Maboudian R 2014 Highly flexible, all solid-state micro-supercapacitors from

vertically aligned carbon nanotubes *Nanotechnology* **25** 055401

[28] Lin J, Zhang C, Yan Z, Zhu Y, Peng Z, Hauge R H, Natelson D and Tour J M 2013 Three-dimensional graphene carbon nanotube carpet-based microsupercapacitors with high electrochemical performance *Nano Lett.* **13** 72–8

Permissions

The contributors of this book come from diverse backgrounds, making this book a truly international effort. This book will bring forth new frontiers with its revolutionizing research information and detailed analysis of the nascent developments around the world.

We would like to thank all the contributing authors for lending their expertise to make the book truly unique. They have played a crucial role in the development of this book. Without their invaluable contributions this book wouldn't have been possible. They have made vital efforts to compile up to date information on the varied aspects of this subject to make this book a valuable addition to the collection of many professionals and students.

This book was conceptualized with the vision of imparting up-to-date information and advanced data in this field. To ensure the same, a matchless editorial board was set up. Every individual on the board went through rigorous rounds of assessment to prove their worth. After which they invested a large part of their time researching and compiling the most relevant data for our readers.

The editorial board has been involved in producing this book since its inception. They have spent rigorous hours researching and exploring the diverse topics which have resulted in the successful publishing of this book. They have passed on their knowledge of decades through this book. To expedite this challenging task, the publisher supported the team at every step. A small team of assistant editors was also appointed to further simplify the editing procedure and attain best results for the readers.

Apart from the editorial board, the designing team has also invested a significant amount of their time in understanding the subject and creating the most relevant covers. They scrutinized every image to scout for the most suitable representation of the subject and create an appropriate cover for the book.

The publishing team has been an ardent support to the editorial, designing and production team. Their endless efforts to recruit the best for this project, has resulted in the accomplishment of this book. They are a veteran in the field of academics and their pool of knowledge is as vast as their experience in printing. Their expertise and guidance has proved useful at every step. Their uncompromising quality standards have made this book an exceptional effort. Their encouragement from time to time has been an inspiration for everyone.

The publisher and the editorial board hope that this book will prove to be a valuable piece of knowledge for researchers, students, practitioners and scholars across the globe.

List of Contributors

Fengniu Lu
International Center for Materials Nanoarchitectonics (MANA), National Institute for Materials Science (NIMS) 1-2-1 Sengen, Tsukuba 305-0047, Japan

Takashi Nakanishi
International Center for Materials Nanoarchitectonics (MANA), National Institute for Materials Science (NIMS) 1-2-1 Sengen, Tsukuba 305-0047, Japan
Warsaw University of Technology, Warsaw 02-507, Poland
Institute for Molecular Science (IMS), 5-1 Higashiyama, Myodaiji, Okazaki 444-8787, Japan

Daniele Pergolesi, Emiliana Fabbri, Christof W Schneider, Thomas Lippert
Paul Scherrer Institut, Department of General Energy Research, CH-5225, Villigen-PSI, Switzerland
International Center for Materials Nanoarchitectonics (WPI-MANA), National Institute for Materials Science (NIMS), 1-1 Namiki, Tsukuba, Ibaraki 305-0044, Japan

Vladimir Roddatis
CIC Energigune, Albert Einstein 48, E-01510 – Miñano (Álava), Spain

Enrico Traversa
Physical Science and Engineering Division, King Abdullah University of Science and Technology (KAUST), Thuwal 23955-6900, Saudi Arabia

John A Kilner
Department of Materials, Imperial College London, London SW7 2BP, UK

Tetsuyuki Ochiai
Photonic Materials Unit, National Institute for Materials Science (NIMS), Tsukuba 305-0044, Japan

Akira Yamamoto
Department of Molecular Engineering, Graduate School of Engineering, Kyoto University, Kyotodaigaku Katsura, Nishikyo-ku, Kyoto 615-8510, Japan

Kentaro Teramura
Department of Molecular Engineering, Graduate School of Engineering, Kyoto University, Kyotodaigaku Katsura, Nishikyo-ku, Kyoto 615-8510, Japan
Elements Strategy Initiative for Catalysts & Batteries (ESICB), Kyoto University, Kyotodaigaku Katsura, Nishikyo-ku, Kyoto 615-8520, Japan
Precursory Research for Embryonic Science and Technology (PRESTO), Japan Science and Technology Agency (JST), 4-1-8 Honcho, Kawaguchi, Saitama 332-0012, Japan

Saburo Hosokawa and Tsunehiro Tanaka
Department of Molecular Engineering, Graduate School of Engineering, Kyoto University, Kyotodaigaku Katsura, Nishikyo-ku, Kyoto 615-8510, Japan
Elements Strategy Initiative for Catalysts & Batteries (ESICB), Kyoto University, Kyotodaigaku Katsura, Nishikyo-ku, Kyoto 615-8520, Japan

Yukinori Yoshimura and Ken-Ichiro Imura
Department of Quantum Matter, AdSM, Hiroshima University, Higashi-Hiroshima, 739-8530, Japan

Koji Kobayashi and Tomi Ohtsuki
Department of Physics, Sophia University, Chiyoda-ku, Tokyo, 102-8554, Japan

Huan Wang, Sen Liu, Yong-Lai Zhang, Jian-Nan Wang, Lei Wang, Hong Xia and Qi-Dai Chen
State Key Laboratory on Integrated Optoelectronics, College of Electronic Science and Engineering, Jilin University, 2699 Qianjin Street, Changchun, 130012, People's Republic of China

Hong Ding
State Key Laboratory of Inorganic Synthesis and Preparative Chemistry, College of Chemistry, Jilin University, 2699 Qianjin Street, Changchun, 130012, People's Republic of China

Hong-Bo Sun
State Key Laboratory on Integrated Optoelectronics, College of Electronic Science and Engineering, Jilin University, 2699 Qianjin Street, Changchun, 130012, People's Republic of China
College of Physics, Jilin University, 119 Jiefang Road, Changchun, 130023, People's Republic of China

Geoffrey Lawrence, Arun V Baskar, Wang Soo Cha and Ajayan Vinu
Australian Institute for Bioengineering and Nanotechnology, The University of Queensland, #75 Corner Cooper and College Road, Brisbane 4072, QLD, Australia

Salem S Al-Deyab and Mohammed H El-Newehy
Petrochemical Research Chair, Department of Chemistry, College of Science, King Saud University, Riyadh 11451, Saudi Arabia

Irene Prencipe, David Dellasega, Alessandro Zani, Daniele Rizzo and Matteo Passoni
Dipartimento di Energia, Politecnico di Milano, Milan, Italy

Ji-Guang Li and Yoshio Sakka
Advanced Materials Processing Unit, National Institute for Materials Science, 1-2-1 Sengen, Tsukuba, Ibaraki 305-0047, Japan

Fu Wang, Dexin Ma and Andreas Bührig-Polaczek
Foundry Institute, RWTH Aachen University, D-52072 Aachen, Germany

Yoshinao Nakagawa, Riku Tamura, Masazumi Tamura andKeiichi Tomishige
Department of Applied Chemistry, School of Engineering, Tohoku University, 6-6-07, Aoba, Aramaki, Aoba-ku, Sendai 980-8579, Japan

Kouji Segawa
Institute of Scientific and Industrial Research, Osaka University 8-1 Mihogaoka, Ibaraki, Osaka 567-0047, Japan

Xingzhong Guo, Jie Song, Yixiu Lvlin and Hui Yang
School of Materials Science and Engineering, Zhejiang University, Hangzhou, 310027, People's Republicof China

Kazuki Nakanishi and Kazuyoshi Kanamori
Department of Chemistry, Graduate School of Science, Kyoto University, Kitashirakawa, Sakyo-ku, Kyoto 606-8502, Japan

Ken Kanazawa, Shoji Yoshida, Hidemi Shigekawa and Shinji Kuroda
Faculty of Pure and Applied Sciences, University of Tsukuba, Tsukuba 305-8573, Japan

Junji Tominaga, Alexander V Kolobov, Paul J Fons, Xiaomin Wang, Yuta Saito, Takashi Nakano
Nanoelectronics Research Institute, National Institute of Advanced Industrial Science & Technology (AIST), Tsukuba Central 4, 1-1-1 Higashi, Tsukuba 305-8562, Japan

Muneaki Hase
Faculty of Pure and Applied Science, University of Tsukuba, 1-1-1 Tennodai, Tsukuba 305-8573, Japan

Shuichi Murakami
Department of Physics & TIES, Tokyo Institute of Technology, 2-12-1 Ookayama, Meguro-ku, Tokyo 152-8551, Japan

Jens Herfort and Yukihiko Takagaki
Paul-Drude-Institut für Festkörperelektronik, Hausvogteiplatz 5-7, 10117 Berlin, Germany

Alexei A Belik
International Center for Materials Nanoarchitectonics (WPI-MANA), National Institute for Materials Science (NIMS), 1-1 Namiki, Tsukuba, Ibaraki 305-0044, Japan

Harvey Amorín, Miguel Algueró, Rubén Del Campo, Eladio Vila, and Alicia Castro
Instituto de Ciencia de Materiales de Madrid, CSIC, Cantoblanco, 28049 Madrid, Spain

Pablo Ramos
Universidad de Alcalá, 28871 Alcalá de Henares, Spain

Mickael Dollé
Département de Chimie, Université de Montréal C.P. 6128, succursale Centre-Ville Montréal, QC, H3C 3J7, Canada

Yonny Romaguera-Barcelay and Javier Pérez De La Cruz
IFIMUP and IN-Institute of Nanoscience and Nanotechnology, Faculdade de Ciências da Universidade do Porto, Rua do Campo Alegre, 687, 4169-007 Porto, Portugal

Benjamin Dargatz, Jesus Gonzalez-Julian and Olivier Guillon
Friedrich Schiller University of Jena, Otto Schott Institute of Materials Research, Löbdergraben 32, D-07743 Jena, Germany

Martin Drábik and Dirk Hegemann
Empa, Swiss Federal Laboratories for Materials Science & Technology, Laboratory for Advanced Fibers, Lerchenfeldstrasse 5, 9014 St.Gallen, Switzerland

Josef Pešička
Charles University in Prague, Faculty of Mathematics and Physics, Department of Physics of Materials, Ke Karlovu 5, 121 16 Prague 2, Czech Republic

Hynek Biederman
Charles University in Prague, Faculty of Mathematics and Physics, Department of Macromolecular Physics, V Holešovičkách 2, 180 00 Prague 8, Czech Republic

Amin M Saleem
Smoltek AB, Regnbågsgatan 3, Gothenburg, SE-41755, Sweden
Micro and Nanosystems group, BNSL, Department of Microtechnology and Nanoscience, Chalmers University of Technology, Gothenburg, SE-41296, Sweden

Sareh Shafie and Vincent Desmaris
Smoltek AB, Regnbågsgatan 3, Gothenburg, SE-41755, Sweden

Peter Enoksson
Micro and Nanosystems group, BNSL, Department of Microtechnology and Nanoscience, Chalmers University of Technology, Gothenburg, SE-41296, Sweden

Ioanna Savva
Department of Mechanical and Manufacturing Engineering, University of Cyprus, Nicosia, Cyprus

Theodora Krasia-Christoforou
Department of Mechanical and Manufacturing Engineering, University of Cyprus, Nicosia, Cyprus
Nanotechnology Research Center (NRC), University of Cyprus, Nicosia, Cyprus

Gert Göransson
Department of Chemistry and Molecular Biology, University of Gothenburg, Gothenburg, SE-41296, Sweden

Index